T0215116

Cambridge Studies in Biological and Evolutionary Anthropology 60

Monkeys on the Edge: Ecology and Management of Long-Tailed
Macaques and their Interface with Humans

Long-tailed macaques (*Macaca fascicularis*) have a wide geographical distribution and extensively overlap with human societies across Southeast Asia,
regularly utilizing the edges of secondary forest and inhabiting numerous
anthropogenic environments, including temple grounds, cities, and farmlands.
Yet despite their apparent ubiquity across the region, there are striking gaps in
our understanding of long-tailed macaque population ecology.

This timely volume, a key resource for primatologists, anthropologists, and
conservationists, underlines the urgent need for comprehensive population
studies on common macaques. Providing the first detailed look at research on
this underexplored species, it unveils what is currently known about the population of *M. fascicularis*, explores the contexts and consequences of human–
macaque sympatry, and discusses the innovative programs being initiated to
resolve human-macaque conflict across Asia. Spread throughout the book are
boxed case studies that supplement the chapters and give a valuable insight
into specific field studies on wild *M. fascicularis* populations.

MICHAEL D. GUMERT is an Assistant Professor in the Division of Psychology
at Nanyang Technological University, Singapore, where he leads a field program investigating the behavioral biology and ecology of *Macaca fascicularis*
in Indonesia, Singapore, and Thailand. Recent research focuses on practical
issues facing long-tailed macaque populations and he has organized international experts in a cooperative group to better understand the conservation
and management needs of long-tailed macaques.

AGUSTÍN FUENTES is Professor of Anthropology and Director of the Institute
for Scholarship in the Liberal Arts at the University of Notre Dame. His current research projects include assessing behavior, ecology, and pathogen transmission in human–monkey interactions in Southeast Asia and Gibraltar and
examining the roles of cooperation, social negotiation, and niche construction
in primate and human evolution.

LISA JONES-ENGEL is a Senior Research Scientist at the Washington National
Primate Research Center, University of Washington. Her current research focuses
on the human–primate interface where she coordinates several multidisciplinary
research projects in Asia, which focus on the role synanthropic macaques play in
the evolution and emergence of infectious diseases.

Cambridge Studies in Biological and Evolutionary Anthropology

Series Editors

HUMAN ECOLOGY
C. G. Nicholas Mascie-Taylor, University of Cambridge
Michael A. Little, State University of New York, Binghamton
GENETICS
Kenneth M. Weiss, Pennsylvania State University
HUMAN EVOLUTION
Robert A. Foley, University of Cambridge
Nina G. Jablonski, California Academy of Science
PRIMATOLOGY
Karen B. Strier, University of Wisconsin, Madison

Monkeys on the Edge

Ecology and Management of Long-Tailed Macaques and their Interface with Humans

Edited by

MICHAEL D. GUMERT
Nanyang Technological University, Singapore

AGUSTÍN FUENTES
University of Notre Dame, Indiana, USA

LISA JONES-ENGEL
Washington National Primate Research Center, Seattle, USA

CAMBRIDGE
UNIVERSITY PRESS

CAMBRIDGE
UNIVERSITY PRESS

University Printing House, Cambridge CB2 8BS, United Kingdom

One Liberty Plaza, 20th Floor, New York, NY 10006, USA

477 Williamstown Road, Port Melbourne, VIC 3207, Australia

314-321, 3rd Floor, Plot 3, Splendor Forum, Jasola District Centre, New Delhi - 110025, India

79 Anson Road, #06-04/06, Singapore 079906

Cambridge University Press is part of the University of Cambridge.

It furthers the University's mission by disseminating knowledge in the pursuit of
education, learning and research at the highest international levels of excellence.

www.cambridge.org
Information on this title: www.cambridge.org/9781108822558

© Cambridge University Press 2011

This publication is in copyright. Subject to statutory exception
and to the provisions of relevant collective licensing agreements,
no reproduction of any part may take place without the written
permission of Cambridge University Press.

First published 2011
First paperback edition 2020

A catalogue record for this publication is available from the British Library

Library of Congress Cataloging in Publication data
Monkeys on the edge : ecology and management of long-tailed macaques and their interface with
humans / [edited by] Michael D. Gumert, Agustín Fuentes, Lisa Jones-Engel.
    p.   cm. – (Cambridge studies in biological and evolutionary anthropology)
    Includes bibliographical references and index.
    ISBN 978-0-521-76433-9 (hardback)
    1. Kra–Ecology.   2. Kra–Conservation.   3. Kra–Effect of human beings on.
    I. Fuentes, Agustin.   II. Gumert, Michael D.   III. Jones-Engel, Lisa.
    IV. Title.   V. Series.
    QL737.P93M657 2011
    599.8´644–dc22
    2010054305

ISBN   978-0-521-76433-9   Hardback
ISBN   978-1-108-82255-8   Paperback

Cambridge University Press has no responsibility for the persistence or
accuracy of URLs for external or third-party internet websites referred to in
this publication, and does not guarantee that any content on such websites is,
or will remain, accurate or appropriate.

Contents

Contributors

Nur Afendi, Karimunjawa National Park Office, Semarang, Java Tengah, Indonesia

I. G. A Arta Purta, Pusat Kajian Primata, Universitas Udayana, Bali, Indonesia

F. Brotcorne, Behavioural Biology Unit, University of Liege, Belgium

Sharon Chan, Central Nature Reserve, Conservation Division National Parks Board, Singapore

Gregory Engel, Washington National Primate Research Center, University of Washington, Seattle, WA, USA

Kaitlyn-Elizabeth Foley, TRAFFIC, Malaysia

Agustín Fuentes, Department of Anthropology University of Notre Dame, Notre Dame, IN, USA

Shunji Goto, Amami Wild-Animal Research Center Kagoshima, Japan

Lisa Guidi, Department of Anthropology, Washington University, St. Louis, MO, USA

Michael D. Gumert, Division of Psychology, School of Humanities and Social, Sciences, Nanyang Technological University, Singapore

Yuzuru Hamada, Evolutionary Morphology Section, Primate Research Institute, Kyoto University, Japan

I. D. K. Harya Putra, Pusat Kajian Primata, Universitas Udayana, Bali, Indonesia

M. C. Huynen, Behavioural Biology Unit, University of Liege, Belgium

Entang Iskandar, Primate Research Center, Bogor Agricultural University, Bogor, West Java, Indonesia

Lisa Jones-Engel, Washington National Primate Research Center, University of Washington, Seattle, WA, USA

Phouthone Kingsada, Department of Biology, National University of Laos, Dong Dok, Vientiane, Lao People's Democratic Republic

Hiroyuki Kurita, Board of Education, Oita City, Japan

Randall C. Kyes, Washington National Primate Research Center, University of Washington, Seattle, WA, USA

Benjamin P. Y-H. Lee, Nature Parks Branch, Parks Division, National Parks Board, Rupublic of Singapore; formerly Durrell Institute of Conservation and Ecology, University of Kent, UK

Suchinda Malaivijitnond, Primate Research Unit Chulalongkorn University, Bangkok, Thailand

Badrul Munir Md-Zain, School of Environmental Science, and Natural Resources, Universiti Kebangsaan Malaysia Bangi, Selangor, Malaysia

Mastura Mohd-Zaki, School of Environmental Science and Natural Resources, Universiti Kebangsaan Malaysia, Bangi, Selangor, Malaysia

Yoshiki Morimitsu, Institute of Natural and Environmental Sciences, University of Hyogo, Japan

Toru Oi, Ph D., Wildlife Ecology Laboratory Forestry and Forest Products Research Institute, Ibaraki, Japan

Nada Padayatchy, Bioculture Ltd., Mauritius

Joko Pamungkas, Primate Research Center, Bogor Agricultural University, Bogor, West Java, Indonesia

Sitideth Pathonton, Department of Biology, National University of Laos, Dong Dok, Vientiane, Lao People's Democratic Republic

Bounnam Pathontone, Department of Biology, National University of Laos, Dong Dok, Vientiane, Lao People's Democratic Republic

Bounthob Praxaysombath, Department of Biology, National University of Laos, Dong Dok, Vientiane, Lao People's Democratic, Republic

Devis Rachmawan, Orangutan Foundation-UK, Pangkalan Bun, Kalimantan Tengah Indonesia

Sandeep K. Rattan, Wildlife Wing, HP Forest Department, Himachal Pradesh, India

Aida L. T. Rompis, Pusat Kajian Primata, Universitas Udayana, Bali, Indonesia

Fong Samouth, Department of Ecology, National University of Laos, Dong Dok, Vientiane, Lao People's Democratic Republic

Aye Mi San, Department of Zoology, University of Yangon, Yangon, Myanmar

Wayan Selamet, Padangtegal Wenara Wana, Bali

Christopher A. Shaffer, Department of Anthropology, Washington University, St. Louis, MO, USA

Chung-Tong Shek, Agriculture, Fisheries and Conservation Department, Hong Kong

Chris R. Shepherd, TRAFFIC, Malaysia

M. Farooq Siddiqi, Aligarh Muslim University, Aligarh, U.P., India

I. G. Soma, Pusat Kajian Primata, Universitas Udayana, Bali, Indonesia

Charles H. Southwick, Department of Ecology and Evolutionary Biology, University of Colorado, Boulder, CO, USA

Rebecca Stephenson, Department of Anthropology, University of Guam, Guam

I. Nyoman Suartha, Pusat Kajian Primata, Universitas Udayana, Bali, Indonesia

Robert W. Sussman, Department of Anthropology, Washington University, St. Louis, MO, USA

Mohamed Reza Tarmizi, School of Environmental Science and Natural Resources, Universiti Kebangsaan Malaysia Bangi, Selangor, Malaysia

Yolanda Vazquez, School of Biology, Faculty of Science, Agriculture and Engineering, Newcastle University, UK

Chanda Vongsombath, Department of Biology, National University of Laos, Dong Dok, Vientiane, Lao People's Democratic Republic

I. Nyoman Wandia, Pusat Kajian Primata, Universitas Udayana, Bali, Indonesia

Ni Luh Watiniasih, Pusat Kajian Primata, Universitas Udayana, Bali, Indonesia

Bruce P. Wheatley, Anthropology Department, University of Alabama, Birmingham, AL, USA

Foreword

Thirteen ways of looking at a monkey

DAVID QUAMMEN

If the world of nonhuman primates offers an emblem of the tensions of modernity, it's *Macaca fascicularis*, the long-tailed macaque. This alert, adaptable Asian species is one of the world's most familiar monkeys, but also among the most sorely taken for granted. Its behavior is flexible and complex. Its intelligence and opportunism are famous, even notorious. It has been called many names, of which "weed" and "ethnotramp" aren't the worst. Its current population status is poorly known but, by reliable accounts, combines the good news of broad distribution with the bad news of declining numbers. Its relations with *Homo sapiens* are close, diverse, ambivalent, and in some cases problematic. Although it has recently been reclassified as a species of "least concern" by the IUCN, concern does remain high among some primatologists, who see the long-tailed macaque facing multiple challenges throughout its distributional range. Some of those challenges (of which habitat loss, habitat fragmentation, culling and other population-control actions intended to reduce conflict with humans, and capture for use in biomedical research are foremost) could lead to local extinctions, disappearance of subspecies, and compromised population viability overall. No wonder, then, that Michael D. Gumert, Agustín Fuentes, Lisa Jones-Engel, and many of their colleagues have felt an urgent need to assess what is known, and to target what isn't known but should be, about *Macaca fascicularis*. This book is an expression of that heightened concern.

The long-tailed macaque is an extraordinary species, much valued (especially as a laboratory test animal) and at the same time much disdained. It's so plastic in its attributes and roles, so various, so shimmery – and human attitudes toward it are so varied too – that I'm put in mind of the Wallace Stevens poem, "Thirteen Ways of Looking at a Blackbird," first published in 1917, when cubism in art and imagism in poetry were the cresting waves. Stevens was getting at the matter of perspective and subjectivity when he wrote those thirteen little haiku-like bits, each bit a vision or a thought of the bird. For instance: "Among twenty snowy mountains/The only moving thing/Was the eye of the blackbird." And: "I do not know which to prefer/The beauty of inflections/Or the beauty of innuendos,/The blackbird whistling/Or just after." In a similar spirit (but far more prosaically), I've made a list of adjectives that have been or could be applied to the long-tailed macaque, holding myself to a canonical limit of thirteen. This creature is: smart, adaptive, widespread, resilient, winsome, pestiferous, synanthropic, variable, exploited, sacred, profane, numerous, and besieged. I won't annotate the list because every one of those topics is treated expertly in the chapters that follow.

My own early impression of *Macaca fascicularis* was skewed by the fact that I encountered it first on the island of Mauritius, far from its native range. On Mauritius, to which it was introduced by Portuguese or Dutch sailors sometime in the sixteenth century, it has thrived as one among many exotic species. Those exotics together have caused severe ecological damage, even helped push some Mauritian endemics to extinction; the exact culpability of the long-tailed macaque, amid the jumble of invaders, is hard to measure. Later I saw it in the monkey temples of Bali, manifesting its typical boldness, its special mojo, and I began to appreciate its influential place within human attitudes (especially Hindu, Buddhist, and Animist attitudes) toward nature. Still later, while tagging along on a field trip to Bangladesh with Lisa Jones-Engel and Gregory Engel, during which their focus was the rhesus macaque, *Macaca mulatta*, I learned more about *Macaca fascicularis* also – enough to confute my early bias. It was Jones-Engel who alerted me to the sad and dangerous paradox of the long-tailed macaque in the twenty-first century: that, because so much of its natural habitat is being destroyed, and because it can compensate somewhat by gravitating to the fringes and interstices of human environments, the species is simultaneously becoming more conspicuous and less abundant. Jones-Engel also reminded me that, despite whatever misguided human actions may have placed *Macaca fascicularis* where it doesn't belong, fundamentally it is a wild animal, part of the fullness of life on Earth.

The logical extension of those two fateful trends – more conspicuous, less abundant – is that we might allow ourselves to be surprised when this "common" monkey becomes badly reduced, scarce, even threatened with extinction from parts or all of its range in the wild. We have made that mistake before, with other seemingly plentiful and misprized species – most famously, the passenger pigeon in North America. It would be a shame to commit the same lazy blunder again.

The long-tailed macaque, as this book's title declares, is a species on the edge. It lives on the edges of forests, along the edges of rivers and seas, at the edges of human settlements, and maybe on the edge of catastrophic decline. Wallace Stevens wrote, as the ninth of his thirteen ways: "When the blackbird flew out of sight/It marked the edge/Of one of many circles." A glimpse of a fleeting arc, and then it was gone.

Preface

A growing concern in Southeast Asian countries is the interface between long-tailed macaques (*Macaca fascicularis*) and human beings (*Homo sapiens*) and the consequences that this sympatric relationship brings to the affected human communities and macaque populations. A variety of potential negative and positive consequences exist in zones of interface for both humans and macaques. As a result, the relationship between *M. fascicularis* and humanity has become a recent focus for academics and NGOs worldwide. In addition, regional governments, whose citizens have expressed concern about the disturbances caused by living closely with macaques, have attempted or are newly initiating programs to manage or reduce populations of macaques living along-side people. The occurrence of human–macaque overlap is not isolated to a few exceptional locations, but rather macaque synanthropy is a widespread phenomenon existing in regions all throughout peninsular and insular Southeast Asia. In this volume, we begin to build a more comprehensive understanding of long-tailed macaque populations and the extent of their overlap with humans.

In several regions of Southeast Asia, governments and NGOs have already initiated management programs to control macaque populations. These programs are a result of an effort to respond to citizen complaints and have set out to reduce what they consider overpopulated and/or nuisance macaque populations. Management programs have occurred or are occurring in Thailand, Malaysia, Hong Kong and Singapore, and to a lesser extent in some regions of Indonesia. The goals of these programs either aim at reducing populations and pushing macaques away from human settlements by whatever means, or are focused on humanely controlling or eradicating the population through means of mass sterilization. Nearly all management programs have been initiated with insufficient or complete lack of information on the population being managed, with action plans often being initiated on only assumptions and non-researched estimates of the size, distribution, density, and other important characteristics of the populations being manipulated. Consequently, there is no way to gauge the impact of these programs on the long-tailed macaque population, and whether the consequences of these programs are indeed pathways to producing the results sought (i.e., reduction of macaque–human conflict).

The human–macaque interface is not only occurring with long-tailed macaques, but also has been an issue for rhesus macaques, Assamese macaques, and bonnet macaques in India, Japanese macaques in Japan, Taiwanese macaques in Taiwan, Tibeten macaques in China, several species of macaques in Sulawesi, and to a lesser extent pig-tailed and stump-tailed macaques in several regions of Southeastern Asia. It is a genera-wide dilemma, but we have chosen to focus on long-tailed macaques in this volume because of the large geographic range they encompass and the number of countries affected by this single species. Issues associated with long-tailed macaques interfacing with humans are truly an international environmental dilemma, affecting sixteen different nations and millions of people. This vast overlap between humans and macaques in Southeast Asia is largely driven by the expansive anthropogenic development and habitat modification that is rapidly occurring in the region, as ASEAN nations race to achieve newfound wealth and status amongst the international community. For these reasons, the large influence this single species has over Southeast Asia warrants a detailed look into its population and overlap with people.

In general, it appears that there has been an increase in conflict between humans and macaques over the last decade in Southeast Asia, although it still remains unclear all the factors driving this change. Likely factors include, human expansion into previously undisturbed macaque habitats, an increase of macaques moving near human settlement, increases in populations of macaques and/or humans in regions of interface, increased conflict as the result of communities gaining more wealth and resources to defend from macaques, or just increased reports and complaints by disgruntled citizens becoming more accustomed to modern lifestyles. The uncertainty of the causal factors and their degree of impact on increasing conflict are a clear indication that better investigation is needed on this issue. Understanding the causes of conflict will certainly be of high importance in mitigating it. With current trends in the report of human-macaque conflict, it is anticipated that the development of national and international-level wildlife management organizations dedicated solely to monitoring, evaluating, and implementing action to develop sustainable human-macaque communities and/or reduce conflict between the two will become a major focus of wildlife management in Southeast Asia over the next decade. Management efforts will need to be paralleled by similarly extensive research programs that will help to support and guide these programs so that they do not blindly manipulate their macaque populations.

Long-tailed macaques have had a very long history of living in close proximity to human activity and their settlements, and they are well adapted to reproductively succeed in human-influenced environments. For example, long-tailed macaques tend to prefer forest edge habitats and regions altered by human activity hence, they are often found along the edges of human-landscaped environments. They also are generalist feeders and are attracted and adjust well to human-based food resources, such as contained refuse, litter, and food directly given to them. Macaques are also attractive to people, and humans keep them for recreational use (i.e., pets and performance), and this has led to humans manipulating and moving macaques around. This close relationship to humankind, as well as their high level of flexibility and generalist dietary habits, has allowed long-tailed macaques to colonize new island environments after being transported by humans to a few islands beyond their core range in Southeast Asia. The level of impact of these exotic macaques varies from highly invasive to little impact in their new insular environment. Accordingly, the impact of long-tailed macaques is not restricted to only Southeast Asia, and wildlife managers now have to consider efforts to control future introductions of macaques to new islands, as well as managing the impact they have had on the flora and fauna of islands already colonized. Moreover, understanding how long-tailed macaques adjust to such a wide variety of environmental circumstance should be of immense interest to evolutionary scientists interested in the natural selection of variability and flexible behavior.

In this volume, we have attempted to compile what is currently known about the long-tailed macaque population and its overlap with human communities in response to the current needs we have highlighted above. This is a highly needed compilation of information because, although this species is a commonly occurring monkey of Southeast Asia and one of the world's most widely geographically distributed, there is a striking gap in our understanding of its population distribution and abundance. Moreover, this volume will be a useful guide as attention toward this species continues to increase over the coming years. More and more organizations arc attempting manipulations of long-tailed macaque populations in an effort to lessen the extent of human–macaque overlap, and therefore it will be imperative to have information readily available about the population-level characteristics of *M. fascicularis*. Furthermore, systematic assessments of the impact and consequences of human–macaque sympatry are needed to guide future actions towards macaques. This volume is an attempt to compile information from experts around the world currently researching wild populations of *M. fascicularis* to gather and to impart what is known about the present population of long-tailed macaques. The volume is intended to be a resource for scientists and researchers interested in the biology of primate populations and the effects of human–wildlife interaction. Moreover, it should become a key source of information for organizations and policy-makers struggling to develop strategies and legislation for developing peaceful human–macaque communities and managing their interface between humans and long-tailed macaques.

The volume is broken into five major parts. Part I deals with the population of long-tailed macaques. Chapter 1 describes the basic population-level characteristics of long-tailed macaques, as well as explaining the features of the human–macaque interface. Chapters 2 and 3 focuses on work from a group led by Dr Yuzuru Hamada from Kyoto University and Dr Aye Mi San from Myeik University, which has been doing extensive work throughout mainland Southeast Asia assessing the distribution and evolution of macaque species. These two chapters look at the first reports generated on the long-tailed macaque populations in Myanmar and the Peoples Republic of Laos. Part II looks in more detail at the human–macaque interface, and in this section we are introduced to situations in Malaysia, Thailand, Singapore and Bali, Indonesia. Chapter 7 closes the section on the human–macaque interface by detailing the concerns of transmission of infectious agents between humans and macaques. Part III focuses attention on the colonization of long-tailed macaques beyond their natural range, and Chapters 8–10 are focused on the exotic populations in Mauritius and Palau. Part IV is a short section looking at the closest relative of long-tailed macaques, the rhesus macaque (*M. mulatta*), and in Chapter 11, Dr Charles Southwick's 50 plus years of research in India is summarized to show how rhesus populations and human–macaque contact have changed over that time. Part V brings the volume to a close by providing various approaches and needs for studying and managing long-tailed macaques and their interface with humans. In Chapter 12, we discuss potential resolutions to lessen human–macaque conflict and develop sustainable human–macaque communities (i.e., communities where macaques and humans can coexist without major conflict and without serious threats to each other's health, wellbeing, or existence). Chapter 13 outlines the future directions for researching macaque populations, provides a discussion on approaches to studying macaque synanthropy, and questions the need for conservation programs for some parts of the population. Throughout the volume, boxes are presented from separate authors to supplement the chapters. These short excerpts provide detailed information of specific activities being carried out in the field to better understand the population or to help resolve human–macaque conflict.

Acknowledgements

This volume is the result of two international workshops on the relationship between humans and long-tailed macaques. In 2007, we held the *Macaca fascicularis Workshop: Resolving Macaque-Human Conflicts* at the 30th meeting of the American Society of Primatologists in Wake Forest, North Carolina. Later, in 2008, we continued our discussions by holding the *Macaca fascicularis Workshop: Understanding and Managing Macaque-Human Commensalism* and an associated roundtable discussion at the XXII Congress of the International Primatological Society in Edinburgh, UK. The products of these gatherings eventually became this volume.

We would like to express our gratitude to the invited panel members and speakers that attended these programs: Nantiya Aggimarangsee, Irwin Bernstein, Antje Engelhardt, Joseph Erwin, Ardith A. Eudey, Yuzuru Hamada, Entang Iskandar, Randall C. Kyes, Badrul Munir Md-Zain, Benjamin P. Y-H. Lee, Suchinda Malaivijitnond, Stephen J. Schapiro, Charles Southwick, Robert Sussman, and Bruce B. Wheatley. We would also like to thank Mathew F. S. X. Novak, Allyson Bennett, Paul Honess, Phyllis Lee, and Alexander Weiss for their assistance in organizing and advertising these events at the conferences. We further extend great thanks to the many others who attended and participated in the open workshops and discussions.

Special thanks are given to Regina Liszanckie and Nurul Haiyu Binti Rosli for providing helpful administrative support during the preparation of this volume. We finally would like to thank Lynette Talbot, Martin Griffiths, Sally Philip, Zewdi Tsegai, Jo Bottrill, Matt Davies, and John Normansell for their assistance and guidance during editing and production. Grant RG95/07 from the Academic Research Fund of Nanyang Technological University, provided by the Ministry of Education, Singapore supported parts of the production of this book.

A stone statue at Sangeh Monkey Forest, Bali, Indonesia depicts an evil giant from the Ramayana, Kumbakarna, being attacked by an army of long-tailed macaques. Kumbakarna is the brother of the epic's evil antagonist, Rawana, and was convinced by him to fight against the hero, Rama. Kumbakarna tried to convince Rawana to stop the unnecessary war, but out of loyalty to his own kind, agreed to fight. In the Ramayana, macaques represented an army of warriors that assisted Rama to fight against a powerful evil and save his true love, Sita. In the real world, macaques live on the edge of our own war to conquer and control our global environment. As in the Ramayana, the macaques fight against a powerful force; and like Kumbakarna, maybe we too are knowingly fighting an unworthy war. Photograph by Michael D. Gumert.

"Ramadewa looked at the numerous troops of monkeys. They were at ease and happy and showed their liveliness. All their movements, their noisy voices, their way of sleeping on branches made him happy just to look at them."

Verse 151:VI from the Ramayana Kakawin,
as translated from Javanese by
Soewito Santoso, 1980.

Part I

The status and distribution of long-tailed macaques

1 The common monkey of Southeast Asia: Long-tailed macaque populations, ethnophoresy, and their occurrence in human environments

MICHAEL D. GUMERT

The long-tailed macaque (*Macaca fascicularis*) population spreads over one of the widest geographical ranges of any primate, trailing only humans (*Homo sapiens*) and rhesus macaques (*M. mulatta*) (Wheatley, 1999) (Figure 1.1). According to Fooden (1995, 2006), the population extends across the majority of mainland Southeast Asia. They occur in the southeastern most part of Bangladesh, spreading south along the coast of Myanmar, east through the southern two-thirds of Thailand, all of Cambodia, the southeastern tip of Laos, and through the southern half of Vietnam. Through Thailand, the population extends past the Isthmus of Kra, and occurs all through Sundaland (i.e., peninsular Malaysia and the Indonesian archipelago west of the Wallace line) and into the Philippines. Long-tailed macaques also occur on smaller islands. For example, long-tailed macaques occur off the northern coast of Sumatra on the most southern Indian Nicobar Islands, as well as occurring on small islands off the west coast, such as Simeulue and Lasia. Other island habitats include Maratua, off Kalimantan, Karimunjawa, off Java, Koh Khram Yai, off Thailand, and Con Son, off Vietnam. They certainly occur on many other small islands, as the region is covered with tens of thousands of islands.

Long-tailed macaques are found predominantly on the western side of the Wallace line and are considered Asian fauna. Despite this, populations in Wallacea exist on the eastern side of the line, which are possibly the result of historical human introductions (e.g., Lombok, Nusa Tenggara, and East Timor) (Kawamoto *et al.*, 1984). Confirmed cases of recent human introductions of macaques across the Wallace line have occurred on an island off Sulawesi (Froehlich *et al.*, 2003) and on West Papua (Kemp and Burnett, 2007). In

Monkeys on the Edge: Ecology and Management of Long-Tailed Macaques and their Interface with Humans, eds. Michael D. Gumert, Agustín Fuentes and Lisa Jones-Engel. Published by Cambridge University Press. © Cambridge University Press 2011.

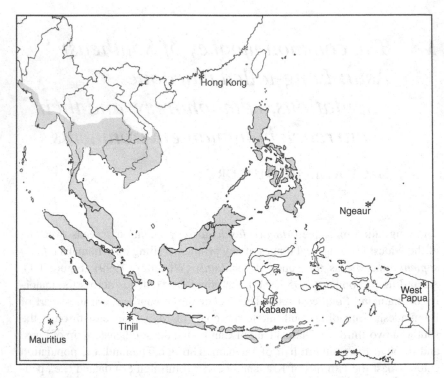

Figure 1.1. A map of Southeast Asia showing the approximate distribution of *M. fascicularis* shaded in gray. The regions they have colonized over the last several centuries are marked by asterisks and labeled by name. Mauritius is inlayed on the map, is located off the east coast of Africa, and is not to scale.

addition, there are also confirmed cases of recently human-introduced populations in other regions far from their natural range such as Mauritius (Sussman and Tattersall, 1986), Hong Kong (Southwick and Southwick, 1983), and Palau (Poirer and Smith, 1974). An intentional introduction to a formerly unpopulated island within the boundaries of their natural range occurred on Tinjil Island to establish a breeding colony (Kyes, 1993).

The population size of *M. fascicularis* is not well known (Southwick and Siddiqi, 1994). In the 1980s, Kathy Mackinnon (1986) provided population estimates for primates in Indonesia and estimated that there were 3,726,860 long-tailed macaques. A year later, it was estimated that there were 309,360 macaques in mainland Southeast Asia (Mackinnon and Mackinnon, 1987). Using these figures, Fooden (1995) estimated that the entire natural population of long-tailed macaques in the 1980s was somewhere around 5 million individuals. Eleven years later, he reassessed his estimation and suggested there are approximately 3 million long-tailed macaques in existence today (Fooden, 2006). If these numbers are accurate, there has been a 40 percent total population decline in about a quarter of a century. Also pointing towards population decline, Mackinnon and Mackinnon (1987) reported that 63 percent of primate habitat in Indonesia was no longer suitable for inhabitation as early as the 80's, and Khan *et al.* (1982) pointed out a 23 percent decline in the population of *M. fascicularis* between 1957 and 1975 in Malaysia. *M. fascicularis* numbers are relatively high compared to other extant non-human primates, but they are continually declining, suggesting monitoring of their population may be necessary in the near future (Wheatley and Putra; 1994b; Eudey, 2008).

Southwick and Siddiqi (1994) pointed out that Mackinnon's estimates may not have taken into account the patchy distribution of long-tailed macaques and so the figures reported may overestimate actual population levels. Mackinnon (1986) admits that population census techniques can be subjective, but that best efforts were taken to accurately depict population densities despite attempting to census a species that is unevenly distributed through a vast range. Therefore, even with the best correction techniques it is still possible that populations were overestimated. Regions that are most easily surveyed, rivers and forest edges, are where long-tailed macaques predominantly range, making it possible that extrapolations into unsurveyed regions, where they may not be, might overestimate the population size. We must also consider that long-tailed macaques are among the most easily observable of wildlife because they are conspicuous and prefer habitats near human settlements and along forest edges. Consequently, it is much easier to physically count entire regional populations than many other species. Population counts based on extrapolations from populations intermixed into human settlements and/or on forest borders may not be accurate, as what is observed may represent the large percentage of their population with little extrapolation needed.

We can predict that long-tailed macaques are decreasing in number in their total population. However, it is possible that a larger proportion of the population is now interfacing with human settlement, as there arc a number of reports highlighting increased occurrence of human-macaque conflict in several regions of South Asia (Eudey, 2008; Fuentes, 2006; Fuentes *et al.*, 2008; Malaivijitnond *et al.*, 2005; Malaivijitnond and Hamada, 2008; Sha *et al.*, 2009a; Sha *et al.*, 2009b; Wong and Chow, 2004). Increased conflict is suspected to be a consequence of human-induced habitat disturbance, which is causing forest loss, expansion of edge habitat, and production of mosaic environments. The rapid development occurring in Southeast Asia presents a challenge to wildlife management programs because it has produced more environments suitable to sustaining long-tailed macaques. As a result, the consequences of human-macaque overlap may become more prominent in more

Table 1.1. *Reported habitat types inhabited by
long-tailed macaques*

Bamboo forest	Mangrove forest
Beach	Mixed forest
Broadleaf forest	Montane forest
Coastal regions	Primary forest
Deciduous forest	Rain forest
Dipterocarpel forest	Riverine
Dry forest	Rocky shores
Edge habitats	Scrub forest
Evergreen forest	Secondary forest
Grassland	Semideciduos forest
Hills	Submontane forest
Islands	Swamp forest
Lowland forest	Tropical forest

and more regions across Asia. It will be imperative to better assess how much
of the long-tailed macaque population is actually residing in human settlements and repair the striking gap in our knowledge about the population-level
characteristics of this supposedly common and well-understood monkey.

Habitat preferences

Long-tailed macaques are found in a wide variety of habitats including mangrove, rainforest, swamp, coastal, tropical, deciduous, evergreen, scrub, riverine, and secondary forests (Fooden, 2006) (Table 1.1). They are not equally
distributed in these environments and long-tailed macaques are most commonly
found along forest edges, especially in swamp forests and riverine habitats or
on the edges of disturbed habitats (Bismark, 1991; Bismark, 1992; Chivers
and Davies, 1978; Crockett and Wilson, 1980; Fittinghoff and Lindburg, 1980;
Gurmaya *et al.*, 1994; McConkey and Chivers, 2004; van Schaik *et al.*, 1996;
Suaryana *et al.*, 2000; Supriatna *et al.*, 1996). The most recent surveys of long-
tailed macaques in Sumatra and Kalimantan indicate they may be less abundant
in lowland and montane forests (Yanuar *et al.*, 2009), and in swamp forests are
difficult to find beyond 1 km from river edges (Gumert *et al.*, 2010).

Anthropogenic land-use generates large amounts of forest edge, and being
an edge species, long-tailed macaques are already well adapted for exploiting
the fragmented forests created by current human land development. Not surprisingly, long-tailed macaques are commonly reported to inhabit the edges of a variety of anthropogenic environments (Fuentes, 2006; Fuentes *et al.*, 2008, Hadi,

Table 1.2. *Types of human-macaque interface zones*

Agricultural land
Cemeteries
Eco-lodges
Metropolitan cities
Plantations
Rural villages
Roads
Small islands
Towns
Temples/Religious grounds
Recreation parks

2005; Malaivijitnond *et al.*, 2005; Malaivijitnond and Hamada, 2008; Sha *et al.*, 2009a; Sha *et al.*, 2009b; Wong and Ni, 2000) (Table 1.2). Long-tailed macaques are also attracted to the edges of human settlements because of the availability of excess food resources, which can keep macaques around villages, towns and cities and lower their dependency on wild food sources (Sha *et al.*, 2009a). Since, anthropogenic habitat alteration is continually developing new forest edges and disturbed forest habitats throughout Southeast Asia, more and more habitats are being generating that can potentially sustain populations of long-tailed macaques at close proximity to human settlements. In contrast, undisturbed habitats suitable to long-tailed macaques, particularly in unprotect forests, may be quickly becoming scarcer, potentially threatening the sustainability of regional populations of macaques in such conditions (Yanuar *et al.*, 2009).

Variation amongst long-tailed macaques

Distribution of subspecies

Congruent with the great geographic distribution of *M. fascicularis*, there is also a great level of variation in physical characteristics across locations. Historically, there have been approximately 50 different scientific names given to this monkey at various times, but the current classification system suggests there are ten different subspecies of long-tailed macaques (Fooden, 1995; Fooden, 2006; Groves, 2001). Fooden's classification system is arguable, but it is adopted by Groves (2001), and his system will be reviewed here as a good point of origin for classifying the variation found in long-tailed macaques. The species group mainly consists of *M. f. fascicularis*, which inhabits all

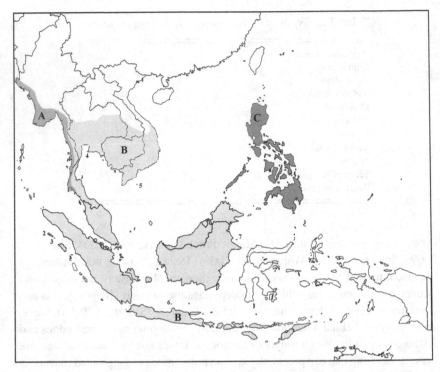

Figure 1.2. A map of the distribution of the ten sub-species of *M. fascicularis*. The core and its fringing subspecies are shaded and labeled by letters. A = *M. f. aurea*, B = *M. f. fascicularis*, C = *M. f. philippinensis*. The isolated island subspecies are labeled with numbers. 1 = *M. f. umbrosa*, 2 = *M. f. lasiae*, 3 = *M. f. fusca*, 4 = *M. f. atriceps*, 5 = *M. f. condorensis*, 6 = *M. f. karimondjiwae*, 7 = *M. f. tua*.

of mainland Southeast Asia east of Thailand, and extends through the major islands of Indonesia and slightly extends into the southwestern region of the Philippines (Figure 1.2).

Two other subspecies lie on the eastern and western extremes of the core species' geographical distribution. The second largest ranging subspecies, *M. f. aurea*, is found northwest of the range of *M. f. fascicularis* in Myanmar, Bangladesh, and west-central Thailand. They have a contact zone with *M. f. fascicularis* along the mountain ranges that cover the border regions of Myanmar and Thailand. *M. f. philippinensis* occupies a similarly sized range on the eastern side, occupying the majority of the Philippine islands except for the western region of Mindanao, which is solely occupied by *M. f. fascicularis*. The ranges of *M. f. fascicularis* and *M. f. philippinensis* overlap in the eastern two-thirds of Mindanao and in the southern Negros Islands. Here the two subspecies are sympatric and long-tailed macaques of mixed phenotype can be found.

All of the other seven subspecies are isolated on small islands. *M. f. atriceps*, *M. f. condorensis*, and *M. f. karimondjiwae* all populate small shallow-water fringing islands. *M. f. atriceps* is found on the island of Koh Khram Yai off the southwestern coast of Thailand. *M. f. condorensis* inhabits Con Son, an island off the southeastern coast of Vietnam. *M. f. karimondjiwae* inhabits Karimunjawa Island, off the north central coast of Java. *M. f. umbrosa*, *M. f. fusca*, *M. f. lasiae*, and *M. f. tua*, all inhabit small deep-water fringing islands. *M. f. umbrosa* is found on the islands of Katchall, Little Nicobar, and Great Nicobar in the Indian Nicobar Island chain north of Sumatra. *M. f. fusca* is located on the small island of Simeulue, which is off the west coast of northern Sumatra. Nearby, *M. f. lasiae* inhabits Lasia Island. Finally, off of the east coast of Kalimantan, *M. f. tua* occupies Maratua Island.

Subspecies characterization

Fooden (1995) assessed the skins of long-tailed macaques from all over its range and found significant morphological variation between regions, and this is the basis for his subspecific classification system. The variation he found was in the dorsal pelage color, crown color, thigh color, lateral crest pattern, head length, body length, and relative tail length. One of the major patterns he observed was that monkeys on fringing islands tended to show a blackish coloration of their pelage, and that large island and mainland monkeys were lighter (Fooden, 1995). Aside from being darker, some island species exhibit other distinct variations. For example, the Philippine subspecies has distinguishing molar variation (Fooden, 1991), while the Simeulue long-tailed macaque has a shorter tail and demonstrates less sexual dimorphism than *M. f. fascicularis* (van Schaik and van Noordwijk, 1985). The Simeulue long-tailed macaque also exhibits distinct behavioral differences. They live in smaller groups that fission and have a distinct loud call that is reported to be similar to the loud call of the Siberut macaque, *M. pagensis* (van Schaik and van Noordwijk, 1985; Sugardjito *et al.*, 1989). Too little is known of the other island subspecies to know if they also exhibit unique characteristics aside from being darker.

Macaca fascicularis aurea, a mainland subspecies, is distinct morphologically appearing to be larger, having darker skin, and a browner-colored pelage (Fooden, 1995; San and Hamada, Chapter 2). Although distinct morphologically, they also exhibit a striking behavioral difference from other populations of long-tailed macaques because they have been observed to customarily use stone tools to crack oysters, as well as other shellfish, snails and nuts (Carpenter, 1887; Malaivijitnond *et al.*, 2007). Although other tool-use behavior has been seen idiosyncratically in *M. f. fascicularis* (Chiang, 1967; Fuentes *et al.*, 2005,

Watanabe *et al.*, 2007; Wheatley, 1988), none of these cases are reported to be customary traditions as in *M. f. aurea* (Gumert *et al.*, 2009).

Some of the distinctions between subspecies concluded from morphological studies have been supported by a small amount of genetic work. The subspecies on the fringing islands off the coast of Sumatra show substantial genetic differentiation from the macaques on mainland Sumatra (Scheffran *et al.*, 1996). There also seems to be a distinct difference in blood proteins between the macaques of the Philippines and others regions, indicating a genetic distinction between *M. f. philippinensis* and *M. f. fascicularis* (Fooden, 1991). Moreover, mitochondrial DNA work has shown that Philippine long-tailed macaques went through a bottleneck after being colonized by Indonesian long-tailed macaques, and have since largely remained genetically isolated from the Indonesian stock. (Blancher *et al.*, 2008) Overall, the subspecific variation of long-tailed macaques needs to be better discerned and it will be important to investigate the differentiation across all the identified subspecies. Such information will help us understand the significance of subspecific differentiation in long-tailed macaques, and will also be a useful guide for conservation and management strategy decisions.

Population status of each subspecies

Long-tailed macaques in the Philippines, *M. f. philippinensis*, were in high abundance prior to 1960. Since that time, the population has been decimated as a result of trapping and forest conversion practices (Fooden, 1991), but no clear population estimates are provided with these claims. The population of Andaman long-tailed macaques (*M. f. aurea*) in Myanmar may not be very large, appears fragmented, and may be threatened by human development and trade (San and Hamada, Chapter 2). In Thailand, *M. f. aurea* are only reported in a few locations around Ranong province (Malaivijitnond and Hamada, 2008), and in Bangladesh they are considered a critically endangered species according to national laws (Khanam *et al.*, 2005). The core subspecies, *M. f. fascicularis* is reported to be widespread but declining because local populations are disappearing in some regions, such as Cambodia (see Box 3.1), while others are expanding into more human-interface zones and thus facing higher levels of conflict (Wheatley and Putra, 1994b; Eudey, 2008). There may now be a larger percentage of the population residing around human settlements, making a large number of monkeys vulnerable to human activity.

The population levels of the small island subspecies are not well understood, but we do have some information. *M. f. umbrosa* has been assessed and is reported to have a population of approximately 3,000 macaques spread across three islands in the Nicobars (Umapathy *et al.*, 2003). This data may be out-dated though as there has been no complete assessment of their loss following the 2004 tsunami (Umapathy, pers. Comm.). The Nicobar macaque is consid-ered a crop pest, but is also listed on Schedule 1 of India's Wildlife Protection Act (Anonymous, 1993). On the island of Simeulue, *M. f. fusca* is reported to have a large population of at least 50,000 individuals, and is also considered a crop raider (Sugardjito *et al.*, 1989). On Karimunjawa, *M.f. karimondjiwae* has a population of around 300 individuals that conflict with the local com-munity (see Box 1.1). Other island populations have no data reported, and *M. f. atriceps*, for example, cannot be observed because they occur on a mili-tary occupied island in Thailand (Malaivijitnond and Hamada, 2008).

Intra-subspecific variation

A high level of morphological and genetic variation occurs within the core subspecies, *M. f. fascicularis*. This high level of variation is greater than that observed within any other macaque species (Fooden, 1995, 2006), which raises questions about the accuracy and usefulness of current subspecies cat-egorizations. Fooden's (1995) extensive research on morphological variation showed that there is high diversity in head, body, and pelage characteristics within *M. f. fascicularis*. The variation found within the core subspecies may be most pronounced in Thailand, where there is a high level of morphological difference between regional populations (Malavijitnond *et al.*, 2005). Island populations of *M. f. fascicularis* are also distinct, such as the Singaporean long-tailed macaque, which tends to be darker, smaller, and have a longer tail–body ratio (Schillaci, *et al.*, 2007). Mangrove-dwelling macaques in Vietnam were found to have different body size and tail-length measurements than monkeys in Indonesia and are reported to have a characteristic "mohawk-like" hair crest pattern (Son, 2003). Studies suggest that the variation seen in body size and tail length within *M. f. fascicularis* can be accounted for by Bergmann's rule, stating body size increases with latitude, and Allen's rule, stating tail length decreases with latitude (Aimi *et al.*, 1982; Son, 2003). The influence of Bergmann/Allen rules can be seen within and between subspecies, although it does not account for all variation observed and its predictions are not always supported (Fooden, 1995; Fooden and Albrecht; 1993; Schillaci *et al.*, 2007). In addition, sexual dimorphism in cranial size appears to vary across geographical regions within the core area of *M. f. fascicularis* (Schillaci, 2010). An anomalous morpho-logical distinction is the yellow long-tailed macaques in Kosumpee Forest Park in Thailand (Hamada *et al.*, 2005), and there also appears to be distinct

Box 1.1   The long-tailed macaques of Karimunjawa (*Macaca fascicularis karimondjiwae*): A small and isolated island subspecies threatened by human–macaque conflict

Nur Afendi, Devis Rachmawan and Michael D. Gumert
In April 2008, the first census of long-tailed macaques on Karimunjawa (*Macaca fascicularis karimondjiwae*) was carried out for four weeks around Karimunjawa National Park (KjNP). The Karimunjawa archipelago is a chain of small islands that is located about 80 km north of Jepara in Central Java, Indonesia. The island chain is located between 5° 49' – 5° 57'S and 110° 04 – 110° 40E, and the three largest islands in this chain are the main island of Karimunjawa (46.19 km²), Kemujan (15.01 km²) and Parang (8.7 km²). Only five islands in the chain are inhabited by humans, and macaques are only reported to inhabit Karimunjawa and Kemujan. The current human population is about 9,054 people on five of the islands, of which 18.5 percent are farmers and 45.4 percent are fishermen (BTNK, 2008).

The islands of Karimunjawa are part of a Marine National Park, and much of the main island of Karimunjawa is part of KjNP. The islands consist of mangrove forest, coastal forest, and lowland tropical forest, and there are protected reefs offshore. The park area on Karimunjawa Island is a hilly area covering most of the central region of the island. The current study occurred only on Karimunjawa, but not Kemujan due to time constraints and limited manpower. The two islands are connected via a mangrove area and a bridge and the majority of the population is thought to be on Karimunjawa, although macaques also occur on Kemujan. During the study, around thirty local people were interviewed and macaque groups were followed daily to count the number of macaques, determine their pattern of distribution, and to assess the level of conflict occurring between the macaques and people of Karimunjawa.

Based on our preliminary survey and interviews conducted, we were able to determine that there were three locations on the island of Karimunjawa where monkeys were frequently observed: Kemloko, Legon Lele, and Nyamplungan (Figure 1.3). After exploring these locations, we counted 23 groups during our study and a total of 269 individuals, of which 82 were fully adult. The average was eleven to twelve individuals per group. About half of the population was found to occur in Legon Lele, and the other half was about equally distributed amongst Kemloko and Nyamplungan. If we consider the macaques as being capable of ranging over the entire 46.19 km² island, the density of macaques on Karimunjawa Island was calculated to be around 5.78 macaques per km². More precisely though, if we attempt

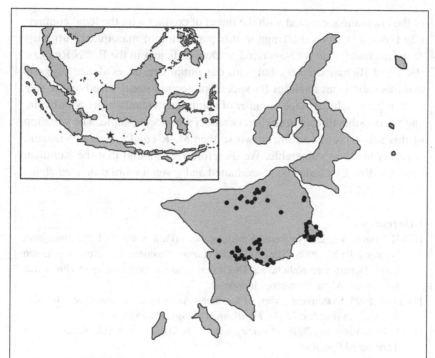

Figure 1.3. The distribution of long-tailed macaques on Karimunjawa Island in Indonesia. Kemujan Island is the northeastern portion of the landmass. The star in the top left inlay shows the location of Karimunjawa. The black dots represent macaque sightings.

to calculate the density of macaques within the range we were able to track them in (23.47 km2), then the actual density in areas they inhabit is 11.46 per km^2. The macaques were observed to range over 50.8 percent of the island, and their density and group size are at the lower end of the spectrum of typical numbers reported in other regions (Fooden, 1995).

Our census only provides numbers from Karimunjawa and therefore it is not possible to count the entire population of these macaques. Despite this, even if the population on the smaller and less forested island of Kemujan were found to have a macaque population of comparable size, the population of this subspecies would still be considered very small. We can therefore safely conclude that the population is well under 1,000 individuals, and quite likely smaller than 500. This population is very small and the entire population overlaps with human activity to some extent. Our questioning of local people found unanimous support that significant conflict was occurring between macaques and humans living around the park. The small size of the population, coupled with the threat of conflict with the local community poses a serious challenge to this population of macaques. Currently, *M. f. karimondjiwae* is categorized as Data Deficient in the IUCN Red List (Ong and Richards, 2009), but with our initial census, evidence suggests that this subspecies qualifies for special protection status based on the size of their geographical range, number of mature individuals in the population, and the possible threats from direct conflict with local people. The situation of this subspecies is as dire or worse than *M. F. condorensis* in Victnam, which is listed as vulnerable. We therefore recommend that the Kemujan macaque Red List status be re-evaluated and given a more protected status from the IUCN.

References

BTNK (Balai Taman Nasional Karimunjawa) (Park Office of Karimunjawa National Park). 2008. Statistik: Balai Taman Nasional Karimunjawa Tahun 2008. Departemen Kehutanan: Direktorat Jenderal Perlingdungan Hutan dan Konservasi Alam. Semarang, Indonesia.

Fooden J. 1995. Systematic review of Southeast Asian long-tail macaques, *Macaca fascicularis* (Raffles, 1821). *Fieldiana: Zoology*, n.s., **81**:v + 206.

Ong P, Richardson M. 2008. *Macaca fascicularis*. IUCN 2010: IUCN Red List of Threatened Species.

variations in patterns of sexual swelling, which have also been observed in Thailand (Malavijitnond *et al.*, 2007a).

The genetic structure of *M. f. fascicularis* varies within and across islands in Indonesia. High levels of genetic variation have been found across the island of Java, and the difference is related to geographical distance (Perwitasari-Farajakkah *et al.*, 1999). Inter-island variation is even more distinct. The Bali macaques are distinct from Javanese and Sumatran macaques, and the variation found between islands has been argued to be the result of genetic drift (Kawamoto and Ischak, 1981; Kawamoto *et al.*, 1984). A small population on Tabuan island between Java and Sumatra has a darker pelage like the fringing island populations, but their genetic variation is still within the range of variation found amongst the lighter-bodied *M. f. fascicularis* of the core regions (Perwitasari-Farajakkah, 2001). Despite all the variation amongst the core subspecies, there has been no well-developed argument for a more refined subspecies classification system to date and we are only beginning to understand the extent of this variation. Recent work on mitochondrial DNA sequences indicates that long-tailed macaques are monophyletic, but that two distinct haplotypes occur, which are separated by oceanic barriers (Blancher *et al.*,

2008). There is an insular clade, constituting the Indonesian and Phillipine macaques, and a continental clade, consisting of the macaques from mainland Southeast Asia.

Ethnophoresy

The close association of long-tailed macaques with humans and their ability to adapt well to differing habitat types has led to the successful colonization of several island habitats beyond their natural range due to ethnophoresy, or the process of an organism being dispersed outside of its range by human transportation (Heinsohn, 2003). Ethnophoresy has caused the colonization of long-tailed macaques on at least five islands in the last several centuries (Figure 1.1) – Mauritius (Sussman and Tattersall, 1981; Sussman and Tattersal, 1986), Ngeaur (a.k.a. Anguar) Island in Palau (Poirier and Smith, 1974; Wheatley *et al.*, 1999), West Papua (Kemp and Burnett, 2003; Kemp and Burnett, 2007), Tinjil Island near Java (Kyes, 1993), and Kabaena island off Sulawesi (Froelich *et al.*, 2003). In addition, humans have also carried long-tailed macaques to the Kowloon Hills of Hong Kong, where they hybridized with rhesus macaques (Southwick and Southwick, 1983; Wong and Ni, 2000). These documented modern cases of colonization provide some evidence that ethnophoresy could have been the cause of earlier pre-historical colonization events of long-tailed macaques to islands east of the Wallace line, such as Lombok, Flores, and East Timur (see Kawamoto *et al.*, 1984 for discussion).

Republic of Mauritius, East Africa

The first recorded event of long-tailed macaques establishing a satellite population on an island outside their range was in Mauritius off the east coast of Africa (Sussman *et al.*, Chapter 8). Portuguese traders from Java or Sumatra probably carried the monkeys during the 1500s along with other forms of exotic wildlife, such as black rats (*Rattus rattus*) and pigs (*Sus scrofa*) (Sussman and Tattersall, 1981; Tosi and Coke, 2007). After this introduction, the macaque population is believed to have exploded and by the early 1600s *M. fascicularis* was well established on the island (Sussman and Tattersall, 1981). Although not documented clearly, the population explosion likely occurred because of a mixture of human-induced habitat destruction and the availability of abundant food resources from the farms of European settlers. An indication that the Mauritian macaques largely relied on the settlers of Mauritius is evident by claims from historians that the early Dutch settlers abandoned Mauritius after suffering severe damage to their plantations and food stocks by macaque raiding (Sussman and Tattersall, 1981).

During the 1980s and 90s the population was estimated to be around 40,000 individuals, but since then macaque numbers have decreased due to trapping and may now be under 10,000 individuals (Sussman et al., Chapter 8). The macaque population on Mauritius has been considered an ecological disaster, and has led to identifying long-tailed macaques as one of the worlds worst invasive species because of the ecological damage and economic loss they are reported to be causing (Lowe et al., 2000). Several authors have claimed macaques have played a predominant role in the decline and extinction of several bird populations on the island (Cheke, 1987; Safford, 1997), including the dodo bird (Quammen, 1996), despite Mauritius having a plethora of other invasive species (Lorence and Sussman, 1986). It has also been claimed that the macaques have helped to disperse the seeds of exotic plants, perpetuating the spread of invasive flora (Cheke, 1987). In addition to creating ecological disturbance, they economically impact Mauritius as a crop pest, and are estimated to be responsible for the loss of up to US $3 million per year in damage to Mauritian farmers (Mungroo and Tezoo, 1999). Despite reports of severe negative impact, Mauritius has significantly prospered from the financial gains of several breeding facilities selling these macaques for biomedical research (Padayatchy, Chapter 9; Maurin-Blanchet, 2006; Stanley, 2003). Therefore, this population is actually the base of a major financial resource to the Mauritian economy.

Ngeaur Island, Republic of Palau

Another location that macaques have colonized through enthnophoresy is Ngeaur Island in the Republic of Palau (Wheatley, Chapter 10). Long-tailed macaques first appeared on Ngeaur Island during German rule in the first decade of the 1900's (Poirer and Smith, 1974; Wheatley, Chapter 10). Although the origin of the monkeys is unclear, the Germans may have brought over several monkeys that were eventually released into the wild. Genetic studies have indicated greater variability than is typical of island populations, such as Bali and Lombok (Matsubayashi et al., 1989), indicating that the monkeys were probably brought from Asia mainland or the Greater Sunda Islands, such as Sumatra or Java, and that there were possibly several repeated introductions. The population has been reported to have fluctuated between 800 and 400 individuals since its introduction, with population losses being the result of eradication efforts by the local communities (Farslow, 1987; Wheatley et al., 1999, 2002). The population has also survived several catastrophic events, including typhoons and WWII bombings on the island (Porrier and Smith, 1974).

Kowloon Hills, Hong Kong Special Administrative Region, China

There are feral groups of poly-specific (i.e., consisting of differing species and their hybrids) monkeys in the Kowloon Hills of Hong Kong, which were released during the 1910s (Southwick and Southwick, 1983). Wong and Ni (2000) have reported that in the 1800s rhesus monkeys were in Hong Kong and thus they are indigenous to the region, but not to the Kowloon Hills. The rhesus macaques were introduced to the Kowloon Hills as the result of an effort to remove a poisonous plant called strychnos (*Strychnos angustiflora* and *S. umbellata*) that had invaded the region after construction of a reservoir system in 1913. During WW II the Kowloon Hills were clear cut and following the war the area was reforested. This presented a highly disturbed habitat along with access to human food resources.

In the early 1950s, a group of five long-tailed macaques were released into the hills and they began to interbreed with the rhesus monkeys already present. In the 1960s, a male and female Tibetan macaque (*M. thibetana*) and their single offspring were released by a group of Chinese acrobats. The Tibetan macaques integrated into the group, but were never observed to interbreed with the rhesus, long-tails, or their hybrids, and the last Tibetan macaque was reported to have died in 1995. Additionally there have been reports of pig-tailed (*M. nemestrina*), Japanese (*M. fuscata*), and possibly Taiwan (*M. cyclopis*) macaques living in these hills (Southwick and Manry, 1987; Burton and Chan, 1989; Burton *et al.*, 1999), but these claims were not reported in later work by Wong and Ni (2000). Long-tailed macaques can produce reproductively viable hybrids with rhesus macaques (Bernstein, 1966; Fooden, 1964), which is why they were able to easily hybridize into this population.

The population remained very small until recently, as there were only around 100 monkeys in the Kowloon Hills in the 1980s (Southwick and Manry, 1987; Southwick and Southwick, 1983). Eventually the Kowloon monkeys became a popular tourist attraction and humans began provisioning the monkeys more and more, which may have been the driver for an increase in their population size. In 1991, census reports indicated the population had increased six-fold to ~600 individuals (Fellowes, 1992). The population increase continued and in 1994 Wong and Ni (2000) observed that the monkey population of Kowloon had risen to ~700 macaques with a composition of 65.3 percent rhesus, 2.2 percent long-tailed, and 32.3 percent rhesus-long-tailed hybrids. The population has continued to grow to around 2,000 individuals and has become a large concern to the Agriculture, Fisheries, and Conservation Department (AFCD) of Hong Kong. They have since begun taking organized measures to manage the population by controlling the level of feeding and initiating a contraception program (Wong and Chow, 2004).

West Papua, Indonesia

M. fascicularis has colonized a small region in West Papua, Indonesia, near the city of Jayapura, (Kemp and Burnett, 2003; 2007). The monkeys have been there for ~30–100 years, but their population has not expanded. In fact, the population consists of only six groups of ~10 macaques each, yielding a total of 60–70 monkeys (Kemp and Burnett, 2003; 2007). They were first reported there in the 1980s, but their introduction may have occurred as early as 1910 during the Dutch administration or they may have been carried by foreigners in the mid 1900s between 1941 and the 1970s (Kemp and Burnett, 2003). Several theories exist for how long-tailed macaques were introduced to Papua. They may have been carried by the armed forces during WW II or possibly were brought during transmigration settlement that moved Indonesian people from Java to Papua (see Kemp and Burnett, 2003 for a list of plausible ideas). However they were carried, it was enthnophoresy. The macaque population on Papua has not become invasive, as Papua seems to be highly resistant to the effects of several typically invasive species that have been introduced to the island (Heinsohn, 2003). The Papua macaques present a case for investigating why this population has not expanded like on Mauritius.

Kabaena Island, Sulawesi, Indonesia

There are general reports that long-tailed macaques have been successfully introduced to Sulawesi (Long, 2003), but there is very little specific information about exotic *M. fascicularis* on this island. It is clearly documented that pet long-tailed macaques exist on the mainland of Sulawesi (Jones-Engel *et al.*, 2004), but free-ranging long-tailed macaques have only been clearly reported to occur on Kabaena Island, a small isle off the southwest coast of the southeastern arm of Sulawesi (Froehlich *et al.*, 2003). Little is known about this population or its introduction, but it is reported that this feral population shows distinct morphological differences from other *M. fascicularis* populations. The females are dwarfed and the males are larger than average, making them more sexually dimorphic than typical (Froehlich *et al.*, 2003). Given the paucity of data on the Kabaena long-tailed macaques, we have little information on their impact. It is reported that the macaques have not overrun Kabaena Island and only range on 25 percent of the island (Froehlich *et al.*, 2003), unlike Mauritius or Ngeaur. Moreover, the unique morphology of these macaques suggests they have been on the island for a long period of time (Froehlich *et al.*, 2003) and thus may not be invasive here.

Tinjil Island, West Java, Indonesia

During 1988 through 1991, 58 male and 420 female wild-captured long-tailed macaques were intentionally released from Sumatra and Java to start a natural habitat breeding facility on Tinjil Island, located off the south coast of Banten Province, Java (Kyes, 1993; Santoso and Winarano, 1992), with the intention of managing a sustainable population (Kyes *et al.*, 1998; Crockett *et al.*, 1996). Since long-tailed macaques have become the predominant primate exported for research purpose (USFW, 2006, see Box 1.2), the breeding colony was started to assist in supplying this demand, while limiting the strain on natural populations caused by capturing wild macaques. From 1970, about 15,000 long-tailed macaques per year were being exported from Indonesia alone (Santosa, 1996; Djuwantoko *et al.*, 1993). When macaques are taken at these large numbers populations can dramatically plummet, as shown by the 90 percent decline in rhesus monkeys through the 1960s and 70s in India (Southwick and Siddiqi, 1994 and Chapter 11). Regulations were passed in 1994 by the Indonesian Forestry Department restricting export of wild-caught primates, and therefore Tinjil became a major source of legal exports for biomedical lab primates out of Indonesia (Pamungkasand Sajuthi, 2003). Furthermore, the colony also allowed for breeders to control infections carried by macaques, which are safer to use in biomedical research than wild-caught animals (Pamungkas *et al.*, 1994). The model was an attempt to make a more environmental-friendly and healthy alternative to the practice of trapping wild primates for the unstoppable demand of laboratory primates.

Since their initial release, the macaque-breeding colony has been provisioned, closely monitored, and managed. In 1990, a survey showed the macaque population was well established and had grown to 760 individuals (Santoso and Winarno, 1992). By 1997, the population was reported to have grown to 1,550 monkeys in eighteen to twenty groups (Kyes *et al.*, 1998). At this time, the island population consisted of very large groups (i.e., 78–86 individuals per group), that was later reduced to 30–40 individuals per group in recent years (Kyes, personal communication). The population is sustained at this manageable level by periodically harvesting offspring for use in biomedical research. Managers remove enough monkeys to prevent overpopulation, but don't remove so many macaques as to disrupt the sustainability of the

---

Box 1.2   Trade in long-tailed macaques (*Macaca fascicularis*)

Kaitlyn-Elizabeth Foley and Chris R. Shepherd

Long-tailed macaques (*Macaca fascicularis*) are heavily traded to supply the demand for biomedical research, food and the pet trade. Presently little is known of the impact this trade has on wild populations. Although population numbers of long-tailed macaques are considered robust throughout their distribution in Southeast Asia, it is a concern that over-harvesting and unregulated trade is leading to population declines. Currently long-tailed macaques are listed as being of Least Concern on the International Union for the Conservation of Nature and Natural Resources (IUCN) Red List of Threatened Species (Ong and Richardson, 2008), which determines the status and conservation needs of a wide variety of species.

There are thirteen macaque species found in Southeast Asia, all of which are listed under Appendix II of the Convention on the Trade of Endangered Species of Wild Fauna and Flora (CITES). Trade of CITES Appendix-II listed species requires an export or re-export permit, stating their origin, and must be in accordance with any national legislation that is in place. This allows countries to control and regulate the trade, and ensure that trade is not a threat to the conservation of the species. Accurate numbers of individuals either wild caught or captive bred in trade are unknown due to illegal trafficking and laundering. It is suspected that many breeding farms export wild-caught animals laundered as captive bred as this is less expensive. Laundering is suspected in all *M. fascicularis* range states, but particularly so in major exporting countries: China, Indonesia, Vietnam and the Philippines.

We evaluated known information on long-tailed macaque trade by collecting data on international trade from the UNEP-WCMC CITES database (www.unep-wcmc.org/citestrade, accessed on 1 February 2010) for 2004–8. At the time of analyses, trade information for 2009 and 2010 was unavailable. The WCMC CITES trade database maintains all records of import and export of CITES-listed species as reported to the CITES Secretariat by participating countries. It is important to note however, that this data is not representative of the entire trade of long-tailed macaques as illegal trade is not documented.

The WCMC CITES trade database shows that the largest exporters of long-tailed macaques are China, Indonesia, Mauritius, Philippines and Vietnam; while the main importers are US, EU and China. Sources of macaques were observed to be from mainly "captive bred," "F1," and "wild-caught" origins (Table 1.3). Over five years, the US, the largest importer of long-tailed macaques, imported over one hundred thousand live macaques

Table 1.3. *Major traders of M. fascicularis with total number of individuals and declared sources (2004–2008).*

| Exporter | C | W | Source | | | |
			F	I	U	R
China	11,585.1	x	1,236	662	20	x
Indonesia	4,350	2,080	10,100	350	x	250
Mauritius	696	10,485.1	58,178.5	386	x	x
Philippines	107,398.1	1,760	x	153	x	x
Vietnam	33,744	32,774	1,670	x	x	x

C-captive bred, W-wild caught, F-f1 or subsequent generations, I-confiscated/seized, U-unknown, R-ranched.

Table 1.4. *Numbers of M. fascicularis imported into the US from 2004–2008.*

Year	Individuals
2004	16,844
2005	24,629
2006	25,878
2007	24,112
2008	26,509

comprising more than 45 percent of the total international trade between 2004 and 2008) (Table 1.4). In addition the US imported large quantities of derivatives, bones and specimens of long-tailed macaques (Table 1.5). In a report by Fooden (2006) it is suggested that there were approximately three million long-tailed macaques in Southeast Asia. If these population estimates are remotely accurate, the trade levels seen from 2004–2008 are extremely unsustainable.

Long-tailed macaques are considered common in Southeast Asia, but local declines have recently been reported by conservation organizations and macaque hunters alike. For example, in the Cambodian province Ratanakiri farmers reported that no macaques were harvested due to drastic population declines (Eudey, 2008). Levels of capture and export are considered to be unsustainable and acquisition is often carried out in violation of national laws and CITES. As such, investigation and population censusing is urgently needed to determine the actual numbers of macaques being harvested from the wild and the quantity being laundered into the international

Table 1.5. *US imports of M. fascicularis by import term.*

Bodies	Bones	Derivatives	Extract	Live	Skin pieces	Skulls	Specimens	2000
1	144			10970		2	19753	2001
	38			13559			89892	2002
		40		14308			70377	2003
1		40		14679		2	49812	2004
	165			16844	60		353955	2005
1	35	350		24629			253337	2006
305	16	15		25878			113268	2007
				24112		1	161427	2008
			20	26509		1	187185	

trade under the guise of captive breeding. Operations which are determined to be illegally catching and trading macaques need to be penalized to the full extent of the law to serve as a deterrent. Moreover, those utilizing macaques need to be made aware that they are potentially consuming a significant portion of the population.

The World Health Organization and the Ecosystem Conservation Group recommends that wild-caught primates should only be used to establish self-sustaining captive breeding colonies (WHO, 1971). Whilst a facility might only obtain animals which have been declared as captive-bred, it is possible that some individuals could have been harvested from the wild. The use of animals of unknown pedigree and disease history as study subjects could undermine the validity of any research conducted on them.

Export countries, still allowing the capture of wild macaques must adhere to Article IV of CITES which requires that a non-detriment finding (NDF) survey be conducted prior to allowing capture for export. Data on wild populations are needed to ensure that harvest is not detrimental to the conservation of this species in the wild. Countries need to ensure that adequate laws are in place and the enforcement of these is essential. CITES is a tool to be used to ensure that trade is not a threat to the conservation of wildlife. Unless this tool is effectively used and enforced it is completely ineffective.

Importing countries should ensure that the import of macaques is conducted in accordance with national legislation in source countries and adheres to CITES regulations. Exporting countries need to work together to ensure their utilization of macaques does not damage natural populations of long-tailed macaques. Illegal and unregulated trade has potentially caused population declines in long-tailed macaques, particularly in Cambodia (Eudey, 2008). Efforts will be needed to ensure that a high volume of uncontrolled trade does not decimate Southeast Asian long-tailed macaque populations.

References

Eudey, M. (2008) The crab-eating macaque (*Macaca fascicularis*): Widespread and rapidly declining. *Primate Conservation*, **23**: 129–132.

Fooden, J. (2006) Comparative review of Fascicularis-group species of Macaques (primates: *Macaca*). *Fieldiana Zoology*, **107**: 1–34.

International Primatological Society (2007) IPS International Guidelines for the Acquisition, Care and Breeding of Nonhuman Primates 2nd Edition. IPS, USA. UNEP-WCMC CITES Trade Database. www.unep-wcmc.org/citestrade, accessed 1 February 2010.

Ong, P and Richardson, M. (2008) *Macaca fuscicularis*. In: IUCN 2008. IUCN Red List of Threatened Species.

World Health Organization (1971) Health Aspects of the Supply and Use of Non-Human Primates for Bio-medical Purposes. Technical Report Series No. 470. World Health Organization, Geneva.

population (Crockett *et al.*, 1996; Kyes *et al.*, 1998). Between 1991 and 1997 there were thirteen harvests and 707 young macaques between the age of one and three were removed from the population (Kyes *et al.*, 1998). These numbers are far below world demand, and current evidence suggests many illegal long-tailed macaques are being brought into legal trading markets from places like Cambodia and Laos (Eudey, 2008; Hamada *et al.*, Chapter 3, see Box 1.2 and 3.1). The Tinjil system provides a more sustainable way of supplying invasive research needs than uncontrolled wild harvesting. Despite this, the system fails to produce enough individuals to meet demand and falls into a legal gray area for trading regulations (i.e., are they wild-caught or captive bred).

The human-macaque interface

Since long-tailed macaques are widely distributed across Southeast Asia and are well adapted for living in human-modified environments (Richard *et al.*, 1989; Wheatley, 1999), they are ecologically associated with humans (i.e., synanthropic) in several different types of environments across a wide geographical range (Aggimarangsee, 1992; Fuentes *et al.*, 2008; Malaivijitnond and Hamada, 2008) (Table 1.2). These regions are referred to as macaque–human interface zones (Patterson, 2005). The relationship between long-tailed macaques and humans in these zones has mostly been referred to as commensal (Wheatley, 1999), meaning the monkeys benefit from the relationship without benefiting or harming their human counterparts. This is true in many cases, however this relationship also has mutual and parasitic components, and therefore cannot simply be viewed as a benign commensalism that solely benefits the synanthrope while leaving humans unaffected (Fuentes, 2006). Rather, there is a whole spectrum of costs and benefits affecting sympatric humans and macaques that consist of competitive, mutual, and neutral components.

There are several positive impacts that long-tailed macaques have on humans that may be considered mutualistic. In many regions, long-tailed macaques are culturally and religiously important to people, particularly where they live in close proximity to Hindu and Buddhist temples throughout Southeast Asia (Aggimarangsee, 1992; Eudey, 1994; Fuentes *et al.*, 2005; Jones-Engel *et al.*, 2002; Wheatley, 1999). Religion plays a role in community cohesion and support in human societies (Atran 2002), and in such communities macaques are peripherally associated to this important institution. A supernatural-based relationship also occurs on some riverine islands in Kalimantan, where local people believe the monkeys contribute to their good fortune (Gumert, 2004). In regions where macaques are associated to religious systems, they often are integrated into festivals and celebrations, which again are important factors in community cohesion and relationship building (Malaivijitnond, *et al.*, Chapter 5). In addition to having a peripheral association to core religious or spiritual beliefs in some communities, monkeys are also associated with the acquisition of resources. In some communities macaques are a significant source of revenue as a tourist attraction. Examples of this are at temple grounds in Bali (Fuentes, 2006; Wheatley and Putra, 1994b; Wheatley and Putra, 1995), Lombok (Hadi, 2005), and Thailand (Malavijitnond *et al.*, 2005, 2008, Chapter 5).

Another benefit for humans from long-tailed macaques is their use as research animals in the bio-medical and technological sciences worldwide (Carlsson *et al.*, 2004; Hagelin, 2004; Kyes *et al.*, 1998). Trends in trade suggest that long-tailed macaques have been and continue to be one of the most used lab primates and constitute greater than 80 percent of the current importation of lab primates in developed nations (MacKinnon, 1986; Djuwantoko *et al.*, 1993; USFWS, 2009, see Box 1.2). In the 70s and 80s world demand reached 35,000 animals per year (Santosa, 1996). Use in research provides one of the greatest benefits any animal has to offer humankind because numerous advancements have been achieved through their use that have aided in the alleviation of human suffering through the development of new medicines and technologies. Despite this significant and costly contribution to humanity by long-tailed macaques, their preferred trade as a major bio-medical model has yet to be mirrored by a similar level of interest or support to the issues facing their natural populations.

Long-tailed macaques have several negative impacts on humans. Macaques compete with humans for food and space, becoming crop-raiders and urban pests, and can be threatening to humans in their proximity. Macaques can parasitize human resources, damage dwellings, and access refuse, causing the spread of debris within a community (Zain *et al.*, Chapter 4). They also occasionally threaten or cause harm to humans by acting aggressively. Aggressive macaque-to-human behavior is mainly the result of direct competition over contestable food sources (Fuentes 2006; Fuentes and Gamerl, 2005; Fuentes *et al.*, 2008; Sha *et al.*, 2009b; Wheatly and Putra, 1994a), but can also occur due to human prompting (i.e., teasing, chasing, approaching to close, etc.), defense of young, and interference with mating activity (McCarthy *et al.*, 2009; Gumert *et al.*, 2009). A potential negative consequence is the spread of infections, which has become a special concern because of the close biological similarity of macaques and humans (Brown, 1997; Jones-Engel *et al.*, Chapter 12).

Human populations significantly affect macaques in negative ways. Human-caused habitat alteration can attract macaques to human settlements and humans are highly interested in feeding macaques, which cause behavioral changes in them. Human-fed macaques learn over time to venture further into human settlement and lose their fear of being in close proximity and contact of people (i.e., loss of flight distance). Sustained feeding by humans can also alter the reproductive success of macaques, by lowering infant and juvenile mortality and thus increasing population growth (Fuentes *et al.*, Chapter 6). In contrast, humans can also limit population growth because they are dangerous to macaques and can harm or kill them through their activity. Humans directly affect the mortality of macaques by hunting, trapping (Louden *et al.*, 2006), culling, and removal of macaques (Di Silva, 2008; Sha *et al.*, 2009a, 2009b; Wheatley and Putra, 1994a), and indirectly through human activity and edifices, such as automobile traffic, (Wong, 1994), electrical wiring, and water tanks. Lastly, contraction of infectious agents by macaques due to human activity, can cause disease, fatality (Jones-Engel *et al.*, 2006) and population collapse (Wheatley, 1999, in this case transmitted from human-kept livestock).

Areas of macaque–human interface range from highly competitive to mutualistic. In regions of great competition, human-macaque conflict is very high causing stress to both species, and generating the general sense that solutions are needed to ameliorate the situation. Consequently, strategies and management plans need to be developed and enacted in order to remedy some of the problems occurring in human-macaque interface zones throughout Southeast Asia. Moreover, effort will need to be taken by wildlife management officials to both protect macaques and humans from the negative consequences of their interface. Solutions will vary depending on the context of the situation, as the macaque–human interface occurs across a wide variety of differing circumstances (Table 1.2).

Types of human-macaque interface zones

Temples and religious sites

The most well known cases of human-macaque interface are at Hindu and Buddhist temples in Indonesia and Thailand (Aggimarangsee, 1992; Eudey, 1994; Hadi, 2005; Fuentes, 2006; Fuentes *et al.*, Chapter 6; Fuentes *et al.*, 2005; Jones-Engel *et al.*, 2002; Malavijitnond *et al.*, 2005; Malaivijitnond and Hamada, 2008; Wheatley, 1999; Wheatley and Putra, 1994b) (Figure 1.4). In Thailand, around 40 temple or shrine areas have been reported to have semi-tame long-tailed macaques (Eudey, 1994; Aggimarangsee, 1992; Malaivijitnond *et al.*, 2005; Malavijitnond and Hamada, 2008), the most famous being Lopburi, or the "City of Monkeys" (Malaivijitnond *et al.*, Chapter 5). The monkey forests of Ubud and Sangeh are well known as tourist spots in Bali, Indonesia, and there are apparently 43 locations in Bali where macaques are associated with temples (Fuentes *et al.*, 2005, Chapter 6). There are also monkey temples in Lombok, Indonesia (Hadi, 2005) and in other countries such as Cambodia, Vietnam and Myanmar (San and Hamada, Chapter 2).

At commercialized (i.e., areas promoted for tourism) monkey forest temples, like at Lopburi, Sangeh, and Ubud, tourists from all over the world visit and feed long-tailed macaques on a daily basis. The benefit of tourism has generated a mutualistic relationship between the local communities and the monkeys. The people feed and attract macaques to the site, while the macaques are responsible for bringing in tourists that provide revenue to the local community (Wheatley and Putra, 1994b; Wheatley and Putra, 1995). Local people make a living as photographers, tour guides, and merchants. The goods and services provided include food for feeding the monkeys, photographs of tourists with monkeys, guided tours, protection from aggressive monkeys, retrieval of items taken by monkeys, local artwork, souvenirs, and refreshments. In Lopburi, annual celebrations occur for the monkeys that bring in large crowds of people and sponsors. Moreover, the local tourist industry utilizes the macaques as their main attraction, having monkey hotels, monkey restaurants, a monkey-themed train station, and even a locally-brewed monkey beer.

The relationship between long-tailed macaques and people at temple sites is not all mutualistic. Macaques are conditioned to see humans as food sources and

Figure 1.4. A common place of human-macaque interface occurs at Hindu and Buddhist temples throughout Southeast Asia. (a) This group of long-tailed macaques walks in the streets near a Buddhist stoopa; (b) in downtown Lopburi, Thailand.

therefore visitors become highly attractive to the macaques. A serious consequence of this is that macaques frequently display aggression, which can result
in bites and scratches (Fuentes, 2006; Fuentes and Gamerl, 2005; Engel *et al.*,
2002). The violence is reciprocal, as humans also show high levels of aggression towards macaques and monkeys, and this can be physically damaging. For
example, some macaques have lost an eye from sling shot pellets or have damaged tails and limbs (Gumert, pers. obs.). Monkeys are fed and taken care of on
temple grounds, but when they move off these grounds they are harassed, hunted,
and killed, generally for pest-related reasons or for food (Aggimarangsee, 1992;
Eudey, 1994). This distinction is formalized legally in Thailand in section 25 of
the 1960 Wildlife Conservation Act, stating that no one can kill or harm animals
at any religious sites, but offers no protection for such animals when they move
outside these regions (Aggimarangsee, 1992). Research shows this is also the
case in Bali. Gunshot pellets were found in the bodies of more than a third of a
sample of temple monkeys X-rayed (Jones-Engel *et al.*, 2002; Schillaci *et al.*,
2010), and in many sympatric regions of Bali they are hunted, eaten and captured
for trade (Louden *et al.*, 2006). The reality is that no formalized legal protection
is offered to long-tailed macaques, and their good fortune while at temple sites is
largely due to the sanctity of the real estate.

Reserves and recreation parks

The macaque-human interface also occurs at recreation parks and nature
reserves, such as in Pangandaran, West Java, (Engelhardt, 1997) where
they are habituated and interact with tourists. They are also found in similar forest parks throughout Thailand (Aggimarangsee, 1992; Malaivijitnond
et al., 2005). In Singapore, macaques frequent several nature parks, including Bukit Timah and MacRitchie Reservoir (Sha *et al.*, 2009a; Sha *et al.*,
2009b). This type of interface occurs all over Southeast Asia and represents
a common interface for macaques and humans. In park and reserve settings,
macaques snatch food from people picnicking in the parks and can be threatening to visitors. People tend to enjoy feeding the macaques in this context
and appear interested and frequently stop to photograph them. The severity
of conflict in these setting ranges greatly.

Monkey islands

Sometimes, dense long-tailed macaque populations are found on small
riverine islands near large human settlements, which are likely the result of

historical translocation efforts to remove the macaques as these settlements developed. Two such cases occur in Kalimantan on Lampahen Island in the Little Kapuas River of Central Kalimantan and Kembang Island in the Barito River of South Kalimantan (Mackinnon *et al.*, 1996; Gumert, 2004). Local legends portray these macaques as the spirits descended from corrupted humans that were abandoned to these islands for bad behavior. According to this folklore, they bring good fortune to the local community if treated kindly. Kembang Island is a tourist location for travelers to Banjarmasin, but it is not commercialized like some of the Thailand or Bali temple locations. People that visit the island provide food to the macaques, which mob the food bearers and assertively take their food (Gumert, per. obs.), and will also chase people onto docked boats in search of food (Southwick, per. comm.). Lampahen Island, on the other hand, is not influenced from tourism and thus modern tourism has not disrupted a traditional macaque-human sympatry. The human interaction that the monkeys receive occurs with the local community. They provide offerings of food to the monkeys after times of good fortune and leave ceremonial flags on the island for each offering they provide (Gumert, 2004) (Figure 1.5).

Human settlements

Long-tailed macaques interface with people along and within human settlements, including villages, towns, and even metropolitan cities. Macaques can be found free-ranging in urban settings, such as Singapore (Fuentes *et al.*, 2008; Jones-Engel *et al.*, 2006; Lee and Chan, see Box 12.2; Sha *et al.*, 2009a; Sha *et al.*, 2009b), Hong Kong (Fellowes, 1992; Southwick and Southwick, 1983; Wong, 1994; Wong and Ni, 2000), and Kuala Lumpur (Chiew, 2007; Di Silva, 2008; Twigg and Nijmin, 2008; Zain *et al.*, Chapter 4). Smaller towns in Thailand, such as Loburi and Petchaburi have macaques that inhabit the downtown regions and walk along the streets as freely as the people living there (Malaivijitnond *et al.*, 2005; Malaivijitnond and Hamada, 2008; Malavijitnond *et al.*, Chapter 5). Numerous towns and villages across Southeast Asia have macaques living on their fringes and thus experience varying levels of interface with them. These situations present threats to humans and macaques and in some villages, such as Alas Dowo in Jawa Tengah, Indonesia, a local macaque pet owner reported very recent disappearance of long-tailed macaques in the area over the last decade. In addition, villages such as Anjir in Kalimantan Tengah have recently initiated systematic trapping efforts to remove macaques from their farmlands (Gumert, per. obs.). Such reports raise the question of whether macaques interfacing with farm-based villages are

Figure 1.5. Long-tailed macaques have been placed on riverine islands in Kalimantan where the local people consider them sacred. Ceremonial flags can be seen in the background as tokens left for the monkeys in a belief that kind treatment of these macaques will bring good fortune to the local communities.

facing population-threatening pressures, as modernization increases the ability of villagers to exterminate monkey pests.

Where macaques live in close proximity to human settlements, their diets are supplemented to varying degrees from the food sources produced from the settlement (Wheatley *et al.*, 1996). In towns and cities, macaques will raid homes, feed off refuse, and wait along roads for handouts from cars. Near villages, macaques also raid homes and trash heaps, as well as farms and gardens. To help alleviate conflict, people in some sites build barriers and use fencing or caging to block access by the monkeys into their homes and yards (Figure 1.6). Other people use different forms of scare tactics such as sticks, slingshots, or firecrackers, and sometimes macaques are culled or removed at times when conflict becomes too great for the affected community (Di Silva, 2008; Sha *et al.*, 2009a; Sha *et al.*, 2009b; Twigg and Nijman, 2008; Wheatley and Putra, 1994a). Macaques also face other dangers specific to human landscapes. They are at risk of human aggression, they move around and utilize electrical wiring for travel, and frequently move on roads (Figure 1.7), presenting the risk of injury and fatality. In Singapore and Hong Kong, automobile accidents represent a large proportion of mortality (Sha *et al.*, 2009a; Wong, 1994).

Eco-tourist lodges

Long-tailed macaques have close contact with humans around some eco-tourist facilities. A case example of this occurs at Tanjung Puting National Park, in Kalimantan where a group of ~50 long-tailed macaques live around the Rimba Orangutan Eco-lodge (Gumert, 2007). These macaques used to take food from the waste piles behind the lodge (Figure 1.8a) and also scavenged around the lodge looking for food (Figure 1.8b), particularly focusing in the kitchen area. They were also often observed waiting above the kitchen each morning, when leftover food was thrown out and while lodge cooks prepared food for the day. In the afternoons, the monkeys often foraged through the lodge searching for food and litter left out by the hotel staff and guests, or doors and windows left open that provided access to food stores. Conflict mainly occurred between the macaques and the staff permanently living at the facility, and only on a few occasions were the tourists disturbed or threatened by them.

Pets and performing macaques

The interface that long-tailed macaques have with humans has led to the problem of a high abundance of macaques being taken into captivity (Eudey, 1994;

Figure 1.6. In towns and cities, caging and fencing are sometimes used to keep long-tailed macaques out of houses, apartments, and shops.

Figure 1.7. Urban long-tailed macaques sometimes congregate along roadsides and cross streets.

Jones-Engel *et al.*, 2004; Schillaci *et al.*, 2005; Schillaci *et al.*, 2006), with long-tailed macaques often representing the most common type of primate found in the animal trade and pet markets (Malone *et al.*, 2003). Pet macaques are caged and chained, and sometimes developed into performing monkeys to bring their owners revenue. However, after they begin to sexually mature they are not good pets, and most of them end up with nowhere to go, leaving wildlife and rescue centers in Southeast Asia to deal with these surplus animals. For example, during 2004 in Indonesia, numerous long-tailed macaques were contained in the animal rescues centers across the country (Smits, personal communication). These macaques had been confiscated by wildlife authorities or given up by pet owners. Because there are few specialized facilities to contain them or standard release programs, facilities are filling up because there are just too many of them.

Agricultural areas and crop-raiding

Macaques are commonly found around agricultural lands and crop-raiding can occur in these zones. Although not much research has been done directly on *M. fascicularis* as a crop raider in their natural range, there have been several reports of raiding in Sumatra, Borneo, Bali, and the Nicobar Islands

Figure 1.8. (a) A group of long-tailed macaques feed in a trash pile behind an eco-tourist lodge in Kalimantan, Indonesia. (b) Macaques turn over an unprotected trash bin in search of food scraps.

(Das and Ghosal, 1977; Fuentes, 2006; Linkie *et al.*, 2007; Richard *et al.*, 1989; Sugardjito *et al.*, 1989; Umapathy *et al.*, 2003;, 1989; 1999). In Bali, macaques frequently raid crops, such as rice, sweet potatoes, cassava, taro, coconuts, bananas, and salak, around temple areas (Wheatley, 1989, 1999; Louden *et al.*, 2006; Fuentes, 2006). In the Nicobar Islands of India, these monkeys raid crops to such a degree that the local people set traps and use dogs to kill the monkeys (Umapathy *et al.*, 2003).

These reports indicate long-tailed macaques are crop raiders, but the actual degree of crop raiding and crop damage they perform has yet to be fully investigated. One study at Kerinci Seblat National Park, Southern Sumatra partially addressed the impact of long-tailed macaques as crop-raiders along with eleven other large mammals (i.e., muntjac (*Muntiacus muntjak*), banded langurs (*Presbytis melalophos*), pig-tailed macaques (*Macaca nemestrina*), porcupines (*Hystrix brachyuran*), and wild pigs (*Sus scrofa*). Long-tailed macaques were not amongst the most serious crop pests in this sample. (Linkie *et al.*, 2007). Perhaps, under less human-altered conditions, long-tailed macaques may not be the most prolific crop raiders, which contrasts to the largely human-affected locations around temples in Bali where macaques are serious crop pests (Wheatley, 1989; Wheatley *et al.*, 1996).

A questionnaire survey conducted in villages just outside Gunung Leuser National Park in North Sumatra showed that primates were the most destructive crop-raiders among vertebrate crop-raiders, with long-tailed macaques being perceived by villagers to be the largest problem, especially for fruit crops (i.e., other primates included orangutans, langurs, and pig-tailed macaques, and other vertebrates included elephants, pigs, deer, porcupines and squirrels) (Marchal and Hill, 2009). It is possible that in places like Gunung Leuser we are seeing the effects of the early stages of human development along the borders of large tracts of forest on wildlife. As more food sources become available, wildlife, especially macaques, can become more accustomed and dependent on the settlement. Perhaps as human development continues in undeveloped forest regions we will observe the development of closer, more proximate human-macaque interfaces.

Conclusions

Long-tailed macaques are one of the world's most populous primates and extend over one of the largest geographical ranges of any primate. They are also an edge species, and prefer to be on the periphery of forest environments. As a result, these monkeys overlap with humans in many locations and thus affect the lives of millions of people. They inhabit most habitat types across Southeast Asia and they have been carried beyond their original geographic distribution over the last several centuries. Population estimates are unclear, but they appear to show a declining number in natural habitats (i.e., where there is no or little human influence) and potentially an increase along human settlements. Since population status is not well known, better knowledge of wild macaque populations will be necessary, as we need to ensure that populations are not damaged by uncontrolled exploitation from their use in the animal trade and pest removal programs. A species like this will require sophisticated organizations to handle and manage their population, similar to the game commissions that manage ungulates and other common large mammals in many developed nations. It is interesting to speculate why an animal that is so conspicuous and has had such a large impact on humanity still remains so poorly understood and unmanaged. Nonetheless, it is clear we can no longer ignore this species' large impact, close relationship and importance to humanity, and during the 21st century, we will likely need to consider them, not as an unimportant pest, but rather as one of the major foci for wildlife management efforts across Southeast Asia.

Acknowledgements

I would like to give special thanks to those participating in the *Macaca fascicularis* workshop at the 30th Meeting of American Society of Primatologist in 2007 and also to those in attendance at the pre-congress workshop and round-table meeting at the 22nd Congress of the International Primatological Society. Discussion with these participants formed the foundation for the writing of this chapter. Research used for the preparation of this paper has been supported by a Fulbright Graduate Fellowship, a Howard-Hughes Postdoctoral Fellowship at Hiram Collage, the HSS Staff Development Fund, and Tier 1 Grant RG 95/07 from the Ministry of Education, Singapore.

References

Aggimarangsee, N. 1992. Survey for semi-tame colonies of macaques in Thailand. *Natural History Bulletin of the Siam Society* **40**: 103–166.
Aimi, M., Bakar, A. and Supriatna, J. 1982. Morphological variation of the crab-eating macaque, *Macaca fascicularis* (Raffles, 1821), in Indonesia. *Kyoto University Overseas Research Report of Studies on Asian Non-Human Primates* **2**: 51–56.
Anonymous. 1993. Wildlife Protection Act 1972 (amended 1993). Government of India, New Dehli.
Atran, S. 2002. *In Gods We Trust: The Evolutionary Landscape of Religion*. Oxford University Press.

Bernstein, I. S. 1966. Naturally occurring primate hybrid. *Science* **154**: 1559–1560.

Bismark, M. 1991. Analisis populasi monyet ekor panjang (*Macaca fascicularis*) pada beberapa tipe habitat hutan. *Buletin Penelitian Hutan* **532**: 1–9.

(1992). Peranan hutan di luar kawasan pelestarian alam dalam konservasi populasi *Macaca fascicularis. Buletin Penelitian Hutan* **549**: 9–18.

Brown, D. W. G. 1997. Threat to humans from virus infections of non-human primates. *Reviews in Medical Virology* **7**: 239–246.

Blancher, A., Bonhomme, M., Crousu-Roy, B., *et al.* 2008. Mitochondrial DNA sequence phylogeny of 4 populations of the widely distributed cynomolgus macaque (*Macaca fascicularis fascicularis*). *Journal of Heredity*, **99**: 254–264.

Burton, F. and Chan, L. 1989. Congenital limb malformations in the free-ranging macaques of Kowloon Hong Kong. *Journal of Medical Primatology* **18**: 397–404.

Burton, F., Bolton, K. and Campbell, V. 1999. Soil-eating behavior of the hybrid macaques of Kowloon. *NAHSON Bulletin* **9**: 14–20.

Carlsson, H.E., Schapiro, S.J., Farah, I., and Hau, J. 2004. Use of primates in research: a global overview. *American Journal of Primatology* **63**: 225–237.

Carpenter, A. 1887. Monkeys opening oysters. *Nature* **36**: 53.

Cheke, A. 1987. An ecological history of the Mascarene Islands, with particular reference to extinction and introduction of land vertebrates. In *Studies of Mascarene Island Birds*. Cambridge University Press. pp. 5–89.

Chiang, M. 1967. Use of tools by wild macaque monkeys in Singapore. *Nature* **214**: 1258–1259.

Chiew, H. 2007. Reimpose export ban on macaque, urges animal rights groups. *The Star*. http://thestar.com.my/lifestyle/story.asp?file=/2007/9/11/lifefocus/20070911 081410&sec=lifefocus

Chivers, D. J. and Davies, A. G. 1978. Abundance of primates in the Krau Game Reserve, Peninsular Malaysia. In *The Abundance of Animals in Malesian Rain Forest*, G. Marshall (ed.). Misc PT. Series no 22, Department of Geography, University of Hull (Aberdeen-Hull Symposium on Malesia Ecology). pp. 9–32.

Crockett C, and Wilson, W. 1980. The ecological separation of *Macaca nemestrina* and *Macaca fascicularis* in Sumatra. In *The Macaques: Studies in Ecology, Behavior and Evolution*. New York, NY: Van Nostrand Reinhold. 148–181.

Crockett, C. M., Kyes, R. C., and Sajuthi, D. 1996. Modeling managed monkey populations: Sustainable harvest of long-tailed macaques on a natural habitat island. *American Journal of Primatology* **40**: 343–360.

Das, P. and Ghosal, D. 1977. Notes on the Nicobar crab-eating macaque. *Newsletter of the Zoological Survey of India* **3**: 264–267.

Di Silva, R. 2008. SpotLight: Culling solution to macaque 'explosion'. *New Straits Times*. 11 February.

Djuwantoko, Thojib, A., and Hasanbahru, S. 1993. Ekologi perilaku kera ekor panjang (*Macaca fascicularis* Raffles, 1821) di hutan tanaman jati. Laparon Penilitian. Yjojakarta, Indonesia: Fakultas Kehutanan, Universitas Gadjah Mada.

Engel, G.A., Jones-Engel, L., Schillaci, M.A., *et al.* 2002. Human exposure to herpesvirus B-seropostive macaques, Bali, Indonesia. *Emerging Infectious Diseases*, **8**: 789–795.

Engelhardt, A. 1997. Impact of tourist contact on the competitive and social behaviour of wild long-tailed macaques (*Macaca fascicularis*). MS Thesis. Berlin, Germany: Free University of Berlin.

Eudey, A. A. 1994. Temple and pet primates in Thailand. *Revue D'Ecologie (Terre et la Vie)* **49**: 273–280.

   2008. The crab-eating macaque (*Macaca fascicularis*) widespread and rapidly declining. *Primate Conservation* **23**: 129–132.

Farslow, D. L. 1987. The behavior and ecology of the long-tailed macaque *(Macaca fascicularis)* on Angaur, Island, Palau, Micronesia. Dissertation, Columbus, OH, Ohio State University.

Fellowes, J. 1992. Hong Kong macaques: Final report to the WWF Hong Kong projects committee. Hong Kong, WWF.

Fittinghoff, N. A., Jr and Lindburg, D. G. 1980. Riverine refuging in East Bornean *Macaca fascicularis*. In *The Macaques: Studies in Ecology, Behavior and Evolution*, New York, NY: Van Nostrand Reinhold. pp. 182–214.

Fooden, J. 1964. Rhesus and crab-eating macaques: Intergradation in Thailand. *Science* **143**: 363–364.

   1991. Systematic review of Philippine macaques (Primates, Cercopithecidae: *Macaca fascicularis* subspp.). *Fieldiana Zoology n.s.* **64**: iv + 44.

   1995. Systematic review of Southeast Asian long-tail macaques, *Macaca fascicularis* (Raffles, 1821). *Fieldiana Zoology n. s.* **81**: 1–206.

   2006. Comparative review of fascicularis-group species of macaques (Primates: *Macaca*). *Fieldiana: Zoologyn.s.* **107**: 1–43.

Fooden, J. and Albrecht, G. 1993. Latitudinal and insular variation of skull size in crab-eating macaques (Primates, Cercopithecidae: *Macaca fascicularis*). *American Journal of Physical Anthropology* **92**: 521–538.

Froehlich, J., Schillaci, M., Jones-Engel, L., Froehlich, D., and Pullen, B. 2003. A Sulawesi beachead by longtail monkeys (*Macaca fascicularis*) on Kabaena Island, Indonesia. *Anthropologie* **41**: 76–74.

Fuentes, A. 2006. Human culture and monkey behavior: assessing the contexts of potential pathogen transmission between macaques and humans. *American Journal of Primatology* **68**: 880–896.

Fuentes, A. and Gamerl, S. 2005. Disproportionate participation by age/sex classes in aggressive interactions between long-tailed macaques (*Macaca fascicularis*) and human tourists at Padangtegal monkey forest, Bali, Indonesia. *American Journal of Primatology* **66**: 197–204.

Fuentes, A., Kalchik, S., Gettler, L., *et al.* 2008. Characterizing human–macaque interactions in Singapore. *American Journal of Primatology* **70**: 879–883.

Fuentes, A., Southern, M. and Suaryana, K. 2005. Monkey forests and human landscapes: Is extensive sympatry sustainable for *Homo sapiens* and *Macaca fascicularis* on Bali. In *Commensalism and Conflict: The Primate–Human Interface*, Norman, OK: The American Society of Primatologists Publications. pp. 168–195.

Groves, C. P. 2001. *Primate Taxonomy*. Washington, DC: Smithsonian Institute Press.

Gumert M. D. 2004. Spirits or Demons? *Tempo*, **52**: 27.

2007. Payment for sex in a macaque mating market. *Animal Behaviour*, **74**: 1655–1667.

Gumert M. D., Kluck, M., and Malaivijitnond, S. 2009. The physical characteristics and usage patterns of stone axe and pounding hammers used by long-tailed macaques in the Andaman Sea region of Thailand. *American Journal of Primatology* **71**: 594–608.

Gumert, M. D., Rachmawan, D., Iskandar, E., and Pamungkas, J. 2010. Preliminary population census of long-tailed macaques at Tanjung Puting National Park, Kalimantan Tengah, Indonesia. Presented at Quest for coexistence with nonhuman Primates, ASIAN-HOPE 2010 IPS Precongress Symposium and Workshop, Japan. September.

Gurmaya, K. J, Adiputra, I. M. W., Saryatiman, A. B., Danardono, S. N., and Sibuea, T. T. H. 1994. A preliminary study on ecology and conservation of the Java primates in Ujung Kulon National Park, West Java, Indonesia. In *Current Primatology: vol 1: Ecology and Evolution*, Strasbourg, France: Université Louis Pasteur. pp. 87–92.

Hadi, I. 2005. Distribution and present status of long-tailed macaques (*Macaca fascicularis*) in Lombok Island (Indonesia). *The Natural History Journal of Chulalongkorn University* (Suppl. **1**): 90.

Hagelin, J. (2004). Use of live nonhuman primates in research in Asia. *Journal of Postgraduate Medicine* **50**: 253–256.

Hamada, Y., Hadi, I., Urasopon, N. and Malaivijitnond, S. (2005). Preliminary report on yellow long-tailed macaques (*Macaca fascicularis*) at Kosumpee Forest Park, Thailand. *Primates* **46**: 269–273.

Heinsohn, T. 2003. Animal translocation: long-term human influences on the vertebrate zoogeography of Australasia (natural dispersal versus ethnophoresy). *The Australian Zoologist* **32**: 351–376.

Jones-Engel L, Engel G, Rompis A, *et al.* 2002. The not so sacred monkeys of Bali: A radiological assessment of human-macaque (*Macaca fascicularis*) commensalism. *American Journal of Primatology* **57** (Suppl. 1): 35.

Jones-Engel, L., Engel, G.A., Schillaci, M.A., *et al.* 2004. Prevalence of enteric parasites in pet macaques in Sulawesi, Indonesia. *American Journal of Primatology* **62**: 71–82.

Jones-Engel, L., Engel, G.A., Schillaci, M.A., 2006. Considering human-primate transmission of measles virus through the prism of risk analysis. *American Journal of Primatology* **68**: 868–879.

Kawamoto, Y. and Ischak, T. M. 1981. Genetic differentiation of the Indonesian crab-eating macaque (*Macaca fascicularis*): I. Preliminary report on blood protein polymorphism. *Primates* **22**: 237–252.

Kawamoto, Y., Ischak, T. M., and Supriatna, J. 1984. Genetic variations within and between troops of the crab-eating macaque (*Macaca fascicularis*) on Sumatra, Java, Bali, Lombok, and Sumbawa, Indonesia. *Primates* **25**: 131–159.

Kemp, N. J. and Burnett, J. B. 2003 (revised, 2006). A biodiversity risk assessment and recommendations for risk management of long-tailed macaques (*Macaca fascicularis*) in New Guinea. Washington, DC: Indo-Pacific Conservation Alliance.

Kemp, N. J. and Burnett, J. B. 2007. A non-native primate (*Macaca fascicularis*) in Papua: implications for biodiversity. In *The Ecology of Papua: Part II*, Singapore: Periplus Editions Ltd. pp. 1348–1364.

Khan, M., Elagupillay, S. and Zainal, Z. 1982. Species conservation priorities in the tropical rain forests of peninsular Malaysia. In *Species conservation priorities in the tropical forests of Southeast: Proceedings of a symposium held at the 58th Meeting of the IUCN Species Survival Commission Asia*, International Union for Conservation of Nature and Natural Resources. pp. 9, 15.

Khanam, S., Sarker, S. U., Hasan, R., Baten, A. 2005. Review of the literature on primates in Bangladesh. *The Natural History Journal of Chulalongkorn University* (Suppl. 1): 95.

Kyes, R. C. 1993. Survey of the long-tailed macaques introduced onto Tinjil Island, Indonesia. *American Journal of Primatology* 31: 77–83.

Kyes, R. C., Sajuthi, D., Iskandar, E., *et al.* 1998. Management of a natural habitat breeding colony of long-tailed macaques. *Tropical Biodiversity* 5: 127–137.

Linkie, M., Dinata, Y., Nofrianto, A., and Leader-Williams, N. 2007. Patterns and perceptions of wildlife crop raiding in and around Kerinci Seblat National Park, Sumatra. *Animal Conservation* 10:127–135.

Long, J. 2003. *Introduced Mammals of the World*. Wallingford, UK: CABI Publishing.

Lorence, D. and Sussman, R. 1986. Exotic species invasion into Mauritius wet forest remnants. *Journal of Tropical Ecology* 2: 1470–162.

Louden J. E., Howells, M. E., Fuentes A. 2006. The importance of integrative anthropology: A preliminary investigation employing primatological and cultural anthropological data collection methods in assessing human-monkey co-existence in Bali Indonesia. *Ecological and Environmental Anthropology* 2: 2–13.

Mackinnon, K. 1986. The conservation status of nonhuman primates in Indonesia. In *Primates: The Road to Self-Sustaining Populations*, New York, NY, Springer-Verlag. pp. 99–126.

Lowe, S., Browne, M. and Boudjelas, S. 2000. One hundred of the worst invasive species: a selection from the global invasive species database. *Aliens* 12: 1–12.

MacKinnon, J. and Mackinnon, K. 1987. Conservation status of the primates of the Indo-Chinese Subregion. *Primate Conservation* 8: 187–195.

MacKinnon, K., Hatta, G., Halim, H., and Mangalik, A. 1996. *The Ecology of Kalimantan*. Singapore: Periplus Editions Ltd.

Malaivijitnond, S. and Hamada, Y. 2008. Current status and situation of long-tailed macaque (*Macaca fascicularis*) in Thailand. *The Natural History Journal of Chulalongkorn University* 8: 185–204.

Malaivijitnond, S., Hamada, Y., Varavudhi, P., and Takenaka, O. 2005. The current distribution and status of macaques in Thailand. *The Natural History Journal of Chulalongkorn University* (Suppl. 1): 29–34.

Malaivijitnond, S., Hamada, Y., Suryobroto, B., and Takenaka, O. 2007a. Female long-tailed macaques with scrotum-like structures. *American Journal of Primatology* 69: 721–735.

Malaivijitnond, S., Lekprayoon, C., Tandavanittj, *et al.* 2007b. Stone-tool usage by Thai long-tailed macaques (*Macaca fascicularis*). *American Journal of Primatology* **69**: 227–233.

Malone N. M., Fuentes, A., Purnama, A. R., and Adi Putra, I. M. W. 2003. Displaced hylobatids: Biological, cultural, and economic aspects of the primate trade in Jawa and Bali, Indonesia. *Tropical Biodiversity* **8**: 41–49.

Marchall V. and Hill, C. 2009. Primate crop-raiding: A study of local perceptions in four villages in North Sumatra, Indonesia. *Primate Conservation* **24**: 107–116.

Matsubayashi, K., Gotoh, S., Kawamoto, Y., Nozawa, K., and Suzuki, J. 1989. Biological characteristics of crab-eating monkeys on Angaur Island. *Primate Research* **5**: 8–57.

Maurin-Blanchet, H. 2006. Données récentes sur les Singes en provenance de I'lle Maurice: Illustration d'une production, de *Macaca fascicularis* (Cynomolgus or Macaque crabier), sous contrôle qualitatif et quantitatif, à des fins de recherche scientifique. *Sciences et techniques de l'animal de laboratorie* **31**: 143–152.

McCarthy, M. S., Matheson, M. D., Lester, J. D., Sheeran, L. K., Li, J-H., and Wagner, S. 2009. Sequences of Tibetan macaque (*Macaca thibetana*) and the tourist behaviors at Mt. Huangshan, China. *Primate Conservation* **24**: 145–151.

McConkey, K. R. and Chivers, D. J. 2004. Low mammal and hornbill abundance in the forests of Barito Ulu, Central Kalimantan, Indonesia. *Oryx* **38**: 439–447.

Mungroo, Y. and Tezoo, V. 1999. Control of invasive species in Mauritius. In *Invasive Species in Eastern Africa: Proceedings of a workshop held at ICIPE*.

Pamungkas, J. and Sajuthi, D. 2003. The breeding of naturally occurring B virus-free cynomolgus monkeys (*Macaca fascicularis*) on the Island of Mauritius. In *International Perspectives: The Future of Nonhuman Primate Resources*, Washington, DC: The National Academic Press. p 20.

Pamungkas, J., Sajuthi, D., Lelana, P. A., *et al.* 1994. Tinjil island, a natural habitat breeding facility of simian retrovirus-free *Macaca fascicularis*. *American Journal of Primatology* **34**: 81–84.

Patterson, J. 2005. *Commensalism and conflict: The Primate–human Interface*, Norman, OK: The American Society of Primatologists Publications.

Perwitasari-Farajallah, D., Kawamoto, Y. and Suryobroto, B. 1999. Variation in blood proteins and mitochondrial DNA within and between local populations of long-tail macaques, *Macaca fascicularis* on the Island of Java, Indonesia. *Primates* **40**: 581–595.

Perwitasari-Farajallah, D., Kawamoto, Y., Kyes, R., Agus Lelana, R., and Sajuthi, D. 2001. Genetic characterization of long-tailed macaques (*Macaca fascicularis*) on Tabuan Island, Indonesia. *Primates* **42**: 141–152.

Poirier, F. E. and Smith, E. O. 1974. The crab-eating macaques (*Macaca fascicularis*) of Angaur Island, Palau, Micronesia. *Folia Primatologica* **22**: 258–306.

Quammen, D. 1996. *The Song of the Dodo: Island Biogeography in an Age of Extinctions*. New York, NY: Simon & Schuster.

Richard, A. F., Goldstein, S. J., and Dewar, R. E. 1989. Weed macaques: the evolutionary implications of macaque feeding ecology. *International Journal of Primatology* **10**: 569–594.

Safford, R. J. 1997. Nesting success of the Mauritius Fody *Foudia rubra* in relation to its use of exotic trees as nest sites. *Ibis* **139**: 555–559.

Santoso, N. and Winarno, G. D. 1992. Studi populasi dan perilaku monyet ekor panjang (*Macaca fascicularis*) di Pulau Tinjil, Jawa Barat. Bogor, Indonesia: Falkutas Kehutanan Institut Pertanian Bogor.

Santosa, Y. 1996. Beberapa parameter bio-ekologi penting dalam pengusahaan monyet ekor panjang (*Macaca fascicularis*). *Media Konservasi* **V**: 25–29.

van Schaik, C. and van Noordwijk, M. A. 1985. Evolutionary effect of the absence of felids on the social organization of the macaques on the island of Simeulue (*Macaca fascicularis fusca*, Miller 1903). *Folia Primatologica* **44**: 138–147.

van Schaik, C. P., van Amerongen, A. and van Noordwijk, M. A. 1996. Riverine refuging by wild Sumatran long-tailed macaques (*Macaca fascicularis*). In *Evolution and Ecology of Macaque Societies*. Cambridge University Press. pp. 160–181.

Scheffran, W., de Ruiter, J. R., and van Hooff, J. A. R. A. M. 1996. Genetic relatedness within and between populations of *Macaca fascicularis* on Sumatra and off-shore islands. In *Evolution and Ecology of Macaque Societies*. Cambridge University Press. pp. 20–42.

Schillaci, M. 2010. Latitudinal variation in cranial dimorphism in *Macaca fascicularis*. *American Journal of Primatology* **72**: 151–160.

Schillaci, M., Jones-Engel, L., Engel, G. A., *et al.* 2005. Prevalence of enzootic simian viruses among urban performance monkeys in Indonesia. *Tropical Medicine and International Health* **10**: 1305–1314.

Schillaci, M., Jones-Engel, L., and Engel, G. A. 2006. Exposure to human respiratory viruses among urban performing monkeys in Indonesia. *American Journal of Tropical Medicine and Hygiene* **75**: 716–719.

Schillaci, M., Jones-Engel, L., Lee, B. P. Y-H., *et al.* 2007. Morphology and somatometric growth of long-tailed macaques *Macaca fascicularis fascicularis* in Singapore. *Biological Journal of the Linnean Society* **92**: 675–694.

Schillaci, M., Engel, G., Fuentes, A., *et al.* 2010. The not-so-sacred monkeys of Bali: A radiographic study of human-primate commensalism. In *Indonesia Primates*. New York, NY: Springer. pp. 249–256.

Sha, J., Gumert, M. D., Lee, B. P. Y-H., *et al.* 2009a. Status of the long-tailed macaque *Macaca fascicularis* in Singapore and implications for management. *Biodiversity and Conservation* **18**: 2909–2926.

Sha, J., Gumert, M. D., Lee, B. P. Y-H., *et al.* 2009b. Macaque-human interactions and the societal perceptions of macaques in Singapore. *American Journal of Primatology* **71**: 825–839.

Son, V. D. 2003. Morphology of *Macaca fascicularis* in a mangrove forest, Vietnam. *Laboratory Primate Newsletter* **42**: 9–11.

Southwick, C. and Manry, D. 1987. Habitat and population changes for the Kowloon macaques. *Primate Conservation* **8**: 48–49.

Southwick, C. H. and Siddiqi, M. F. 1994. Population status of nonhuman primates in Asia, with emphasis on rhesus macaques in India. *American Journal of Primatology* **34**: 51–59.

Southwick, C. H. and Southwick, K. L. 1983. Polyspecific groups of macaques on the Kowloon Peninsula, New Territories, Hong Kong. *American Journal of Primatology* **5**: 17–24.

Stanley, M. A. 2003. The breeding of naturally occurring B virus-free cynomolgus monkeys (*Macaca fascicularis*) on the Island of Mauritius. In *International Perspectives: The Future of Nonhuman Primate Resources*. Washington, DC: The National Academic Press. pp. 46–48.

Suaryana, K., Rompis, A. and Sibang, I. 2000. Status dan distribusi monyet ekor pan-jang (*Macaca fascicularis*) di Bali. Denpasar, Indonesia: Pusat Kajian Primata Lembaga Penelitian, Universitas Udayana.

Sugardjito, J., van Schaik, C., van Noordwijk, M., and Mitrasetia, T. 1989. Population status of the Simeulue monkey (*Macaca fascicularis* fusca). *American Journal of Primatology* **17**: 197–207.

Supriatna, J., Yanuar, A., Martarinza, *et al.* 1996. A preliminary survey of long-tailed and pig-tailed macaques (*Macaca fascicularis* and *Macaca nemestrina*) in Lampung, Bengkulu, and Jambi provinces, southern Sumatra, Indonesia. *Tropical Biodiversity* **3**: 131–140.

Sussman, R. W. and Tattersall, I. 1981. Behavior and ecology of *Macaca fascicularis* in Mauritius: A preliminary study. *Primates* **22**: 192–205.

   1986. Distribution, abundance, and putative ecological strategy of *Macaca fascicularis* on the island of Mauritius, southwestern Indian Ocean. *Folia Primatologica* **46**: 28–43.

Tosi, A. J. and Coke, C. S. 2007. Comparative phylogenetics offer new insights into the biogeographic history of *Macaca fascicularis* and the origin of the Mauritian macaques. *Molecular Phylogenetics and Evolution* **42**: 498–504.

Twigg, I. C. and Nijman, V. 2008. Export of wild-caught long-tailed macaques from Southeast Asia. 22nd Congress of the International Primatological Society. Edinburgh.

Umapathy, G., Singh, M., and Mohnot, S. M. 2003. Status and distribution of *Macaca fascicularis umbrosa* in the Nicobar Islands, India. *International Journal of Primatology* **24**: 281–293.

USFWS (United States Fish and Wildlife Service). 2009. Office of Law Enforcement, Information Service, Branch of Planning and Analysis.

Watanabe, K., Urasopon, N., and Malaivijitnond, S. 2007. Long-tailed macaques use human hair as dental floss. *American Journal of Primatology* **69**: 940–944.

Wheatley, B. 1988. Cultural behavior and extractive foraging in *Macaca fascicularis*. *Current Anthropology* **29**: 516–519.

   1989. Diet of Balinese temple monkeys, *Macaca fascicularis*. *Kyoto University Overseas Research Report Studies on Asian Non-Human Primates* **7**: 62–75.

   1999. *The Sacred Monkeys of Bali*. Prospect Heights, IL: Waveland Press, Inc.

Wheatley, B., Stephenson, R., and Kurashina, H. 1999. The effects of hunting on the long-tailed macaques of Ngeaur Island, Palau. In *The Nonhuman Primates*. Mountain View, CA: Mayfield Publishing Company. pp. 159–163.

Wheatley, B., Stephenson, R., Kurashina, H., and Marsh-Kautz, K. 2002. A cultural-primatological study of *Macaca fascicularis* on Ngeaur Island, Republic of Palau.

In *Primates Face-to-Face: Conservation Implications of Human and Nonhuman Primate Interconnections.* Cambridge University Press. pp. 240–253.

Wheatley, B. and Putra. D. K. H. 1994a. Biting the hand that feeds you: Monkeys and tourists in Balinese monkey forests. *Tropical Biodiversity* **2**: 317–327.

1994b. The effects of tourism on conservation at the monkey forest in Ubud, Bali. *Revue D'Ecologie (Terre et la Vie)* **49**: 245–257.

1995. Hanuman, the monkey god, leads conservation efforts in Balinese Monkey Forest at Ubud, Indonesia. *Primate Report* **41**: 55–64.

Wheatley. B. P., Putra, D. K., and Gonder, M. K. 1996. A comparison of wild and food-enhanced long-tailed macaques (*Macaca fascicularis*). In *Evolution and Ecology of Macaque Societies.* Cambridge University Press. pp. 182–206.

Wong, C. L. 1994. Studies on the feral macaques of Hong Kong. Masters thesis, Biology. Hong Kong: Hong Kong University of Science and Technology.

Wong, C. L. and Chow, G. 2004. Preliminary results of trial contraceptive treatment with SpayVac™ on wild monkeys in Hong Kong. *Hong Kong Biodiversity: AFCD Newsletter* **6**: 13–16.

Wong, C. L. and Ni, I. H. 2000. Population dynamics of the feral macaques in the Kowloon Hills of Hong Kong. *American Journal of Primatology* **50**: 53–66.

Yanuar, A., Chivers, D. J., Sugardito, J., Martyr, D. J., and Holden, J. T. 2009. The population distribution of pig-tailed macaque (*Macaca nemestrina*) and long-tailed macaque (*Macaca fascicularis*) in West Central Sumatra, Indonesia. *Asian Primates Journal* **1**: 2–11.

2 Distribution and current status of long-tailed macaques (Macaca fascicularis aurea) in Myanmar

AYE MI SAN AND YUZURU HAMADA

Introduction

Myanmar is situated in the west of the Indochina Peninsula and geographically ranges from 9°58' N to 28°29'N and from 92°10'E to 101°10'E, with a land area of 676,553 km² and a coastline of 2,832 km (Bird Life International, 2005). The wide variation in topography and climate has produced a rich diversity of wildlife in Myanmar, which is a component of the Indo-Myanmar Hotspot for biodiversity (Bird Life International, 2005). New mammal species have been discovered quite recently, such as the leaf deer (*Muntiacus putaoensis*, Amato et al., 1999) and the Kachin woolly bat (*Kerivoula kachinensis*, Bates et al., 2004). Primate fauna are also rich in Myanmar (Tun Yin, 1967; FAO, 1985; Kyaw Nyunt Lwin, 1995; Parr and Tin Than, 2007), including one species of slow loris (*Nycticebus coucang*), five species of macaques (*Macaca assamensis*, *M. arctoides*, *M. fascicularis*, *M. mulatta*, and *M. nemestrina*), five species of leaf monkeys (*Trachypithecus obscurus*, *T. phayrei*, *T. cristatus*, *T. pileatus*, and *Presbytis femoralis*), and two species of gibbons (*Hylobates hoolock* and *H. lar*).

The long-tailed macaque in Myanmar is classified as a distinctive subspecies (*Macaca fascicularis aurea*), with parts of this subspecies population also occurring in Thailand and Bangladesh. For the most part, the biology and evolution of this subspecies remains unknown. Evolutionary scenarios on the origins of long-tailed macaques (*Macaca fascicularis*) have suggested that proto-*fascicularis* expanded north from Sundaland into continental Southeast Asia (Delson, 1980). After this dispersal, proto-*fascicularis* diversified into the rhesus macaque (*M. mulatta*) and the ten subspecies of *M. fascicularis* (Fooden, 1995). The Myanmar subspecies (*Macaca fascicularis aurea*) is considered to have arisen from such northern colonizers, perhaps after becoming

Monkeys on the Edge: Ecology and Management of Long-Tailed Macaques and their Interface with Humans, eds. Michael D. Gumert, Agustín Fuentes and Lisa Jones-Engel. Published by Cambridge University Press. © Cambridge University Press 2011.

isolated from the rest of Southeast Asia by the Bilauktaung Mountain Range (i.e., the Dawna Range), which runs from north to south along the national border between Myanmar and Thailand. This hypothesis on the origin of *M. f. aurea* needs to be tested. Consequently, a detailed understanding of their distribution is necessary in order to reconstruct their evolutionary history.

Fooden (1995) stated that the Myanmar subspecies *M. fascicularis aurea* has an infrazygomatic lateral facial crest pattern, a relative tail length (tail length/crown-rump length x 100%) of > 90%, a frequent appearance of crested hair at the crown, a pelage color of grayish brown without patterning in juveniles and adults (cf., bi-partite pattern in rhesus and assamese macaques), a black pelage in infants under three months of age, and whiskers and other hairs around the face. The direction of hair at the cheek is a major subspecific key character, demonstrating an infrazygomatic pattern (*M. f. aurea*; Fooden, 1995) rather than the transzygomatic pattern seen in *M. f. fascicularis*. Possible hybrids have been found in Myanmar that appears to be either inter-specific with rhesus macaques (*Macaca mulatta*) or inter-subspecific with the nominotypical subspecies (*M. f. fascicularis*; Fooden, 1995, 2000).

The status and distribution of long-tailed macaques in Myanmar is still only poorly known. There are rough reports on the distribution of the long-tailed macaque in Myanmar (Tun Yin, 1967; Fooden, 1995), but these may not capture the full extent of their distribution, nor are they necessarily valid today. These reports show that the long-tailed macaques of Myanmar range in lower and southern parts of Myanmar. However, their status and population levels have not been well reported and significant environmental changes have occurred in Myanmar since these surveys. Consequently, adequate surveys are needed to complete our understanding of their distribution.

Long-tailed macaque populations are reported to be widespread but rapidly declining due to habitat alteration and the animal trade (Eudey, 2008). In Myanmar, there are several threats facing long-tailed macaques and other species. Myanmar is still one of the most forested countries in mainland Southeast Asia, but the forests are declining and have continued to decline by 0.3 percent annually since early 1990 (Leimgruber *et al.*, 2005). Forest loss has been due to logging, construction of infrastructure, and conversion to agricultural and aquacultural lands. Consequently, forest habitat and quality have been significantly reduced in Myanmar, and this is especially so in Southern Myanmar (Molur *et al.*, 2003). The lowland, coastal, and mangrove forests, which are the primary natural habitats of long-tailed macaques, have been significantly affected by forest conversion. As a result, the habitats available to them are deteriorating and becoming more fragmented. Furthermore, hunting and the wildlife trade also threaten long-tailed macaques in Myanmar. Therefore, a better understanding of their population is needed in order to adequately assess how land use and utilization of macaques as a resource are affecting their population levels.

Long-tailed macaques have been regarded as a "weed species," (Richard *et al.*, 1989), meaning that they are resistant to habitat deterioration, live near or inside human settlements, and easily exploit resources from these settlements. Therefore, it is possible that long-tailed macaques will become more weed-like as destruction of natural forest habitat continues, and human development expands. This will place these macaques into a vulnerable position, because although they can exploit human habitat, their continued existence in regions where they live sympatrically with humans will depend on the attitude of humans towards macaques. If the human communities become intolerant of macaques, they could begin to exterminate local populations, and this type of population pressure may have already begun in Myanmar. Consequently, studies are needed to assess how living near human settlements are impacting macaque populations in Myanmar.

In this study, we have surveyed Myanmar to assess the distribution and status of the long-tailed macaque population. Since 2004 we have been conducting interviews and have carried out pet observations as well as direct field observation. In this manuscript, we report the preliminary results of our research. Long-tailed macaques were found in Rakhine, Ayeyarwady Delta, Bago Yoma, and Tanintharyi Biogeographical regions in Myanmar, and we described the present status and estimate population of long-tailed macaques in each of these regions.

Materials and methods

Study region and interviews

Based on factors of physical geography, rainfall, and forest cover (FAO, 1985), Myanmar is divided into ten biogeographic regions, and we travelled through seven of these between July 2004 and March 2009 to assess the presence or absence of long-tailed macaques (Figure 2.1). In these regions, we interviewed 380 local people in 184 of villages along the highways and seven protected areas. In villages, we identified either the head of each village or persons who knew the forests and wildlife and could describe the presence and abundance of non-human primates. Photographs illustrating the typical morphology of the species were shown to the participants, so they could identify which species they had observed in the region. We recorded the following information at each location where we conducted interviews: address of the village, the geographical coordinates (latitude, longitude) and altitude with GPS (Global

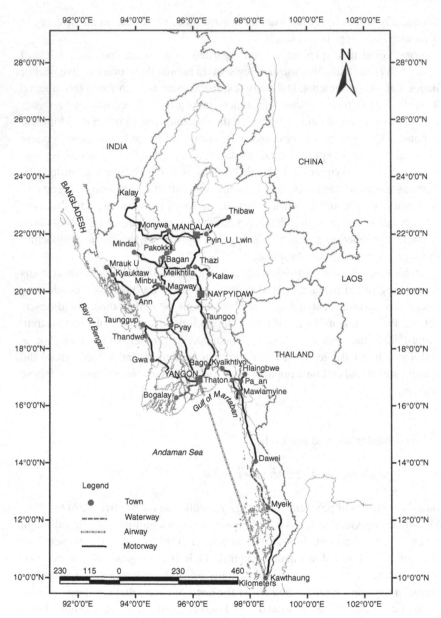

Figure 2.1. Myanmar map showing interview survey routes (2004–2009).

Positioning System), ethnicity of residents, habitat conditions, impacts on wildlife (hunting, consumption, and trading), conflicts between humans and primates, and the local names of primates. In protected areas, we interviewed forest rangers and staff about species diversity, numbers of troops, habitat conditions, and the range of forest protection law. The interview survey form is presented in Appendix 2.1.

Pet observation

During the interview survey, we actively inquired about the location of pet long-tailed macaques in each village. After locating pets, we interviewed the pet owners about the origin of the monkey, the hunting method used to capture the pet, how they obtained the monkey, the price of the monkey, and the route of trade. We observed their morphological characteristics and assessed if there was any evidence of hybrid characteristics (i.e., interspecies or intersubspecies), based on descriptions from Fooden (1995).

Field survey

The region of the field survey ranged from 12°27' N to 17°17'N and from 95°14' E to 99°06'E in southern Myanmar, and covered mangrove forests and isolated limestone mountains which free-ranging and habituated troops of long-tailed macaque were known to inhabit. We observed and classified the *Macaca fascicularis aurea* subspecies by their morphological traits, that is, infrazygomatic pattern of cheek hairs and no hair-crest at the vertex of the head (Fooden, 1995). Individuals were classified into sex and age classes based on their morphological characteristics and population sizes were counted by direct observation. The background history of the troop and condition of the habitat were recorded by interviews with local residents, mostly from the monks living at the monasteries.

Population estimation

We estimated the population of long-tailed macaques in Myanmar using several assumptions to assess each habitat region. In the Mon and Kayin States of the northern part of Tanintharyi region, long-tailed macaques troops were found in isolated mountains. The minimum estimation was made by multiplying the number of troops by an average size of 30, which is an average macaque group size that is in accordance with a review of numerous studies in free-ranging conditions (Fooden, 1995). Although we made an exhaustive survey in these regions, we may have missed as many troops as we found. Therefore we generated a maximum estimate that was two times are count to factor in the likelihood that we missed up to 50 percent of the groups occurring in the region.

In the Rakhine and southern Tanintharyi regions, we estimated the total population in lowland forests by using a strip transect method. The parameters obtained from trip surveys were the distance we covered (L) and the number of sites (n) where long-tailed macaques were reported. Forest cover area was obtained from the most recent existing statistics (Leimgruber *et al.*, 2005). Our suppositions were as follows: the strip width was 10 km, troop size was 30 individuals, and 50 percent and 33 percent of the forested areas were habitable for long-tailed macaques in the southern Tanintharyi and Rakhine regions, respectively. The proportions of habitable area were determined by assessing the development of forest and influence of human activities. The density was established as $5n/L$ individuals/km^2. The minimum estimate was then calculated as $3n/L*A$ for the southern Tanintharyi region and $1.5n/L*A$ for the Rakhine region. Since the forest cover is decreasing from the year of the report, the estimate was corrected by average decreasing rates; in ten years, equaling 9 percent in southern Tanintharyi and 5 percent in Rakhine. The maximum estimate was established at ten times the minimum to provide a large buffer to avoid an underassessment of the population. However, mangrove forests do not appear as heavily exploited by human activity as the other types of lowland forests, therefore the populations of long-tailed macaques in mangrove forests are estimated from the area of forest and a density calculated at ten times greater than other lowland forests in each region.

Results

Interview reports on distribution

We found positive records for long-tailed macaque in four of the seven biogeographic regions that we surveyed; Rakhine, Ayeyarwady Delta, Bago Yoma and Tanintharyi (Table 2.1). We obtained reports on the occurrence of long-tailed macaques at 98 villages (53.26 percent) of the 184 villages we visited (Table 2.2 and 2.3). Overall, we found that the range of long-tailed macaques in Myanmar stretches all along the coastal regions from northwestern 20°32'N, near the Bangladesh border, to the southernmost area, 9°58'N (Figure 2.2). The ranges of long-tailed macaques between the Rakhine, Ayeyarwady, and

Table 2.1. *Physical features of the biogeographic regions in the survey areas*

No.	Biogeographic Region	Political Divisions	Mean Rainfall (mm)	Vegetation Types
1	Chin Hill	Chin State	1750–3750	Hill evergreen, pine forest and bamboo forest
2	Shan Plateau	Shan State	1250–3750	Evergreen, mixed deciduous forest and sub-tropical pine forests
3	Rakhine *	Rakhine State	2500–6250	Evergreen and deciduous forest. Bamboo forest and tidal swamps, beach forest, islands
4	Dry Zone	Mandaly, Magway and Sagaing Divisions	625–1000	Dry forest and scrub
5	Bago Yoma*	Bago Division	1250–3500	Mixed deciduous and semi-evergreen forest
6	Ayeyarwady Delta*	Ayeyarwady Division	2500–5000	Tidal swamp forest
7	Tanintharyi*	Mon, Kayin States and Tanintharyi Division	3750–5000	Evergreen tropical rainforest Tidal swamp forest Beach forest Offshore islands and coral reef

* Biogeographic regions where long-tailed macaques were found

Table 2.2. *Information of long-tailed macaque in biogeographic regions*

Biogeographic Region	Date of Survey	Total Interview Sites	Long-tail positive	Frequency (%)
Rakhine	19 to 22 Nov 2004	45	21	46.7
	14 to 22 Jan 2007	6	6	100
	16 to 20 Jan 2009	43	31	72.1
Ayeyarwady Delta	18 to 22 Jul 2004	1	1	100
Bago Yoma	1 Dec 2004	3	1	33.3
Tanintharyi (North)	20 to 24 May 2008	34	10	29.4
Tanintharyi (South)	1 to 8 Mar 2009	52	28	53.8
TOTAL		**184**	**98**	**53.26**

Table 2.3. *Positive records of long-tailed macaque in different regions*

Site	Date	Region	Village	Latitude	Longitude	Altitude(m)
1	19-Nov-04	Rakhine	Hlay-lone-taung	16°57′44.1″	94°30′25.6″	+
2	19-Nov-04	Rakhine	Seit Gyi	16°57′19.7″	94°3′14.6″	+
3	19-Nov-04	Rakhine	Chaung Tha	16°58′13.7″	94°27′02.6″	+
4	19-Nov-04	Rakhine	Chaung Tha	16°58′14.9″	94°27′04.7″	+
5	19-Nov-04	Rakhine	U-To Chaung	16°57′32.4″	94°28′35.3″	+
6	19-Nov-04	Rakhine	Chaung-khwa	17°28′18.9″	94°56′23.0″	+
7	19-Nov-04	Rakhine	16 miles camp	17°30′54.7″	94°34′26.2″	+
8	19-Nov-04	Rakhine	Mya-yar-pin	17°32′50.6″	94°49′40.0″	+
9	19-Nov-04	Rakhine	Baw-di Camp	17°34′08.8″	94°43′48.1″	+
10	20-Nov-04	Rakhine	Gwa Township	17°35′19.9″	94°41′42.2″	+
11	20-Nov-04	Rakhine	Dawn Chaung Kwin	17°35′08.9″	94°38′23.7″	+
12	21-Nov-04	Rakhine	Old Myay Kwin	17°39′03.1″	94°35′43.1″	+
13	21-Nov-04	Rakhine	Ye-thit-kone	17°40′24.4″	94°36′05.7″	+
14	21-Nov-04	Rakhine	Tie-kyoe	17°49′05.0″	94°29′23.1″	+
15	21-Nov-04	Rakhine	Sar-chet	17°57′42.3″	94°30′14.3″	+
16	21-Nov-04	Rakhine	Boga-lay	18°07′11.2″	94°29′06.6″	+
17	21-Nov-04	Rakhine	Thit Ngot Toe	18°11′17.8″	94°28′52.7″	+
18	21-Nov-04	Rakhine	Kyauk-gyi	18°14′48.5″	94°28′36.5″	+
19	21-Nov-04	Rakhine	Nat-taung	18°33′21.0″	94°20′15.9″	+
20	21-Nov-04	Rakhine	Ah Bay	18°41′40.1″	94°15′47.0″	+
21	21-Nov-04	Rakhine	Tha Phan Shwe	18°51′20.0″	94°14′36.0″	+
22	10-Jan-07	Rakhine	Gat Gyi	19°51′22.1″	94°26′30.8″	+
23	10-Jan-07	Rakhine	Lay Dan Ku	19°48′12.3″	93°58′16.7″	+
24	12-Jan-07	Rakhine	Kyay Taw	20°32′48.7″	92°58′44.0″	+
25	12-Jan-07	Rakhine	Tan Pauk Chaung	20°20′02.3″	93°20′10.6″	+
26	13-Jan-07	Rakhine	Sa Nyin	19°58′49.6″	93°43′11.9″	+
27	13-Jan-07	Rakhine	Kha Maung Taw	19°51′30.0″	93°54′23.9″	+
28	16-Jan-09	Rakhine	Sein-taung-kone	16°49.107′	94°34.096′	12
29	16-Jan-09	Rakhine	Maw-tin Junction	16°48.789′	94°33.635′	3
30	16-Jan-09	Rakhine	Elephant Camp	16°48.630′	94°29.560′	42
31	16-Jan-09	Rakhine	Shaut pin chaung	16°49.340′	94°27.709′	30
32	16-Jan-09	Rakhine	Ya-mon-nar-oo Hotel	16°49.874′	94°23.759′	10
33	17-Jan-09	Rakhine	Tha-latt-khwa	16°49.328′	94°36.545′	18
34	17-Jan-09	Rakhine	Nga-thaing-chaung	17°23.396′	95°04.002′	20
35	17-Jan-09	Rakhine	Chaung-kwa	17°28.331′	94°56.407′	20
36	17-Jan-09	Rakhine	Naung-ta-kha	17°30.611′	94°53.499′	209
37	17-Jan-09	Rakhine	Nyaung-ta-kha	17°30.725′	94°53.144′	173
38	17-Jan-09	Rakhine	25 mile camp	17°32.701′	94°49.146′	286
39	17-Jan-09	Rakhine	31 miles 4 farlon	17°34.121′	94°46.526′	459
40	17-Jan-09	Rakhine	33 miles	17°33.458′	94°45.088′	213
41	18-Jan-09	Rakhine	Kan-thar-yar beach	17°43.696′	94°32.526′	10
42	18-Jan-09	Rakhine	Zi-kone	17°45.676′	94°31.351′	15
43	18-Jan-09	Rakhine	Sat-twar-kone	17°46.569′	94°30.360′	10
44	18-Jan-09	Rakhine	Maw-shwe-chaing	17°47.981′	94°29.169′	3
45	18-Jan-09	Rakhine	Tai-kyoe	17°49.947′	94°29.430′	10

Table 2.3. (*cont.*)

Site	Date	Region	Village	Latitude	Longitude	Altitude(m)
46	18-Jan-09	Rakhine	Chaung-tha	17°50.673'	94°29.839'	6
47	18-Jan-09	Rakhine	Tha-pyu-chaung	17°53.223'	94°30.175'	3
48	18-Jan-09	Rakhine	Kyein-ta-li	18°00.298'	94°29.424'	4
49	18-Jan-09	Rakhine	Kyauk-khaung-kwin	18°05.591'	94°28.686'	21
50	18-Jan-09	Rakhine	Kha-ye-tan	18°08.652'	94°28.764'	18
51	18-Jan-09	Rakhine	Thit-gnot-to	18°11.283'	94°28.909'	15
52	18-Jan-09	Rakhine	Me-ne-kwin	18°22.310'	94°25.563'	10
53	19-Jan-09	Rakhine	Kway-chaung	18°35.444'	94°21.052'	11
54	19-Jan-09	Rakhine	Tha-ka-pyin	18°42.418'	94°19.658'	17
55	19-Jan-09	Rakhine	Tha-la-ku	18°48.550'	94°15.618'	10
56	19-Jan-09	Rakhine	Kyauk-ta-gha camp	18°30.131'	94°20.393'	+
57	19-Jan-09	Rakhine	Sa-lu	18°44.603'	94°30.002'	646
58	19-Jan-09	Rakhine	Ye-paw-gyi	18°40.475'	94°34.780'	699
59	19-Jul-04	Ayeyarwady	Meinmahla Kyun	15°51' to 16°05'	95°14' to 95°21'	+
60	1-Dec-04	Bago	Day-son-par	17°32'50.7"	96°32'30.2"	+
61	21-May-08	N.Tanintharyi	Bayin Nyi Naung	16°58.2'	97°29.6'	28
62	21-May-08	N.Tanintharyi	Taung-ga-lay	16°53'09.2"	97°32'04.2"	44
63	21-May-08	N.Tanintharyi	Kaw-kun-gu	16°49'21.6"	97°35'9.2"	15
64	21-May-08	N.Tanintharyi	Ya-yhae-pyan-gu	16°50'6.5"	97°34'14.8"	21
65	22-May-08	N.Tanintharyi	Mt. Zwe-ka-pin	16°49'27.7"	97°40'05.4"	726
66	22-May-08	N.Tanintharyi	Kaw-ka-thaung-gu	16°49'42.6"	97°42'21.9"	23
67	22-May-08	N.Tanintharyi	Shwe-pyi-tahung	16°44'21.5"	97°45'30.7"	36
68	22-May-08	N.Tanintharyi	Kha-yon-gu	16°32'0.5"	97°42'53.5"	13
69	23-May-08	N.Tanintharyi	Indian Single Rock	16°19'19.1"	97°42'33.3"	73
70	23-May-08	N.Tanintharyi	Mahar-kotthein-nar-yon	17°17'9"	97°13'0.1"	45
71	1-Mar-09	S.Tanintharyi	Aung-thu-kha	09°59'04.2"	98°32'53.7"	36
72	3-Mar-09	S.Tanintharyi	10 miles	10°04'55.4"	98°32'00.8"	58
73	3-Mar-09	S.Tanintharyi	Ban-ka-chun	10°08'59.5"	98°35'35.5"	9
74	3-Mar-09	S.Tanintharyi	Ban-ka-chun	10°08'59.7"	98°35'33.2"	10
75	3-Mar-09	S.Tanintharyi	Ma-li-wun	10°15'11.8"	98°36'06.6"	9
76	3-Mar-09	S.Tanintharyi	San-thida	10°28'0.12"	98°37'45.5"	31
77	3-Mar-09	S.Tanintharyi	Kha-maut-gyi	10°21'05.3"	98°37'24.2"	20
78	3-Mar-09	S.Tanintharyi	Shwe-pyi-thar	10°00'48.8"	98°33'46.3"	42
79	3-Mar-09	S.Tanintharyi	kyay-mar-thiri	10°01'13.6"	98°33'49.8"	42
80	4-Mar-09	S.Tanintharyi	Karathuri	10°55'59.2"	98°45'34.7"	20
81	4-Mar-09	S.Tanintharyi	Bokpyin Town	11°15'39.9"	98°45'27.9"	10
82	4-Mar-09	S.Tanintharyi	Khe-mine	11°13'04.2"	98°47'33.9"	19
83	4-Mar-09	S.Tanintharyi	Shwe-bon-thar	11°15'48.3"	98°45'28.8"	4
84	5-Mar-09	S.Tanintharyi	Lenya	11°26'58.7"	98°59'40.2"	9
85	5-Mar-09	S.Tanintharyi	Pyi-gyi-mine-dine	11°28'24.9"	99°00'35.0"	7
86	5-Mar-09	S.Tanintharyi	Htin-mei-ywa	11°31'17.1"	99°03'17.5"	15
87	5-Mar-09	S.Tanintharyi	Chaung-naut-pyan-ywa	11°43'56.6"	99°06'27.5"	64

Table 2.3. *(cont.)*

Site	Date	Region	Village	Latitude	Longitude	Altitude(m)
88	5-Mar-09	S.Tanintharyi	Tanintharyi, Orgyi	12°06'14.1"	98°59'10.1"	48
89	5-Mar-09	S.Tanintharyi	Kaw-ma-pyin	12°06'27.8"	98°58'28.6"	22
90	5-Mar-09	S.Tanintharyi	Pa-nan-nge	12°09'13.9"	98°57'34.2"	8
91	5-Mar-09	S.Tanintharyi	San-thit	12°13'27.2"	98°53'36.1"	18
92	5-Mar-09	S.Tanintharyi	Ah-thar	12°21'19.6"	98°47'18.6"	minus 1
93	6-Mar-09	S.Tanintharyi	Pa-htaw-taung	12°27'24.7"	98°34'36.9"	127
94	6-Mar-09	S.Tanintharyi	Shin-ma-kan	12°27'33.5"	98°34'54.7"	minus 4
95	8-Mar-09	S.Tanintharyi	Zet-her	14°04'54.2"	98°13'53.5"	minus 1
96	8-Mar-09	S.Tanintharyi	Ta-laing-taung	14°04'54.6"	98°14'16.9"	66
97	8-Mar-09	S.Tanintharyi	Tha-bya-ywa	14°04'21.8"	98°16'20.6"	minus 9
98	8-Mar-09	S.Tanintharyi	Pa-kar-yi	14°06'21.7"	98°18'10.6"	0

+ Elevation data was not recorded
N.Tanintharyi (North Tanintharyi)
S.Tanintharyi (South Tanintharyi)

Tanintharyi regions were found to be separated by human settlements during our surveys.

The mountains of Rakhine Yoma are covered by patches of primary forest within a landscape dominated by secondary vegetation (largely bamboo) that has resulted from logging and shifting cultivation. Lowland and coastal forests tend to be inhabited by long-tailed macaques. On the other hand, mountainous forests tend to be inhabited by other cercopithecoid species such as pig-tail and rhesus macaques and dusky langurs.

Most of the mangrove forest in the Ayeyarwady Delta region had already been converted to human settlements and agricultural fields, and intensive human population pressures, agriculture, and fishing have destroyed mangroves in most of the surrounding areas. The Meinmahla Kyun Wildlife Sanctuary (MKWS, site No. 59) is the only remaining protected mangrove forest in Ayeyarwady Delta, and affords some protection to the wildlife of this region. We recorded two troops of wild long-tailed macaques inhabiting in MKWS during our surveys.

The Bago Yoma region is believed to be inhabited by a small number of long-tailed macaques. According to interviews, local populations of long-tailed macaques do exist here, but have been almost entirely exterminated from this region, by both human activities and the harsh environment, the dry and deciduous vegetation does not suit the long-tailed macaque. We could only confirm one positive report of pet (Site No. 60, Table 2.3) as evidence of long-tailed macaques in this region.

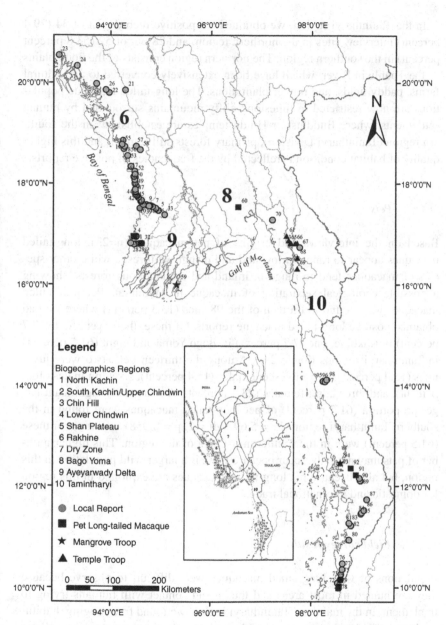

Figure 2.2. Myanmar map showing distribution of long-tailed macaque along coastal area.

In the Tanintharyi region, we obtained ten positive records out of 34 (29.4 percent) interview sites in the northern region, and 28 out of 52 (53.8 percent percent) in the southern region. The northern region consists of the flood plains of the Thanlwin River, which have been extensively converted to agricultural farms, paddy fields, and rubber plantations. The long-tailed macaque populations are now restricted to limestone cliffy mountains surrounded by human settlements, where Buddhist or Hindu temples are established. In the southern region (Tanintharyi Division), primary forests still remain; and this higher quality of habitat condition is reflected by the frequencies of positive reports.

Pets

Based on the interviews and origins of pet macaques (n=23), long-tailed macaques appear to range in lowland and mangrove forests, while other species of macaques tend to range in inland and mountainous forests, showing a possible ecological separation of macaque in Myanmar. Pet long-tailed macaques were found in thirteen of the 98 sites (13.3 percent) where we had obtained positive long-tailed macaque reports. Of these, there were four (30.7 percent) in Rakhine, one (7.7 percent) in Bago Yoma and eight (61.5 percent) in Tanintharyi regions (Table 2.4). Among the thirteen pets, two were juveniles (15.4 percent), two adolescent males (15.4 percent), six adult males (46.2 percent), and three adult females (23.1 percent). A marginally significant larger proportion (61.5 percent) of pet long-tailed macaques were found in the southern Tanintharyi region ($X^2 = 5.692$, df = 2, p = 0.058), and most of these (61.5 percent) were in the southernmost area of the region. This higher number of pets may not only occur because there is a larger wild population in this region, but also because pet long-tailed macaques are exploited in this region for domestic and international trade.

Field survey results

Populations of wild long-tailed macaques were difficult to observe because they are hunted in most areas and thus avoid contact with humans trying to track them. In the mainland Tanintharyi region, we found free-ranging, habituated long-tailed macaque troops: eight in the northern region (Mon and Kayin States) and two in the southern region (Table 2.5). The habitats they were found in varied, and included vegetation that was disturbed (secondary) forests, riverine or coastal mangrove forests, and steep limestone mountains or coastal hills. Nine of the groups we found live in the vicinity of Buddhist or Hindu temples

Table 2.4. *Pet long-tailed macaques*

No	Date	Sex/Age	G.P.S	Region	Origin (Forest)	Captured Method	Purpose
1	19-Nov-04	Ad F	16°58'13.7" 94°27'02.6"	Rakhine	Mangrove	Bow and Arrow	Street-exhibitor
2	19-Nov-04	Ad F	16°58'13.7" 94°27'02.6"	Rakhine	Mangrove	Bow and Arrow	Street-exhibitor
3	19-Nov-04	Juv M	16°58'13.7" 94°27'02.6"	Rakhine	Mangrove	Snare	Trade
4	16-Jan-09	Ad M	16°49.874' 94°23.759'	Rakhine	Mangrove	Snare	Pet
5	1-Dec-04	Ad M**	17°32'50.7" 96°32'30.2"	Bago Yoma	Unknown	Unknown	Pet
6	1-Mar-09	Sad M	9°59'04.2" 98°32'53.7"	South Tanintharyi	Mangrove	Snare	Trade
7	3-Mar-09	Ad M	10°08'59.7" 98°35'33.2"	South Tanintharyi	Mangrove	Unknown	Trade
8	3-Mar-09	Ad M	10°00'48.8" 98°33'46.3"	South Tanintharyi	Mangrove	Unknown	Pet
9	3-Mar-09	Ad F	10°00'48.8" 98°33'46.3"	South Tanintharyi	Mangrove	Unknown	Pet
10	3-Mar-09	Ju M	10°00'48.8" 98°33'46.3"	South Tanintharyi	Mangrove	Unknown	Pet
11	3-Mar-09	Ad M	10°01'13.6" 98°33'49.8"	South Tanintharyi	Unknown	Unknown	Pet
12	5-Mar-09	Sad M	12°13'27.2" 98°53'36.1"	South Tanintharyi	Mangrove	Snare	Trade
13	5-Mar-09	Ad M*	12°21'19.6" 98°47'18.6"	South Tanintharyi	Mangrove	Snare	Trade

* Crest at the vertex
** Mixed characters between long-tail and pig-tail

Table 2.5. *Long-tailed macaque troops encountered in Myanmar*

Biogeographic Region	Locality	Habitat	G.P.S (N, E)	Group Size
North Tanintharyi	Bayin Nyi Naung	Isolated mountain	16°58.2', 97°29.6'	ca 40–50
	Taung-ga-lay	Isolated mountain	16°53.15',97°32.1'	ca 40–50
	Kaw-kun	Isolated mountain	16°49.4', 97°35.2'	ca 51–60
	Ya-thae-pyan	Isolated mountain	16°50.1', 97°34.2'	ca < 10
	Mt.Zwe-ka-pin	Isolated mountain	16°49,7', 97°40.5'	ca 51–60
	Kaw-ka-thaung	Isolated mountain	16°49.7', 97°42.4'	ca < 10
	Indian Single Rock	Isolated mountain	16°19.3', 97°42.6'	ca 70–80
	Kha-yone-cave	Isolated mountain	16°32.0', 97°42.9'	Not recorded
South Tanintharyi	Pa-htaw-taung	Isolated mountain	12°27.4', 98°34.6'	ca 40–50
	Shin-ma-kan	Mangrove forest	12°27.6', 98°34.9'	ca 40–50
Ayeyarwady Delta	Meinmahla Kyun Wildlife Sanctuary	Mangrove forest	15°52'–16°05' 95°14' – 95°21'	ca 40 – 50* ca 40–50*

* Meinmahla Kyun Wildlife Sanctuary Office reported on 19 July 2004

on steep limestone mountains. Although natural resources appear poor in these limestone mountain habitats, these troops are provisioned to varying degrees from their interactions with monks and pilgrims. The groups we observed averaged about 50 individuals in size, but had a wide range of distribution (range: 10–100; Table 2.5).

Population estimate

We estimated the population of long-tailed macaques throughout the regions that we interviewed and surveyed. In the Bago Yoma and Ayeyarwady Delta regions, the population of long-tailed macaques is estimated to be between 90 and 300 individuals, based on the records from Meinmahla Kyun Wildlife Sanctuary. In Mon and Kayin States, we estimated a population between 350 and 700, and these figures are based on our sighting of eight groups. In the Rakhine region, we traveled 300 km, and found 31 sites with macaques. We estimated the density of macaque in this region to be 0.155 individuals/ km², and estimated the population to be between 1,300 and 13,000 individuals. The population of long-tailed macaques in mangrove forests (decreasing by 8 percent from 1996) was estimated to be between 2,250 and 22,500 individuals. Therefore, the total population estimated for the Rakhine region was between 3,550 and 35,500. In the Southern Tanintharyi region, from Kawthaung to Myeik, we traveled about 390 km, and on this route found 28 sites with positive reports for long-tailed macaques. We estimated a density of 0.215 individuals/km², and estimated the population to be between 2,760 and 27,600. We separately calculated the population in mangrove forests, which has an area of 2,600 km² (Leimgruber *et al.*, 2005). We estimated the population of long-tailed macaques in mangrove forests to be between 4,380 and 43,800 individuals. In the southern Tanintharyi region, where villages are sparse and the forests are more intact, the density of long-tailed macaques may be higher. We therefore estimated the total population for the southern Tanintharyi region to be between 7,140 and 71,400 individual.

Based on these figures in the various regions, we calculated the total population of long-tailed macaques in Myanmar to be somewhere between 11,130 and 107,900 individuals. This is a wide range, but we prefer to be conservative in our estimation, and further census work will be needed to better refine these numbers. The southern Tanintharyi region holds the greatest population (64.1–66.4 percent), and the Rakhine region the second greatest (31.9–33.0 percent). The northern Tanintharyi region has the second-smallest population (3.14–0.65 percent), and Ayeyarwady Delta and Bago Yoma have the smallest (0.81–0.28 percent).

Human-macaque relationships

Conflict between humans and macaques was found to occur in Rakhine State, Bago Yoma and Tanintharyi Division (Table 2.6), and the most common conflict reported was by farmers having macaques raid their crops. In the Rakhine region (site No. 48, Table 2.3), long-tailed macaques were reported to have damaged nipa-palm fruits by drinking nipa-palm juice. In the Ayeyarwady Delta regions, conflict between humans and long-tailed macaques was not reported, perhaps because long-tailed macaques have been extensively exterminated, excepting two troops in the Meinmahla Kyun Wildlife Sanctuary.

Table 2.6. *Local name and conflicts of long-tailed macaque*

			Conflicts and threats			
Region	Ethnicity	Local Name	Damage crop	Hunting	Eating	Trade
Rakhine	Rakhine, Chin, Bamar	de-kyin-myauk	Yes	Yes	Yes	Yes
Bago Yoma	Bamar, Kayin	myauk-ta-nga	No	Yes	Yes	No
Ayeyarwady Delta	Bamar, Kayin	myauk-ta-nga	No	No	No	No
North Tanintharyi	Mon, Kayin, Bamar	myauk-mie-shay	Yes	No	No	Yes
South Tanintharyi	Dawei, Myeik	za-yet-taw-myauk	Yes	Yes	Yes	Yes

   Hunting long-tailed macaques for food and trade was observed in the Rakhine and southern Tanintharyi regions (Table 2.6), hunters use snare, bow with poisoned arrows or gun (fusil). Two wildlife meat restaurants in the Rakhine region and four in southern Tanintharyi region were found. In these restaurants, myauk-chay-kha (cooked monkey's meat and digestive tract) is a popular meal for local people. Monkeys (macaques and langurs) were sold at the price of 15,000 kyats (equivalent to $15 USD) by restaurants in Tanintharyi. In the southern part of Tanintharyi, hunting pressure on long-tailed macaques appears heavy for international trade, smuggling through Kawthaung to Ranong, although neither the quantity of macaques traded nor the sources of the trading were assessed.

   In Tanintharyi, in 2004–2005, Chinese entrepreneurs ran monkey farms to collect *M. fascicularis aurea* and 3,000 monkeys went through this facility in 2005 for export to China or to developed countries via China (Shwe Pyi Thar Report, 2006). They constructed cages, which they falsely called "breeding sites" at the base of Pa-htaw-taung hill (Site No.93). They purchased long-tailed macaques at 5,000 kyats to 20,000 kyats (about $5 to $20 USD) per individual from villagers around the Myeik Archipelago. According to interview, at least 1,000 long-tailed macaques were exported during 2005 and 2006. However, the Forest Ministry of Myanmar banned this trade in 2006 (Myanmar Wildlife Protection Law, 1994). The Chinese company abandoned its business and released macaques at Shin-ma-kan mangrove forest (Site No. 94) and Thandar Island in 2006. In the present study, a troop of semi-wild long-tailed macaques (40–50 individuals) was found in site No. 94.

Discussion

Current distribution of the long-tailed macaque in Myanmar

The distribution of the Myanmar subspecies of long-tailed macaques (*Macaca fascicularis aurea*) extends from the southernmost (Kawthaung, 9°58'N, Tanintharyi) to the northwestern most parts of Myanmar (Kyay Taw, 20°32'N, Rakhine) near Bangladesh along the coastal regions (Figure 2.2). The mangrove forests and riverine lowland forests that they inhabit are continuous along the coasts of the Bay of Bengal and the Andaman Sea. Although we conducted field surveys only in the mainland of Tanintharyi, the Myeik Archipelago, which includes more than 800 islands, also harbors for long-tailed macaques (Tun Yin, 1967; Fooden, 1995), and this will need to be studied further in order to fully assess the population of long-tailed macaques in Myanmar. The population of *M. f. aurea* also extends into southeastern Bangladesh (Khan and Ahsan, 1986) and southwestern Thailand (Malaivijitnond *et al.*, 2005).

The Myanmar long-tailed macaques' distribution appears restricted to the coastal regions and was mainly found in the Rakhine, Ayeyarwady Delta, and Tanintharyi biogeographic regions, with a small population in the Bago Yoma region near the coast. Long-tailed macaques were not found in other biogeographic regions, such as Dry zone, Shan Plateau and Chin Hill. This is likely because long-tailed macaques have not adapted to these regions of higher latitude with drier, seasonal climates and dry forests with mixed deciduous forest vegetation, or that they are outcompeted by other macaque species that live in those regions.

The total mainland population of long-tailed macaque in Myanmar is broadly estimated to be between 11,130 and 107,900. However, these estimates depend on just a preliminary survey and several assumptions and therefore are only a rough estimate of the mainland population. Future work will need to confirm and refine the population census. About 64.0–66.0 percent of the population occurs in the southern Tanintharyi region, 32.0–33.0 percent in the Rakhine region, 0.28–0.81 percent in the Ayeyarwady Delta and Bago Yoma regions, and 0.65–3.20 percent in the northern Tanintharyi region. Thus, Rakhine and southern Tanintharyi are the major ranges of long-tailed macaques in Myanmar. In addition to these two ranges, a small number of troops were found scattered in the Ayeyarwady Delta, Bago Yoma, and northern Tanintharyi regions.

Habitat degradation and fragmentation

It is possible that the distribution of long-tailed macaques in Myanmar has been reduced by human activities. At present, the two major ranges of long-tailed macaques in Myanmar, the Tanintharyi and Rakhine regions, are separated by the Ayeyarwady Delta and northern Tanintharyi regions. These areas, which were previously mangrove and lowland forests around the estuaries of the Ayeyarwady, Sittaung and Thanlwin Rivers have been converted to human settlements and this has likely impacted long-tailed macaque populations. In the northern Tanintharyi and eastern Ayeyarwady Delta regions, long-tailed macaque habitats have been becoming degraded for the last 30 years. Wide areas of land were converted to agricultural fields (rice paddy fields) and human settlements, and forests were cut for timber, fuel and constructed of the country's infrastructure. Thus, in these areas, the troops we found were restricted to steep limestone mountains with temples. These populations appear to be isolated from each other by human settlements, and this may be causing there to be a higher ratio of males in some groups than is typical, such as in the Bayin Nyi Naung Mountain troop, (1:1.2 of male and female ratio) (Aye Mi San, 2007). Male dispersal may now be restricted and therefore, the gene flow between troops will be an important area for future research in Myanmar.

In the Ayeyarwady Delta, vast areas of mangrove forest have been destroyed in the last 30 years by deforestation for the production of fuel wood and the expansion of agriculture and aquaculture (Nay Win Oo, 2002). The annual deforestation rate has been as high as 5.6 percent, and the total forest cover declined from 24 percent of the total area in 1989 to 12 percent in 1998 (Oo, 1998). The total decrease has been from 3,860 km^2 in the early 1900s, to 1,770 km^2 in the 1990s (Oo, 1998). In our survey, two troops of wild long-tailed macaques were identified in the Meinmahla Kyun Wildlife Sanctuary (area of 136 km^2), which is protected by the Myanmar Wildlife Protection Law (1994), and therefore long-tailed macaques appear restricted to protected areas in this region of Myanmar.

There are misconceptions by the human inhabitants of macaque territory that long-tailed macaques are quite resilient to the impact of human activities. Since long-tailed macaques are easily noticed in temples and small, private zoos, and since many forested areas are inhabited by some macaques, the wild local long-tailed macaque population is thought to be large by the local people. However, evidence we found suggests that the population has been reduced over the last few decades by habitat degradation and hunting. Therefore, we suggest that the risk of localized extinctions of populations may be rather high.

The two major ranges of long-tailed macaques, the Rakhine and southern Tanintharyi regions, have undergone significant environmental degradation. In the Rakhine region, the forest cover was 62 percent in 1989, and the annual deforestation rate was rather low at 2.6 percent (NCEA, 2006), because there were few big cities, road conditions were poor, and the coastal area of the

Rakhine region was less densely populated. However, the lowland forest that long-tailed macaques inhabit had been deforested much more than other types of forest, because it is along forest edges, close to human settlements and is good for exploitation. Moreover, the coastal mangrove forests have been encroached upon for paddy cultivation and shrimp farming. In the southern Rakhine area, the habitat conditions have rapidly degraded. Bamboo forests have expanded, and erosion and gullies have been observed (Geissmann *et al.*, 2009). A number of villages have been established by immigrants of Bamar, Rakhine, and Chin ethnicities, and lowland forests have been cultivated. A considerable portion of immigrants subsists on timbers and non-timber forest products such as bamboo, bamboo shoot, mushrooms, and wildlife, including long-tailed macaques, which are hunted by snares, poisonous arrows and guns.

In the southern Tanintharyi region, 9 percent of the tropical rainforest was lost between 1990 and 2000, and 6,350 km^2 were degraded from closed forest to degraded forest (NCEA, 2006). Both illegal and legal logging has had a heavy impact on non-human primate populations (Htin Hla *et al.*, 2003). Habitat loss has also resulted from the conversion of forests to agricultural uses (i.e., wide areas of plantation development for commodity crops), aquacultural farms (i.e., prawn, shrimp, and soft-shell crabs) and construction of roads and other infrastructure. Increased employment opportunities are likely to encourage human immigration, which will put additional pressure on natural resources and habitat. Recently, a government project of an oil palm plantation was realized and rubber and betel nut plantations are increasing in scale. These plantations were established in lowland forests, destroying the habitats of long-tailed macaques and much other wildlife.

Hunting pressure is also high in the southern Tanintharyi region. In the majority of areas in Tanintharyi, primates appear to be hunted for village-scale consumption. They are also hunted for trading. We encountered wildlife meat restaurants between Kawthaung (9°58'N) and Tanintharyi town (12°06'N) that purchase monkeys at a considerably high price (equivalent to $15 USD/individual). Living animals and wildlife products are also internationally smuggled through the border towns, Kawthaung (Myanmar) to Ranong (Thailand). Wildlife products fetch a higher price across the border in Thailand via Kawthaung, and thus there is strong incentive to trade. A living monkey was reported to bring 50,000 kyats ($50 USD) in the area between Bokpyin and Kawthaung, but it was reported to be priced three times higher in Thailand. Between Kawthaung and Bokpyin, 247 km apart, there are few villages along the graveled road ("Tanintharyi highway"). One or two public buses a day connect Kawthaung, Bokpyin, and Myeik. If transport were easier, the wildlife trade would be worse.

Human-macaque conflict

Crop raiding by macaques has not become a serious problem across Myanmar, although in some regions it presents a significant challenge for farmers. Conflict has been reported with long-tailed macaque populations living close to or inside human settlements in other Southeast Asian countries (Aggimarangsee, 1992; Wheatley, 1999; Cortes and Shaw, 2006; Zhao, 2005; Fuentes *et al.*, 2008; Sha *et al.*, 2009), and macaques are commonly found inhabiting temples or city parks where they are provisioned and protected from hunting and predators (Aggimarangsee and Brockelman, 2005; Malaivijitnond *et al.*, 2005). Because of human land development, long-tailed macaques have been forced to live in increasing proximity to human settlements in Java (Iskandar *et al.*, 2008). In Thailand, Indonesia, Gibraltar, China, and Singapore macaque populations can easily increase in size (see Box 6.1), and in Thailand have been reported to reach group sizes of over 200 animals (Malaivijitnond and Hamada, 2008). In contrast, long-tailed macaque troops inhabiting temples in Myanmar tend to be smaller than they are in Thailand, and maybe this is because long-tailed macaques inhabiting temples in Myanmar suffer more from human activity (see Gumert, Chapter 1).

Conclusions

M. fascicularis aurea is distributed from the southern to northwestern borders in Myanmar along the coastal regions. Habitat loss and degradation, hunting, and the wildlife trade may be having negative impacts on the long-tailed macaque, and future work will need to monitor the effect of human activities on long-tailed macaques in this region. Human land-use is causing forest habitats to shrink, and there are, therefore, fewer habitats for long-tailed macaques and their populations appear to be becoming isolated from each other. We will need more extensive surveys of Myanmar long-tailed macaques in the future to fully assess their population and the effects of human activity on them. In particular, surveys will need to be conducted throughout the Myeik Archipelago to determine the extent of Myanmar's island populations of long-tailed macaques. This chapter presents the first census showing the population of long-tailed macaques on mainland Myanmar and thus we provide the first set of data on the conservation status of a data deficient sub-species (*M. fascicularis aurea*) (Ong and Richardson, 2008). We provide evidence that this subspecies may be facing several threats from habitat conversion, habitat fragmentation, hunting pressure, and international trade. The future effects of human activity on their population remain uncertain and needs to be closely monitored.

Box 2.1 Preliminary survey of the long-tailed macaques (*Macaca fascicularis*) on Java, Indonesia: Distribution and human–primate conflict

Randall C. Kyes, Entang Iskandar and Joko Pamungkas

Despite presumed abundance and widespread distribution, little recent data exists on the status of the long-tailed macaque (*Macaca fascicularis*) population in Indonesia. Currently, the long-tailed macaque is categorized as Least Concern (ver 3.1) in the IUCN Red List, a designation based in part on "...its wide distribution..." and "...presumed large population..." (Ong, P. and Richardson, M. 2008). In an effort to provide current information on the distribution of the long-tailed macaque and assess increasing media reports of growing human–primate conflict on Java, Indonesia, we conducted a preliminary survey of the island from 6–12 January 2009.

The survey originated in Bogor, West Java and involved a west-to-east loop of the island covering a total of 2,160 km. Time and funding constraints limited our ability for a more extensive survey of the island. We visited several target sites based on reported macaque sightings by media and forestry officials, and searched for additional sites based on leads from local people along the way. Travel and observation occurred from 7 am until 8 pm daily and involved the use of secondary roads to allow for frequent stops to query local people. We stopped on average, every 15–20 km (i.e., approximately 100 stops along our route), and surveyed more than 250 people to inquire about monkey sightings/conflict in the area. When we received a report of monkeys in the area, we traveled to the location (often into remote village and forest areas) to investigate the report. At each site where monkeys were reported, we walked around the immediate area to permit observation and confirmation of monkey presence and possible human-primate conflict. In cases where we were not able to confirm the presence of monkeys via our personal observation (i.e., "confirmed sighting"), we coded the location as "reported sighting," defined as independent reports of monkeys by at least three individuals who were not associated with one another.

Over the seven-day period, we identified and visited a total of 22 sites along our route where wild, free-ranging long-tailed macaques were reported. The sites included nature reserves (*cagar alam*), agricultural areas, villages, local tourist areas (e.g., parks, picnic areas, recreation

Figure 2.3. (a) A Javanese woman feeding a resident group of long-tailed macaques in the village of Cikakak, in Wangon, Central Java. Macaques were observed in the village and around its cemetery, and they frequently raided crops and homes. (b) Long-tailed macaques at a cemetery in Tulung Agung, East Java. The macaques were fed rice daily by cemetery caretakers and some conflict was reported.

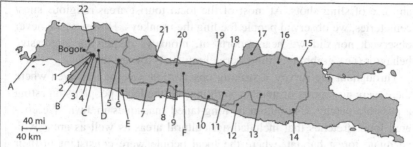

Figure 2.4. The distribution of long-tailed macaques on Java, Indonesia. The line indicates the survey route during the January 2009 survey. Dots with numbers represent the 22 sites where wild populations of macaques were reported/confirmed. Dots with letters represent additional locations where wild populations of long-tailed macaques have been confirmed by the authors within the past two years (since 2008). Key: 1, Taman Safari *(C)*; 2, Gunung Mas *(R)*; 3, Cagar Alam Telaga Warna *(C)*; 4, Puncak Pass *(R)*; 5, Maribaya *(R)* [30], 6, Curug Omas *(C)* [20]; 7, Cisarua, Garut *(R)* [50]; 8, Cimanggu *(R)*; 9, Cikakak *(C)* [150]; 10, Kaligondang *(R)*; 11, Maja Singi *(R)*; 12, Wanagama *(R)*; 13, Tawangmangu *(C)* [150]; 14, Tulung Agung *(C)* [110]; 15, Bektiharjo *(C)* [7]; 16, Sumber Semen *(R)*; 17, Colo *(R)* [100]; 18, Goa Kreo *(C)* [300]; 19, Kutosari *(R)*; 20, Jati Barang *(R)*; 21, Cirebon *(C)* [100]; 22, Cibubur *(C)*; A, Taman Nasional Ujung Kulon; B, Taman Nasional Gunung Halimun; C. Cagar Alam Gunung Simpang; D, Taman Nasional Gunung Gede Pangrango; E, Cagar Alam Leuweung Sancang; F, Gunung Slamet. In the key, *(R)* represents "reported sighting" and *(C)* represents "confirmed sighting." The numbers in brackets indicate the "reported" population size at sites where a number was provided by the people surveryed. Map adapted from Google maps.

sites), religious sites, and cemeteries (Figure 2.3). We confirmed the presence of long-tailed macaques at 10 of the locations, and recorded "reported sightings" at the other twelve sites (Figure 2.4). Due to time constraints, we were not able to generate estimates of abundance at each location. The "reported" population sizes at the 22 sites ranged from "many monkeys" to an approximate number. Among ten sites where a reported estimate was given (Figure 2.4), the average "reported" population size was 102 monkeys (range: 7–300).

Reports of human-primate conflict were noted at 20 of the 22 sites and typically involved various forms of crop raiding (e.g., corn, papaya, sweet potatoes, coconuts) from private gardens and agricultural areas or stealing food (e.g., produce, snacks, drinks) from vendor stands and small restaurants. The typical response from the farmers and merchants included, hollering at/chasing the monkeys, throwing rocks, waving/banging a stick, and use of sling shots. At most of the local tourist areas/religious sites/ cemeteries, we observed people feeding the monkeys. However, we never observed, nor did we hear reports of, monkeys displaying aggressive behavior (i.e., grabbing, biting, scratching, etc.) toward people.

Throughout the survey, a striking observation was the vast area where there were no reports of monkey sightings by the local people, suggesting a possible patchy distribution of long-tailed macaques in Java. We covered long stretches that included agricultural areas as well as areas with adequate forest habitat, where the local people were consistent in their responses of "no monkeys in the area." The fact that long-tailed macaque populations are often located in areas of human habitation, where sightings and conflict occur daily, may lead to assumptions of over-abundance in regions where actual population size may be much smaller then perceived. As such, we believe efforts should be made to conduct thorough population surveys of the long-tailed macaques throughout their range in Indonesia. Our preliminary survey, reported here, is just the first step in an ongoing effort to confirm the locations of long-tailed macaque presence thereby helping to "fill-in-the-blanks" regarding their distribution as we move ahead with plans to conduct an island-wide population survey of the long-tailed macaque on Java.

References
Ong, P. and Richardson, M. (2008). *Macaca fascicularis*. In: IUCN 2009. IUCN Red List of Threatened Species. Version 2009.2. www.iucnredlist.org. Last accessed on 18 January 2010.

Acknowledgements
We thank Erik McArthur and Christine Howard for their expert assistance with the map design and GIS graphics. This study was supported in part by NIH Grant RR-00166.

Acknowledgements
We would like to thank all interviewees for their kind responses to our survey. Our special thanks go to the reverend of Bayin Nyi Naung Mountain, "Bat-dan-ta Pyin-nyar-won-tha" and the forest staff of the Wildlife Sanctuaries for their permission to do this research. This study was supported by the Japanese Society for the Promotion of Science (nos. 16405017 and 20255006.

References

Aggimarangsee, N. 1992. Survey for semi-tame colonies of macaques in Thailand. *Natural History Bulletin of the Siam Society* **40**: 103–166.

Aggimarangsee, N. and Brockelman, W. Y. 2005. Monkey-human interactions in Thailand. *American Journal of Physical Anthropology* **Suppl.128(41)**: 62–63.

Amato, G., Egan, M., and Rabinowitz, A. 1999. A new species of muntjac *Muntiacus putaoensis* (Artiodactyla: Cervidae) from northern Myanmar. *Animal Conservation* **2**:17.

Aye Mi San. 2007. Distribution status of long-tailed Macaques (*Macaca fascicularis aurea*) in some areas of Myanmar and its behavioral study in Bayin Nyi Naung Mountain, Kayin State. Ph. D dissertation, University of Yangon.

Bates, P. J. J., Struebig, M. J., Rossiter, S. J., Kingston, T., Sai Sein Lin Oo, and Khin Mya Mya. 2004. A new species of *Kerivoula* (Chiroptera: Vespertilionidae) from Myanmar (Burma). *Acta Chiropterologica* **6(2)**: 219–226.

Bird Life International. 2005. Myanmar Investment opportunities in biodiversity conservation. www.birdlife.org.

Cortes, J. and Shaw, E. 2006. The Gibraltar macaques: management and future. In *The Barbary Macaque: Biology, Management and Conservation*, Hodges J. K., Cortes J (ed.). Nottingham University Press. pp 199–210.

Delson, E. 1980. Fossil macaques, phyletic relationships and a scenario of deployment. In *The Macaques: Studies in Ecology, Bhavior and Ecology*. New York, van Norstrand. pp. 10–30.

Eudey, A. 2008. The crab-eating macaque (*Macaca fascicularis*): Widespread and rapidly declining. *Primate Conservation* **23**: 129–132.

FAO (Food and Agriculture Organization of the United Nations). 1985. Nature conservation and national parks. Burma, survey data and conservation priorities. Technical report 1.

Fooden, J. 1995. Systematic review of Southeast Asian long-tail macaque, *Macaca fascicularis* (Raffles, 1821). *Fieldiana Zoology* **81**: 1–205.

2000. Systematic review of the rhesus macaque, *Macaca mulatta* (Zimmermann, 1780). *Fieldiana Zoology* **96**: 1–180.

Fuentes, A., Kalchik, S., Gettler, L., Kwiatt, A., Konecki, M., and Jones-Engel, L. 2008. Characterizing human-macaque interactions in Singapore. *American Journal of Primatology* **70(9)**: 879–883.

Geissmann, T., Grindley, M., Momberg, F., Lwin, N., and Moses, S. 2009. Hoolock gibbon and biodiversity survey and training in southern Rakhine Yoma, Myanmar. *Gibbon Journal* **5**: 7–27

Hla, H., Sein, Myo Aung, Moses, S., Eames, J., and Nyunt Tin, Saw. 2003. Gurney's Pitta Survey and Biodiversity Conservation Assesment in Tanintharyi Division, Myanmar, unpublished.

Iskandar, E., Randall, C.K., and Joko, P. 2008. Long-tailed macaques (*Macaca fascicularis*) as an agricultural threat in Java, Indoneisa. 22nd Congress of the International Primatological Society Abstracts.

Khan, M. A. R. and Ahsan, M. F. 1986. The status of primates in Bangladesh and a description of their forest habitats. *Primate Conservation* 7(April): 102–109.

Kyaw Nyunt Lwin. 1995. *Mammals of Myanmar*. Rangoon.

Leimgruber, P., Daniel, S. K., Marck, S., Jake, B., Thomas, M., and Melissa, S. 2005. Forest cover change patterns in Myanmar (Burma) 1990–2000. *Environmental Conservation* 32(4): 356–364.

Malaivijitnond, S., Hamada, Y., Varavudhi, P., and Takenaka, O. 2005. The current distribution and status of macaques in Thailand. *Natural History Journal of Chulalongkorn University* Suppl 1: 35–45.

Malaivijitnond, S. and Hamada, Y. 2008. Current situation and status of long-tailed macaques (*Macaca fascicularis*) in Thailand. *Natural History Journal of Chulalongkorn University* 8(2): 185–204.

Molur, S., Brandon-Jones, D., Dittus, W., *et al.* (eds). 2003. Status of South Asian primates: Conservation and management plan (C.A.M.P.) Workshop Report. Zoo Outreach Organisation/Conservation Breeding Specialist Group, South Asia, Coimbatore, India.

NCEA (National Commission for Environmental Affairs). 2006. National performance assessment and subregional strategic environment framework in the Greater Mekong Subregion. ADB/TA No. 6069-REG,prepared by Project Secretariat, UNEP Regional Resource Centre for Asia and the Pacific. 323 pp.

Ong, P. and Richardson, M. 2008. *Macaca fascicularis*. IUCN 2010: IUCN Red List of Threatened Species. www.iucnredlist.org.

Oo, N. W. 2002. Present state and problems of mangrove management in Myanmar. *Trees—Structure and Function* 16(2–3): 218–223.

Oo, T. P. 1998. Integraded Coastal Zone management. Proceeding of the UNESCO regional seminar Ecotone VII integraded coastal zone management in Southeast and East Asia. 15–19 June 1998, Yangon, Myanmar.

Parr. J. and Tin Than. 2007. *A Guide to the Large Mammals of Myanmar*. Yangon.

Richard, A. F., Goldstein, S. J., and Dewar, R. E. 1989. Weed macaques: The evolutionary implications of macaque feeding ecology. *International Journal Of Primatology* 10(6): 569–594.

Sha C. M., Gumert, M., Lee P. Y.-H., Fuentes, A., Rajathurai, S., and Chan, S. 2009. Status of long-tailed macaque *Macaca fascicularis* in Singapore and implications for management. *Biodiversity and Conservation* 18(11): 2909–2926.

Shwe Pyi Thar. 2006. Myauk-ta-nga breeding project. Project report by Shwe Pyi Thar Co-operative Ltd.

Tun Yin. 1967. Wild animals of Burma. *Rangoon Gazette*, Rangoon.

Wheatley, B. P. 1999. *The sSacred Monkeys of Bali*. Prospect Heights, IL, Waveland Press Inc.

Zhao Q. K. 2005. Tibetan macaques, visitors, and local people at Mt. Emei: Problems and countermeasures. In *Commensalism and Conflict: The Human-Primate Interface*. J. D. Paterson and J. Wallis (ed.) Norman, OK: American Society of Primatologists. 376–399.

Appendix 2.1 INTERVIEW SURVEY FORM (own format)

1. **Interview No. ( )**   **Date** ....
2. **Interviewee Name**..
3. **State / Division**...
3. **Name of village** .... **Ethnicity**..........................
4. **G.P.S. (Latitude, Longitude, Altitude)**.......................................
5. **Species confirmation with the help of macaque's photograph**.....................
    (1) Long-tailed   (2) Rhesus   (3) Pig-tailed   (4) Stump-tailed
    (5) Assamese macaque
6. **If, we found pet long-tailed macaque**
    (a) *Where did they catch?* ....
    (b) *How far forest from here?* ....
    (c) *When/ How did they catch?*..
    (d) *What's purpose for keeping?*...
7. **Observation on Morphology**
    (a) *Pelage color and infant's color* ....
    (b) *Tail Length (TL)* ....
    (c) *Crown-rump Length (CRL)* ....
8. **Conflicts between human and long-tailed**
    (a) Damage field?   (b) Hunting ?   (c) Eating ?   (d) Trading?
9. **Hunting Method** *(a) Snare   (b) Gun   (c) Arrow   (d) Poisonous leaf*
10. **Livelihood of villagers** *(a) Forest   (b) Agriculture   (c) Aquaculture*

3 Distribution and present status of long-tailed macaques (Macaca fascicularis) in Laos and their ecological relationship with rhesus macaques (Macaca mulatta)

YUZURU HAMADA, HIROYUKI KURITA, SHUNJI GOTO, YOSHIKI MORIMITSU, SUCHINDA MALAIVIJITNOND, SITIDETH PATHONTON, BOUNNAM PATHONTON, PHOUTHONE KINGSADA, CHANDA VONGSOMBATH, FONG SAMOUTH AND BOUNTHOB PRAXAYSOMBATH

Introduction

Lao Peoples' Democratic Republic (i.e., Laos), situated in the center of the Indochina Peninsula and encompassing 14–22.5°N, consists of diverse environments, including Xay Phou Louang (Annamite Cordirella) in the east, the Mekong River in the west, and plains in between. More than 15 percent of the national land area has been designated as National Protected Areas (NPAs) by the government of Laos since 1993. Compiling museum data and literature, Fooden (1980, 1995) sketched the distribution of long-tailed macaques in Laos. In addition, assessments were carried out in the 1990 on Laos's wildlife in these NPAs and these surveys also reported the distribution of macaques (Duckworth *et al.*, 1999). These reports suggested that long-tailed macaques were distributed only in southern-most Laos, which is a region consisting of mountainous areas (i.e., Bolaven plateau, Xay Phou Louang) and tributaries of

Monkeys on the Edge: Ecology and Management of Long-Tailed Macaques and their Interface with Humans, eds. Michael D. Gumert, Agustín Fuentes and Lisa Jones-Engel. Published by Cambridge University Press. © Cambridge University Press 2011.

the Mekong River. However, the present distribution and current population status of long-tailed macaques are not known.

Primate fauna is rich in southern-most Laos, which includes prosimians (i.e., lorises), cercopithecids, colobines, and lesser apes (i.e., gibbons) (Duckworth *et al.*, 1999). Laos contains highly endangered species, such as red-shanked douc langurs (*Pygathrix nemaeus*) and yellow-cheeked gibbons (*Nomascus leucogenys gabriellae*; Duckworth *et al.*, 1999), as well as typically common monkeys such as the rhesus (*Macaca mulatta*) and long-tailed macaque (*M. fascicularis*). All non-human primates in Laos are currently under threat of extinction because their habitats have been lost to commercial logging, hydraulic power development, and agriculture, all of which have been driven by increases in human population, economical development dependent on natural resources, and foreign capital (Duckworth *et al.*, 1999). Hunting pressure on non-human primates, although banned by regulation (i.e., Decree of the Council of Ministers N185/CCM, 21 October 1986), is still high for foods, trading, and pest control against agricultural crops (Hamada pers. observation). If this situation continues unchecked, non-human primate populations will be locally or entirely exterminated. To ensure their sustainability, their distribution and diversity should be studied and delineated.

Five species of macaques are distributed in southern Laos, and they have shared habitats by utilizing different ecological strategies, e.g., use of forest types (broad-leaf evergreen forest vs. others) and arboreality vs. terrestriality (Fooden, 1982). Long-tailed macaques range widely throughout Southeast Asia (Fooden, 1995) and appear tough and adaptable to various conditions. Their exact phylogeographic histories have not yet been elucidated, but they are important components in their natural eco-systems. Long-tailed macaques are thought to ecologically compete with rhesus macaques because they belong to the same species group (*fascicularis* group) (Fooden, 1976), share similar biological characteristics, are phylogenetically closely related, and therefore have overlapping biological needs. In this study, we sought to investigate how they are distributed and assessed the degree of their interspecific competition.

The region of overlap between these two species was reported to occur between 15°N and 20°N (Fooden, 1995, 2000), although the region of overlap has never been assessed well in Laos. It is also difficult to document past distribution patterns of the two species to reconstruct their level of historical overlap and how environmental changes might have affected it (e.g., the flowing course of the Mekong River (Meijaard and Groves, 2006) the climates of glacier and inter-glacier alterations in Plio-Pleistocene).

It is well documented that hybridization between long-tailed and rhesus macaques occurs in the eastern half of the Indochinese Peninsula, including Laos (Fooden, 1964, 1995, 2000; Tosi *et al.*, 2002; Hamada *et al.*, 2006,

2008). However, hybridization has yet to be fully studied in Laos, or in many of the regions where it occurs. Recently, artificial disturbance of macaques has occurred in many SE Asian countries, including translocation of populations and release of pet macaques, and sometimes this movement causes human-induced inter-mixing of species (Hamada *et al.*, 2006a). For example, in Vietnam, where long-tailed macaques were distributed in the area of 16°N or lower (Dang *et al.*, 2008), their range has widened to latitudes higher than 18°N (Nhan, 2004), perhaps due to transfer by humans, and rhesus macaques were also artificially transferred to lower latitudes (Cat Tien National Park, around 11.5°N; Polet *et al.*, 2004), which may be expanding their region of overlap and thus increasing the potential for hybridization to occur.

Tail length is one indicator of hybridization. The Indochinese long-tailed macaque population, distributed to the north of the Isthmus of Kra (ca. 10.5°N), tends to have shorter tails (100–120 percent or crown-rump length; Hamada *et al.*, 2008) than its counterpart distributed to the south (120–130 percent; Hamada *et al.*, 2008), which suggests the introgression from rhesus macaques. Rhesus macaques in the eastern half of the Indochina Peninsula, on the other hand, have longer tails (50–80 percent; Hamada *et al.*, in preparation) than their conspecifics to the north (China and vicinity, 35 percent, Hamada *et al.*, 2005) and west (India and vicinity, 45 percent, Hamada *et al.*, 2005), which indicates the Indochinese rhesus population has received introgression from long-tailed macaques. Aside from morphological evidence, hybridization of the two species has also been suggested from molecular analyses (Tosi *et al.*, 2002; Denduangboripant *et al.*, 2005; Malaivijitnond *et al.*, 2008). Rhesus macaque populations in Laos are of importance to delineate the history and mechanism of hybridization because of their natural overlap with long-tailed macaques here.

Based on field surveys, we report here the distribution, tail-length variation, and present status of the rhesus and long-tailed macaques in southern Laos. Taking the distribution patterns of macaques in Thailand (Malaivijitnond *et al.*, 2005) and Vietnam (Hamada *et al.*, 2010) into consideration, we provide implications for macaque phylogeography, including impact of hybridization on long-tailed and rhesus macaques in Laos. We also estimate the national population of long-tailed macaques and discuss the conservation issues and the status of long-tailed macaques in Laos.

Methods

We carried out field surveys to cover southern Laos (Figure 3.1) and interviewed people along several routes. From 8–13 July 2005, 18–26 January 2007, and 6–14 September 2008, we surveyed from Bolikhamxay to the southern

Figure 3.1. Location of survey in Lao PDR.

provinces of Laos (Champasak and Attapeu Provinces, Figure 3.2), from Thakhek to the border to Cambodia along National Route No 13 (NR-13), along NR-9 from Savannakhet to Den Sawan, national boundary to Vietnam; along NR-23 and NR-16, from Pakse, via Thateng and Sekong to Attapeu; and then from Attapeu to the west on NR-18 to Sanamxai and to the east on the new NR to Ban Xe Xou (Figure 3.3). In the survey area, the Mekong River flows in the west and borders Thailand, and the Xay Phou Louang (Annam Mountains) runs north to south in the east. Between them are plains and gently sloping areas with dry forest. These geographical conditions contain three environmental types: the mountainous area, the Mekong River side, and the plains in between (Duckworth *et al.*, 1999). In the plains area, there are low profile hilly areas, e.g., Phou Xang He NPA, which are covered by deciduous dipterocarpus forests. In the South from about 15.25°N, the Mekong River flows inside Laos and plains are located west of the Mekong. In the far south, the Bolaven Plateau is at the center and Xay Phou Louang is in the east. Moreover, there are some low-altitude hilly areas (Xe Pian NPA) to the south of the Bolaven Plateau that are continuous to the forested area in north-eastern Cambodia. Along these routes, we interviewed 91 people at a total of 83 sites.

We interviewed people to gain the following information: interviewee's name, address, age, ethnicity, primate species inhabiting forests near village, abundance of primates, changes in abundance over time, possible reason for changes, human population changes, year village established, agricultural crops, primate damage to crops, counter-actions to such damages, hunting, wildlife trade, and economic conditions (Hamada *et al.*, 2007). We also

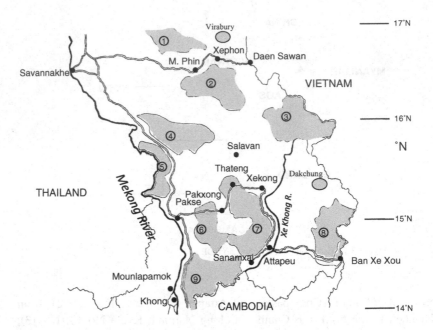

Figure 3.2. Survey routes in southern Lao PDR, from Savannakhet to Daen Sawan through Muang Phin and Xephoe along National Route (NR-) 9; from Savannakhet, through Pakse to the southern border along NR-13; from Pakse to Attapeu through Pakxong, Thateng, and Xekong along NR-23 and NR-16; from Attapeu to Sanamxai along NR-18; and from Attapeu to Ban Xe Xou along a new road. Hatched areas stand for National Protected Areas: 1, Phou Xang He; 2, Dong Phou Viang; 3, Xe Xap; 4, Xe Ban Nuan; 5, Phou Xiang Thong; 6, Dong Hua Sao; 7, Bolaven; 8, Dong Amphan; and 9, Xe Pian. Virabury and Dakchung are the popular localities for wildlife trade.

collected a record on the GPS coordinates and altitude of each location. We chose village heads or other persons who were well versed in the local wildlife as interviewees. By using photos and brochures with morphological and behavioral characteristics of primates that are supposedly distributed in Laos, we determined the species inhabiting each area.

We observed pet primates, and recorded their morphology. We then interviewed pet owners about the pet's origin, how it was acquired, whether hunting was involved, and the reason why they chose to keep a pet. We measured the crown-rump length (i.e., from the vertex of the head to the caudal end of the ischial callosity) and tail length in twelve macaques using a tape measure. Relative tail length, which is the significant subspecific and inter-locality characteristic in macaques (Fooden, 1997; Fooden and Albrecht, 1999; Hamada *et al.*, 2006), was calculated by dividing the tail length by the crown-rump

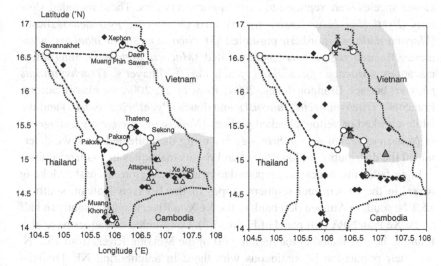

Figure 3.3. Left Panel: Distribution of rhesus (*Macaca mulatta*, painted diamond) and long-tailed macaques (*Macaca fascicularis*, triangle), and hatched area is distribution area of the latter and also contact zone between the two species. Right Panel: Distribution of pig-tailed (*Macaca nemestrina leonina*, diamond), stump-tailed (*Macaca arctoides*, asterisk), and assamese macaques (*Macaca assamensis*, triangle, hatched area is distribution area).

length and multiplying by 100 (percent). For rhesus macaques, we calculated the regression between latitude and relative tail length. We used Excel (Microsoft Co. Ltd.) for statistical analyses.

Bushmeat markets were visited to record wildlife sold and to interview shop-keepers about species, abundance, and frequency of trade. We visited monkey farms in Bolikhamxay and Champasak Provinces, and interviewed animal caretakers or observed primate species kept, number of individuals, age classes, and acquisition manner. Additionally we interviewed village people for monkey transportation or live macaque trade for monkey farms.

Results

Distribution and present status of long-tailed and rhesus macaques in southern Laos

We interviewed 91 people at 83 locations, inspected 20 pet long-tailed and rhesus macaques, and observed one troop of free-ranging rhesus macaques. The diversity of primate fauna was found to be rich in southern Laos and eleven species were reported to exist in our interviews. These included slow loris (*Nycticebus bengalensis*), pygmy loris (*Nycticebus pygmaeus*), rhesus (*Macaca mulatta*), northern pig-tailed (*Macaca nemestrina leonina*), assamese (*Macaca assamensis*), stump-tailed (*Macaca arctoides*), long-tailed macaques (*Macaca fascicularis fascicularis*), Phayer's (*Trachypithecus phayrei*; but see Brandon-Jones, 2004; Roos *et al.*, 2008 for classification), Francois' (*Trachypithecus francoisi*), and doucs (*Pygathrix nemaeus*) and the white-cheeked or yellow-cheeked gibbons (*Nomascus leucogenys leucogenys* or *Nomascus leucogenys gabriellae*). By using the collected data, we determined the distribution of long-tailed and rhesus macaques in southern Laos.

Long-tailed macaques were reported to inhabit the region east of the Mekong River, in the eastern and southern slopes of the Bolaven Plateau, south of 15.2°N; south of Attapeu (left bank of the Xe Xou River); and the southern half of the Xe Pian NPA (left panel of Figure 3.3). Long-tailed macaques were also found to be distributed in the region west of the Mekong River, from 15.2°N, and their population is continuous with those in neighboring NE Thailand (Malaivijitnond *et al.*, 2005; Malaivijitnand and Hamada, 2008). Most of the long-tailed macaque pets observed in the Khong Island were originated from this area (Table 3.1), Mounlapamok District, Champasak Province.

The population of long-tailed macaques was estimated for the two regions of the east and west of the Mekong River on the basis of density and habitable area. We supposed that the average population in a troop is thirty (range 10–48 in Wolfheim, 1983) and minimum densities of troop are the same as those in southern and northeastern Thailand (Malaivijitnond *et al.*, 2005), which are considered to have comparable quality of habitats with eastern and western regions of southern Laos, respectively. The maximum density was supposed to be ten times of the minimum.

The habitable area of the region east of the Mekong River was estimated to be about 27,000 km^2, and the minimum density of 0.012 individual/km^2 found in southern Thailand was used (Malaivijitnond *et al.*, 2005). The total population was estimated to be between 324 and 3,240 individuals.

The total area of the region west of the Mekong River is about 15,500 km^2. The density of this region is supposed to be the same as that in northeastern Thailand (Malaivijitnond *et al.*, 2005), 0.006 individuals/km^2. The total population is estimated to be 93–930. Thus, the national total population of long-tailed macaques in Laos lies within the range of 420–4,200.

Rhesus macaques appeared to be distributed across all of southern Laos including the mountainous area in the south of Pakse (Figure 3.3), except in areas west of the Mekong River (<15.2°N). The recorded southern-most locality in Laos was 14.0°N, however, Xay Phou Kiou, a forest in the Xe Pian NPA that is continuous to the forest in Cambodia across the border, is

Table 3.1. *Pet long-tailed and rhesus macaques observed in southern Laos*

No.	Date	Species	Sex	Age	Location (village, District, Province)	Latitude (°N)	Longitude (°E)	Altitude	Provenance
1	2005/7/17	*M. mulatta*	M	2 yrs.	Ban Vat Loang, Lamam, Sekong	15.35	106.73	no data	Sekong or Attapeu Province
2	2005/7/17	*M. mulatta*	?	Juv.	Ban Kamkok, Thateng, Sekong	15.44	106.48	no data	Tayeun, Thateng District
3	2005/7/19	*M. mulatta*	M	2 yrs.	Ban Sevang Noy, Laongam, Salavan	15.53	106.27	no data	Laongam District
4	2005/7/19	*M. mulatta*	M	2 yrs.	Ban Nakok Pho, Salavan, Salavan	15.72	106.42	no data	Phou Kate, Salavan District
5	2007/1/19	*M. mulatta*	?	?	Thamuang, Champhone District, Savannakhet	16.54	105.25	158	Free-ranging residents
6	2007/1/19	*M. mulatta*	?	?	Dong Mouang, Champhone District, Scavannakhet	16.54	105.25	192	Free-ranging residents
7	2007/1/21	*M. mulatta*	F	?	Houayangkham, Pakse, Champasak	15.12	105.85	168	Moungmon District, Champasak
8	2007/1/23	*M. mulatta*	M	?	Bounleum, Samaxeexay, Attapeu	14.81	106.83	110	Mt. Phou Vong, Phou Vong District, Attapeu
9	2007/1/25	*M. mulatta*	M	?	Panemyxay, Sanexay, Attapeu	14.93	107.05	134	Mt. Phou Xang (locating ca. 8 km from village), Sanexay, Attapeu
10	9/8/2008	*M. mulatta*	F	Y-Ad	Ban Nathong, Phin District	16.61	105.79	196	Phou Xang He NBCA

	Date	Species	Sex	Age	Location				Nearby Mountain
11	9/8/2008	M. mulatta	M	Infant	Bus Station, Ban Densavan, Xephone, Savannakhet	16.63	106.65	212	
12	9/9/2008	M. mulatta	F	Infant	Ban Lakhonesy (109 km to Pakse)	15.89	105.57	136	Mt. Phu Tei (ca. 10km to the west)
13	9/10/2008	M. mulatta	F	Y-Ad	Wat Haisok, Pakse City	15.11	105.84	117	30–40km South from Pakse City
14	9/10/2008	M. fascicularis	F	Y-Ad	Wat Haisok, Pakse City	15.11	105.84	117	from Attapeu
15	9/11/2008	M. fascicularis	F	Adult	Wat Khang, Khong Island	14.12	105.85	92	Ban Done Kamthon, Mounlapamok District*
16	9/11/2008	M. fascicularis	F	Adult	Wat Khang, Khong Island	14.12	105.85	92	Ban Done Kamthon, Mounlapamok District*
17	9/11/2008	M. fascicularis	F	Adult	Wat Khang, Khong Island	14.12	105.85	92	Ban Done Kamthon, Mounlapamok District*
18	9/11/2008	M. fascicularis	F	Adult	Center of Khong Island	14.10	105.78	86	Cambodia
19	9/11/2008	M. fascicularis	M	0.75 yrs.	Center of Khong Island	14.10	152.08	86	Cambodia
20	9/13/2008	M. mulatta	F	0.5 yrs.	Ban Hat Xan	14.77	107.10	153	Mt. Phou Kaphon (Near the village)

Main land west of the Mekong River.

also reported to be inhabited by rhesus macaques. This indicates that rhesus macaques are distributed south of 14.0°N, and thus may occur in northeastern Cambodia. Rhesus macaques were also reported in the high-altitude mountainous forests of the Bolaven Plateau, west of Attapeu City. They appear to be distributed in the Dong Amphan NPA, located in the Xay Phou Louang (Annamite Mountains). However, their absence was reported in the southern bank of the Xe Xou River, which is a tributary of Xe Khong River.

Our results show that long-tailed and rhesus macaques coinhabited the region between 15.2°N to 14.0°N in southern Laos. Long-tailed macaques tended to range in lower altitude forests, such as Xe Pian NPA, and riverine forests along the Xe Khong River and its tributaries, including those in the outskirt of the Bolaven Plateau. Rhesus macaques tended to inhabit higher altitudinal or secondary forests apart from riverside. The two species were not found in the primary forests where pig-tailed and stump-tailed macaques exist.

Khong Island in the Mekong River (Figure 3.2), spanning about 18km north to south and 8km east to west at around 14.2°N, used to be inhabited by gibbons and other non-human primates. However, most of the primate populations have become extinct. At the present time, a small population(s) of macaques with a tail of medium length is reported to occur in the patchy forests there, and is presumed to be rhesus or pig-tailed macaque.

Pet macaques

We inspected six long-tailed and fourteen rhesus pet macaques (Table 3.1). According to pet-owner reports, some of them were brought from a distant locality, but many of them were caught in forests in nearby villages where they were found. Pet long-tailed macaques were found only in Champasak Province, in and to the south of Pakse City. They were reported to have been brought from Attapeu, Mounlapamok (i.e., west of the Mekong River at around 14.5°N) and Cambodia.

The possession of pet rhesus macaques ranged widely in southern Laos, and the southern-most locality one was from the forest at 14.8°N in Attapeu Province. Some of the pet rhesus macaques were probably caught in NPAs close to the village (e.g., Phou Xang He or Xe Pian). There were some localities that were popular for the supply of wildlife trade, including non-human primates. These locations were, Virabury, about 50 km northeast of the Xephon City (Figure 3.2), and Dakchung in Sekong Province, north of Sekong City and near the Vietnam border. Both areas are under development. In Virabury copper and gold mines have recently started operations, and in Dakchung roads and dams were under construction and commercial logging was in operation.

These localities were popular for wildlife trade because of the frequent trans-portation between these localities and cities and also to supply bushmeat to workers there.

Geographic variation in relative tail length of rhesus macaques

Although tail lengths of long-tailed macaques could not be measured because they were not accessible for measurement, they did not appear to vary between individuals and localities, and were similar to those inhabit-ing north-east Thailand (i.e., ca. 115 percent of crown-rump length). On the other hand, relative tail length appeared to vary in rhesus macaques accord-ing to habitat latitude. We measured tail and crown-rump lengths on pet and free-ranging rhesus macaques in Laos (Hamada *et al.*, 2007; Hamada *et al.*, in preparation). The relative tail length (standardized by crown-rump length, percent) showed a significant negative geographical cline with latitude (r^2 = 0.791, n=10, p<0.01; Hamada *et al.*, in preparation). That is, northern individuals tended to have relatively shorter tails, and southern individ-uals had relatively longer tails. The tails of northern-most rhesus macaques (20°N or higher) were comparable in length to those of the eastern group of rhesus macaques (i.e., China and vicinity, average 35.3 percent of crown-rump length, Hamada *et al.*, 2006), those at middle latitude (18–16°N) were of the intermediate range of 45–55 percent, and those at lower latitude (16–14.8°N) were very long, in the range of 75–80 percent. Thus, the degree of hybridization found in rhesus macaques was geographically clinal, mean-ing that it depended on the distance from the long-tailed macaque range (<15.2°N).

Bushmeat Markets

At least four bushmeat markets were found along national routes, in Bolikhamxay, Savannakhet, and Champasak Provinces along NRs-13 and 9, in Attapeu Province along the newly constructed and asphalted NR connect-ing Attapeu to Dakto, Kontum Province, Vietnam via Ban Xe Xou (Figure 3.3). Macaques were not found to be for sale in any of the markets visited and therefore hunting for food does not appear to be a significant threat to Laotian macaques. In contrast, rodents, ungulates (e.g., muntjacs), civ-ets, fowls, lizards and turtles were frequently observed. It was reported that non-human primates, mainly macaques and colobines, were less frequently traded (i.e., once a week). Loris bones were found at traditional medicine shops in the market, though we found no evidence of bones of macaques. Along the NRs running east to west connecting Vietnam and Laos, factories, mines, and large-scale farmlands have recently been established or are being constructed, and they are expected to lead to an increase in bushmeat consumption via trading and by new workers who are mainly Vietnamese. Further study is necessary to assess if these developments will affect macaque populations (see Box 3.1).

Box 3.1   A possible decline in population of the long-tailed macaque (*Macaca fascicularis*) in northeastern Cambodia

Benjamin P. Y-H. Lee

This note reports on the possible decline of the long-tailed macaque (*Macaca fascicularis*) in northeastern Cambodia. After a recent research trip to the northeastern part of the country to investigate the wild meat trade (31 May–2 July 2008), it appeared evident that long-tailed macaques were not easily observed. The survey investigated the wild meat trade in the central market of Ban Lung, in the capital town of Ratanakiri, as well as conducting field trips to rural provinces for interviews with hunters. Throughout the travelling period, the occurrence of all wildlife species, including macaques, was noted.

Being assigned "Least Concern" IUCN conservation status (Ong and Richardson, 2008), long-tailed macaques have been deemed "not in danger" in northeastern Cambodia by Timmins and Soriyun (1998) who reported seeing groups of *M. fascicularis* on five different days in mid 1998 in a survey of wildlife in Tonle San and Tonle Srepok in northeastern Cambodia. They cautioned that many groups may have been undetected during their survey, and according to the limited field surveys in eastern and northeastern Cambodia, the troop size of long-tailed macaques did not seem to exceed ten individuals (Timmins and Soriyun, 1998, Walston *et al.*, 2001). This may be a sign that the monkeys were subjected to heavy hunting pressure as Wheatley *et al.* (2002) reported that high hunting rates in Palau resulted in small groups.

It has been reported that a large-scale harvest of wild *M. fascicularis* is happening throughout Cambodia (Campbell *et al.*, 2006, Rawson, 2007) and began sometime in 2006 (Pollard *et al.*, 2007) in response to heightened demand from "monkey farms" in Cambodia, Vietnam and China (Campbell *et al.*, 2006, Pollard *et al.*, 2007). The wild-caught macaques are used mainly for bio-medical testing. A monkey farm in Cambodia has publicly admitted that wild macaques are acquired from villagers in various areas throughout the country (McFadden and Chandara, 2005). The market forces driving this harvest are intense as each live wild macaque fetches USD$50 (200,000 Riels) (Rawson, 2007), an amount which is at least twice the monthly wage of a rural worker. Hence, it is understandable that no efforts are spared by hunters and villagers to trap this species throughout Cambodia. In the village of I Tub in northern Ratanakiri Province, villagers are actively involved in the harvest of long-tailed macaques for commercial trade, and there seems to be an organized trade structure in dealing macaques (Rawson, 2007). Being a commensal that is tolerant of degraded habitats, many macaque groups occur close to human settlements. Such groups are at severe risk of being hunted or trapped, given the current practice of village hunters to harvest them to supply their demand. Rawson (2007) has reported the *modus operandi* of the village farmers in hunting macaques (Figure 3.4). The target troop of macaques is isolated in a large tree by cutting down all the surrounding trees within a 25–30 m radius. Several nets then encircle the lone tree and a hunter climbs the tree to force the monkeys to drop to the ground and they are captured. Such a hunting method involves wanton forest destruction as about 50–100 trees with a *dbh* (diameter at breast height) above 10 cm are removed from each site with chainsaws and axes, resulting in large forest gaps and cleared land. This has severe implications for the conservation of other wildlife species.

During this study, long-tailed macaques were not observed during a five-day survey to villages in the districts of Veun Sai, Ou Chum and Lumphat covering about 200 km, despite the presence of suitable habitats for *M. fascicularis*. The banks of Tonle San and Tonle Srepok were also visited, and no macaques were observed. Within the vicinity of Ban Lung town, a community-protected area in Yeak Laom Commune was visited on two separate days but no long-tailed macaques were seen. None of the farmer-hunters (n=4) interviewed harvested long-tailed macaques in recent times, although one recalled hunting them and other primates in the 1980s with guns. There was a general consensus that all wildlife species have declined drastically in the past five to ten years. The villagers at I Tub presented a very different view and claimed that long-tailed macaques were "very abundant", even though trapping was occurring (Rawson, 2007). In the central market of Ban Lung where wild meat is traded openly, no long-tailed macaques, dead or alive, were on sale despite carrying out an entire month of daily market monitoring.

To ascertain the population trend of *M. fascicularis* with some degree of reliability, it would be instructive to get conservation staff, researchers,

Figure 3.4. Long-tailed macaques are systematically trapped in Cambodia. Trees are cut down to isolate the group (a), and then the animals are captured and bagged by hunters (b and c). Most of the animals are laundered to monkey farms, but some die during the incident (d) or are consumed for meat (e). (Photographs courtesy of Hugo Rainey, Teth Setha, and Kep Bunna of WCS.)

Figure 3.4. (*cont.*)

(e)

Figure 3.4. (*cont.*)

and forest rangers throughout Cambodia involved in regular monitoring of these monkeys during the course of other wildlife work. Special attention should be paid to riverine and other habitats close to water where these macaques are usually associated (Pollard *et al.*, 2007; Walston *et al.*, 2001). Protected areas (e.g., Seima Biodiversity Conservation Area) and temples where *M. fascicularis* were recently documented should be checked regularly to monitor their populations. If their survival in such areas cannot be guaranteed, it is highly unlikely that they will persist in the rest of the landscape in Cambodia if such high-intensity exploitation continues.

To conclude, the current levels of trapping and trade pose a great threat to *M. fascicularis* in Cambodia and throughout its range if it goes unchecked (Pollard *et al.*, 2007). As the long-tailed macaque is listed on Appendix II of CITES, it is imperative that the relevant country authorities determine if the current trade levels are detrimental to the survival of *M. fascicularis*, before allowing their continued harvest and export. Finally, there is a need for tighter controls on the operations of monkey farms to ensure that monkeys are harvested in a sustainable manner for breeding purposes, and not caught from the wild and laundered through monkey farms.

References

Campbell, I.C., Poole, C., Giesen, W., and Valbo-Jorgensen, J. 2006. Species diversity and ecology of Tonle Sap Great Lake, Cambodia. *Aquatic Sciences* **68**: 355–373.

McFadden, D. and Chandara, L. 2005. Cambodian monkeys bred for testing, exported. *The Cambodia Daily*, 25 November 2005.

Ong, P. and Richardson, M. 2008. *Macaca fascicularis*. In: IUCN 2010. IUCN Red List of Threatened Species. Version 2010.1. www.iucnredlist.org. Accessed on 7 May 2010.

Pollard, E., Clements, T., Hor, N. M., Ko, S., and Rawson, B. 2007. Status and conservation of globally threatened primates in the Seima Biodiversity Conservation Area, Cambodia. Wildlife Conservation Society.

Rawson, B. 2007. Surveys, trade and training in Voensei Division, Ratanakiri Province, Cambodia. Conservation International, Hanoi. Unpublished report.

Timmins, R. J. and Soriyun, M. 1998. A wildlife survey of the Tonle San and Tonle Srepok river basins in northeastern Cambodia. Flora and Fauna International, Indochina Programme, Hanoi and Wildlife Protection Office, Department of Foretsry, Cambodia, Phnom Penh.

Walston, J., Davidson, P., and Soriyun, M. 2001. A wildlife survey of southern Mondulkiri Province, Cambodia. Wildlife Conservation Society (WCS) Cambodia Program.

Wheatley, B., Stephenson, R., Kurashina, H., and Marsh-Kautz, K. 2002. A culturalprimatological study of *Macaca fascicularis* on Ngeaur Island, Republic of Palau. In *Primates face-to-face: Conservation implications of human and non-human primate interconnections*. Cambridge University Press.

Monkey farms

In September 2008, two monkey farms were discovered during this survey of southern Laos. A monkey farm was found to be in operation and undergoing enlargement at Pakkham Village, Bolikhamxay Province along NR-13 (16.32°N, 108.85°E). About 2,000 long-tailed macaques were found to be kept there, and a new cage-house may soon accommodate another 1,000 monkeys. This facility was managed by the same Chinese company that runs another facility at Ban None-Sengchan village, near Vientiane City. It was hard to estimate the number of macaques traded because we could not access.

Another monkey farm was found to be under construction along NR-13, 36 km south of Pakse City (14.81°N, 105.95°E). The Vietnamese company managing this farm has other farm(s) in Bien Hoa State, southern Vietnam. Construction workers told us that 10,000 macaques will be housed in the farm upon its completion within six months.

The majority of the long-tailed macaques kept at the Bolikhamxay farm were either juveniles or young adults. No infants were observed in the farm, suggesting that the macaques were not bred here, but rather wild-caught and temporarily housed here until traded to dealers. On the basis of our data on the wild population, which indicated that long-tailed macaques were restricted to the southern region of Laos, the source of the macaques in these farms was unlikely to be Laos. Rather, they were likely to be from Thailand and/or Cambodia. According to animal caretakers, long-tailed macaque populations inhabiting Buddhist temples and City Parks in Thailand that were sometimes in very large numbers (> 1,000; Malaivijitnond *et al.*, 2005; Malaivijitnond and Hamada, 2008) were caught (most probably illegally, as the Thai government bans hunting and exportation of primates) and sold at $50–80 USD each. We have also heard that in a Buddhist Temple in the Nakhon Sawan Province of Thailand, where more than 1,000 monkeys were counted, that on some days hunters came and caught several hundred monkeys, and brought them to Vientiane by truck. Knowing this, the chief monk and residents chased the truck and confiscated monkeys. It is possible that such illegal hunting and smuggling of long-tailed macaques could be occurring in other places as well.

Animal caretakers even reported that macaques from Cambodia were mainly wild-caught, and thus not habituated, and were healthier than those from Thailand, meaning that considerable numbers of macaques died in transportation or while living at the farm. Residents in southern Laos (<15°N) along NR-13 reported that they tried to catch long-tailed macaques but had little success. According to the animal caretaker, after recovering from transport and reception to the farm, the macaques were tested with drugs (e.g., insulin of ten times the treatment dose), and the survivors were sent to Ho Chi Minh City or Hong Kong. They would then be sent to developed countries (such as Japan, USA and Europe) and sold for bio-medical experiments or for drug testing for $1,000 USD or more each.

Threats to long-tailed macaques in Laos

The population of long-tailed macaques in Laos is potentially under the threat of extinction for several reasons. The major cause of concern is human population growth, which had ca. 2.3 percent annual growth rate in 2005 (http://globalis. gvu.unu.edu). Furthermore, economic development had an annual increase of 5 percent GDP in 2008 (www.globalpropertyguide.com/Asia/Laos/gdp-per-capita-growth.1-year), and will need to continue exploiting natural resources to maintain growth. Some examples of development include, dams for hydraulic power were constructed in the Bolaven Plateau (i.e., tributary of Xe Khong River) to inundate wide areas of riverine forests, and agricultural farm lands are being widely developed in low altitudinal forests near rivers. Particularly, large-scale farmland for the production of commercial commodities such as sugarcane, rubber, coffee, corn, cassava, etc., has been established. Moreover, hunting and trade of live wildlife and its products are prevalent. These threats will continue to increase with the construction of roads and bridges connecting Laos and neighboring countries. A detailed assessment will be necessary to evaluate the impact of these human activities on macaque populations.

Discussion

Distribution of long-tailed and rhesus macaques in southern Laos

The population and distribution of long-tailed macaques in southern Laos was determined using the data obtained in the present and previous studies (Duckworth *et al.*, 1999). Long-tailed macaques were found to be distributed in southern-most Laos, 15.2°N or lower, in the forests of Attapeu and Champasak Provinces. In Attapeu, they distributed in the Xay Phou Louang range and probably in forests south of the Xe Xou River, while in Champasak they are found in the Bolaven Plateau, Xe Pian NPA, and in the region west of the Mekong River. The total national population of Laotian long-tailed macaques was estimated to be in the range of 420–4,200 individuals.

Rhesus macaques were found distributed in almost all areas of Laos (>14°N). The southern limit of rhesus macaque distribution has not yet been determined because studies have yet to be completed in the forested area on the eastern bank of the Mekong River (i.e., Xe Pian NPA), areas along the Xe Khong River, and the forested area in northeastern Cambodia (i.e., Rattanakiri and Stoeng Treng Provinces). In central Laos (i.e., middle latitude), rhesus macaques were found only in the low-profile mountainous and plain areas. That is, they preferred nonevergreen secondary or disturbed forests. They did not appear to have adapted to the dry hilly forests in southern-most Laos, to which northern pig-tailed macaques adapted, which may be why rhesus macaques were not distributed to the western bank of the Mekong River continuous to northeast Thailand.

Sympatric relationship between long-tailed and rhesus macaques

Long-tailed and rhesus macaques are generally allopatrically distributed (Fooden, 1982) to areas of lower and higher latitudes, respectively. This pattern of separation may occur because the two species are closely related and compete for habitats with similar ecological conditions. These two species of macaques tend to prefer forests other than the broadleaf evergreen forests that pig-tailed, stump-tailed, and assamese macaques prefer (Fooden 1982). Consequently, rhesus and long-tailed macaques can be found to be distributed sympatrically with other species of macaques because they use different micro-habitats in the region.

Long-tailed and rhesus macaques were found to share the area east of the Mekong River, in southern Laos (15.2–14.0°N), which may be the zone where the two species interface. Sympatry of the two species was also reported in the Huai Kha Khaeng Wildlife Sanctuary in Thailand, which is located at 15.2°N inside the Dawna Range (Eudey, 1980). The topography of southern Laos is highly varied. It consists mainly of mountainous or peripheral forests that are continuously spread across the country, separated by tributaries of the Mekong River and flooding plains. The habitat preferences the two species share may facilitate them to range in the same area, but interviews in the present study revealed that long-tailed macaques prefer the riverside forest, similar to insular long-tailed macaques in Indonesia (Fittinghoff and Lindburg, 1980). The distribution patterns of long-tailed and rhesus macaques in northeast Thailand (Malaivijitnond *et al.*, 2005) showed that long-tailed macaques inhabit dry and patchy forested areas near bodies of water (i.e., river or pond), suggesting they are adapting to riparian environments, to which rhesus macaques do not compete. The similarities and differences in ecological preferences of the two species will need to be the subject of future study to better discern the degree of their ecological overlap.

Hybridization between rhesus and long-tailed macaques

Long-tailed macaques in Laos are classified as the subspecies, *Macaca fascicularis fascicularis* (Fooden, 1995). Long-tailed macaque populations in the Indochina Peninsula ranging north of the Isthmus of Kra (ca.10.5°N) have shorter tails (ca. 115 percent of the head and body length or crown-rump length) than those in the south or islands (ca. 125 percent or more) (Fooden, 1997; Fooden and Albrecht, 1999; Hamada *et al.*, 2008). Long-tailed macaque pets from Attapeu and Mounlapamok, west of the Mekong River, were found to have tails of about 110–115 percent (Hamada *et al.*, 2008). Mitochondrial DNA analysis showed that Indochina (i.e., northern) and insular Southeast Asian (i.e., southern) long-tailed populations are differently clustered (Blancher *et al.*, 2008). Molecular phylogeographic studies suggested that ancestors of northern long-tailed macaques have been introgressed by male rhesus macaques (Tosi *et al.*, 2002, 2003; Bonhomme *et al.*, 2008).

Rhesus macaques are classified as being either of eastern (i.e., China and vicinity) or western groups (i.e., India and vicinity) without being given sub-specific status (Fooden, 2000). The eastern group has much shorter tails (35.3 percent) than the western group (42.5 percent) (Hamada *et al.*, 2005). The grouping has been supported by molecular phylogenetic analyses (Smith and McDonough, 2005, Smith *et al.*, 2007; Satkoski *et al.*, 2008). In the eastern half of the Indochina Peninsula, including Laos, around the border zone of the two species (14–20°N), supposed hybrid rhesus macaques with tails longer than 45 percent were found (Fooden, 1964, 1995, 2000; Hamada *et al.*, 2006, 2008). Introgression from long-tailed macaques was suggested for these populations (Malaivijitnond *et al.*, 2007, 2008). However, definite molecular analyses have not been reported because of the difficulty in accessing "hybrid" rhesus in the Indochina Peninsula.

Our inspection of rhesus macaque pets in central and southern Laos showed that relative tail length significantly correlated with latitude (Hamada *et al.*, in preparation). Although this geographical cline could be explained by the cli-mate cline, i.e., cold adaptation (i.e., Allen's rule), this is not the case because both the eastern and western groups of rhesus macaques do not show intra-group clinal variation (Fooden, 2000) and because rhesus macaques from northern Laos did not have short enough tails to conform to the trend of length against latitude found in rhesus macaques from central and southern Laos. Therefore, hybridization would be the major cause of tail length variation in rhesus macaques, and the introgression intensity would be clinal in central and southern Laos. On the other hand, geographical variation in tail length or other phenotypic characteristics was not depicted in northern long-tailed macaques. Thus, conditions of the ancestral populations that received introgression may be different in northern long-tailed and rhesus macaques in the contact zone.

Clinal variation in tail length of rhesus macaques could have been produced by either one-way (i.e., introgression) or two-way (i.e., hybridization) genetic flow. It is not impossible that the ancestors of rhesus, or rhesus-like, macaques with longer tails could have been long-tailed macaques that received heavy introgression from rhesus macaques. The heavy introgression altered the phenotypic characteristics into those of rhesus macaques (e.g., tail length, pel-age color pattern, and cranio-facial morphology). On the other hand, it is pre-sumed that the ancestor of northern long-tailed macaques was once confined to refugia in the glacier period with smaller populations and received intro-gression from rhesus macaques. The introgression replaced the Y-chromosome of the long-tailed macaque with that of rhesus macaque and slightly changed its morphological characteristics. Although single nucleotide polymorphism analysis showed complicated genetic relationships between the two species (Street *et al.*, 2007), there is no data on the direction or strength of introgression between the two species. Molecular phylogenetic study is necessary to assess the genetic exchange between these two species.

The hybridization may have occurred through climate and environmental changes during the Plio-Pleistocene epoch with alterations of glacial and inter-glacial climates and accompanying temperature fluctuation, such as in the hypsithermal period of the Holocene (i.e., 5,000–9,000 years ago). Ancestors of long-tailed and rhesus macaques may have been confined to refugia in the glacial period(s), which are thought to have been located in Indochina Peninsula and other areas in Asia, such as southern China to Xay Phou Louang mountainous area, northeast India (Kasi hill, Eudey, 1980), Dawna Range (Eudey, 1980), the northeast Thai mountains, including the present-day Phu Khieow Mountains (Koenig *et al.*, 2003), and in Sundaland (Gathorne-Hardy *et al.*, 2002). Climate change could have influenced the distribution of the two species, and genetic exchange would have occurred between them.

Influences from human activity

Threats to long-tailed macaques appear incipient in southern Laos. It is because the human population is still low and the economic development is limited. However, threats are present and will accelerate in the near future. Several macaque species, including the long-tailed macaque, are considered common wildlife that are tolerant of habitat degradation and hunting and considered by some to be a "weed" (Richard *et al.*, 1989). However, the scale of habitat degradation in Laos may be beyond their ability to accommodate.

Apart from self-consumption cultivation, commodity crop cultivation consisting of rubber, ginger, sugarcane, and corn is driven by capital from foreign countries. Wide areas of coffee and vegetable plantations were established in the Bolaven Plateau. Wide areas of forest in the less steep lands near the river and close to the roads (e.g., Attapeu), which are habitats for long-tailed macaques, have been cut away and turned into large-scale farmlands for production of commodity crops. Industrial afforestation of acacia, eucalyptus or other trees, is expanding rapidly for the supply of paper materials and for the control of carbon discharge (Clean Development Mechanism, Kyoto Protocol). These plantations will be established in similar regions as the farmlands, and will replace or block regeneration of natural forests, separating historic macaque habitat into small patchy forests. These trends will be accelerated as a result of the construction of corridor roads and bridges connecting Laos with southern China, Thailand, Myanmar, Vietnam, and Cambodia. After the roads are constructed or improved in less inhabited regions of southern Laos, human settlements will be established along the roads and people will immigrate to the localities where presently intact landscapes remain. The impact on non-human primates from such human movements is the subject of future study. Mining and factory construction and establishment may increase the hunting pressure for bushmeat trade. Although guns were confiscated by the government, villagers either hid or manufactured guns to hunt animals due to poor gun control (Hamada *et al.*, 2007).

The monkey farm business is now expanding in Laos and is managed by companies from foreign countries. It was reported that 10^3–10^4 long-tailed macaques were annually exported to developed countries but it did not appear that many monkeys were bred at the farms, as this far exceeds our estimated national population. Rather they were thought to have been brought from elsewhere (i.e., neighboring countries) and then subsequently exported. However, the presumed origin countries have banned exportation of wild-caught monkeys according to ratification of the Washington Convention (CITES), which Thailand ratified in 1983, Vietnam in 1994, Cambodia in 1997, and Laos in 2004 (www.cites.org/eng/disc/parties/alphabet.shtml), and so it is still unclear exactly what is happening. More work is needed to uncover what proportion of the wild population is threatened by utilization in trade, as the large volume of macaques being traded from Laos cannot all be bred in farms.

Another concern of monkey farms is that it is suspected that monkeys would escape from the farms, as the cages looked neither secure nor tough. Village areas near monkey farms that are located farther north than the natural distribution area of long-tailed macaques are now being found to be inhabited by these macaques (SP, personal observation). This indicates that macaque farms are responsible of ethnophoresy (see Part III), as some long-tailed macaques are being artificially carried and released beyond their natural range. This needs to be monitored, as these monkeys may distort the macaque distribution and may result in hybrid production with native macaques of other species.

Conclusion

The present surveys did not thoroughly delineate the exact distribution and present status of long-tailed and rhesus macaques in Laos. However, we did find that long-tailed macaques were found to be distributed to the area at 15.2°N or lower in Laos, both in the west and east of the Mekong River. Although the range appears wide, the total population is estimated to be only 420–4,200 individuals. Rhesus macaques were found to range in almost all areas of Laos, except in the area west of the Mekong River in southern Laos. Populations of the two species were found to have hybridized with each other in southern Laos, and we do not yet understand the consequences of this genetic mixing. Several other threats face macaques in southern Laos. Hunting pressure does occur, but it does not appear to have a great impact. However, habitat loss and degradation by dam construction, large-scale farmland, and industrial afforestation (ie., plantation of acacia, eucalyptus, etc.) does appear a significant threat. Soon these land developments will be advanceing more rapidly after the construction of international roads and bridges. These developments will likely destroy low altitude, secondary and riverine forests that are preferred habitats of the rhesus and long-tailed macaques, and thus could be great threats to their future sustainability.

A nationwide assessment of distribution, density and abundance of macaques, together with habitat evaluation is urgently needed to better understand the conservation needs to sustain their populations. Macaques are neither rare nor endemic in Laos. However, studies of local populations in Laos are important for the reconstruction of the phylogeny of this species and for better understanding their population ecology. Moreover, as human land changes bring people and macaques into greater contact, management of human-macaque conflict will also need to be considered. We would like to suggest that international or domestic transport and trade should be strictly controlled, as we do not yet know how this is affecting macaque the size and gene flow of populations. Secondly, land-use for agriculture of large-scale commodity crops should be controlled so the forested areas will be kept continuous with each other. Finally, hunting of wildlife should be controlled, and an alternative protein source should be supplied, along with education for people regarding the fact that wildlife does not have any special nutrition over other foods.

Acknowledgements

We thank staff of the Department of Biology, Faculty of Science, National University of Laos for their help and encouragement to the field survey. This study was financially supported by the Japanese Society for the Promotion of Science (JSPS, Fund Nos. 16405017 and 20255006).

References

Blancher, A., Bonhomme, M., Crouau-Roy, B., Keiji. T., Kitano, T., and Saitou, N. 2008. Mitochondrial DNA sequence phylogeny of 4 populations of the widely distributed cynomolgus macaque (*Macaca fascicularis fascicularis*). *Journal of Heredity* **99**(3): 254–264.

Bonhomme, M., Cuartero, S., Blancher, A., and Crouau-Roy, B. 2009. Assessing natural introgression in 2 biomedical model species: The rhesus macaque (*Macaca mulatta*) and the long-tailed macaque (*Macaca fascicularis*). *Journal of Heredity* 100(2): 158–169.

Brandon-Jones, D. 2004. A taxonomic revision of the langurs and leaf monkeys (Primates: Colobinae) of South Asia. *Zoos' Print Journal* 19(8): 1552–1594.

Dang, N. C., Endo, H., Nguyen, T. S., Oshida, T., Le, X. C., Dang, H. P., Lunde, D. P., Kawada, S., Hayashida, A., and Sasaki, M. 2008. *A Checklist of Wild Mammal Species of Vietnam*. Kyoto: Shoukadoh Book Sellers.

Denduangboripant, J., Malaivijitnond, S., Hamada, Y., Varavudhi, P., and Takenaka, O. 2005. Genetic diversity and phylogeography of long-tailed macaques in Southeast Asia. *Natural History Journal of Chulalongkorn University*, supplement 1: 89.

Duckworth, J. W., Salter, R. E., and Khounboline, K. (compilers). 1999. Wildlife in Lao PDR: 1999 status report. Vientiane: IUCN-The World Conservation Union/Wildlife Conservation Society/Centre for Protected Areas and Watershed Management. Samsaen Printing, Bangkok.

Eudey, A. A. 1980. Pleistocene glacial phenomena and the evolution of Asian macaques. In *The Macaques: Studies in Ecology, Behavior and Evolution,* D. G. Lindburg (ed.). New York: Van Nostrand Reinhold.

Fittinghoff, N. A., Jr., and Lindburg, D. G. 1980. Riverine refuging in East Bornean *Macaca fascicularis*. In *The Macaques: Studies in Ecology, Behavior and Evolution* D. G Lindburg (ed.). New York, Van Nostrand Reinhold.

Fooden, J. 1964. Rhesus and crab-eating macaques: Intergradation in Thailand. *Science* 143: 363–365.

   1976. Provisional classification and key to living species of macaques (Primates: *Macaca*). *Folia Primatologica* 25(2–3): 225–236.

   1980. Classification and distribution of living macaques (*Macaca* Lacepede, 1799). In *The Macaques: Studies in Ecology, Behavior and Evolution*. D. G. Lindburg (ed.). New York: Van Nostrand Reinhold.

   1982. Ecogeographic segregation of macaque species. *Primates* 23 (4): 574 – 579.

   1995. Systematic review of Southeast Asian longtail macaques, *Macaca fascicularis* (Raffles, 1821). *Fieldiana Zoology* 81: 1–206.

   1997. Tail length variation in *macaca fascicularis* and *M. mulatta*. *Primates* 38(3): 221–231.

   2000. Systematic review of the rhesus macaque, *macaca mulatta* (Zimmermann, 1780). *Fieldiana Zoology* 96: 1–180.

Fooden, J. and Albrecht, G. H. 1999. Tail length variation in *fascicularis*-group macaques (Cercopithecidae: *Macaca*). *International Journal of Primatology* 20(3): 431–440.

Gathorne-Hardy, F. J., Syaukani, Davies, R. G., Eggleton, P., and Jones, D. T. 2002. Quarternary rainforest refugia in south-east Asia: using termites (Isoptera) as indicators. *Biological Journal of the Linnean Society* 75: 453–466.

Hamada, Y., Watanabe, T., Chatani, K., Hayakawa, S., and Iwamoto, M. 2005. Morphometrical comparison between Indian- and Chinese-derived rhesus macaques (*Macaca mulatta*). *Anthropological Science* 113(2): 183–188.

Hamada, Y., Urasopon, N., Hadi, I., and Malaivijitnond, S. 2006. Body size and proportions and pelage color of free-ranging *Macaca mulatta* from a zone of hybridization in northeastern Thailand. *International Journal of Primatology* **27**(2): 497–513.

Hamada, Y., Malaivijitnond, S., Kingsada, P., and Bounnam, P. 2007. The distribution and present status of primates in the northern region of Lao PDR. *Natural History Journal of Chulalongkorn University* **7**(2): 161–191.

Hamada, Y., Suryobroto, B., Goto, S., and Malaivijitnond, S. 2008. Morphological and body color variation in Thai *macaca fascicularis fascicularis* north and south of the isthmus of Kra. *International Journal of Primatology* **29**: 1271–1294.

Hamada, Y., Kurita, H., Goto, S., *et al.* 2010. Distribution and present status of macaques in Lao PDR. In *Conservation of Primates in Indochina*. T. Nadler, B. M. Rawson, and V. N. Thinh (eds.). Hanoi: Frankfurt Zoological Society and Conservation International. pp. 27–42.

Kanthaswamy, S., Satkoski, J., George, D., Kou, A., Erickson, B. J.-A., and Smith, D. G. 2008. Hybridization and stratification of nuclear genetic variation in *Macaca mulatta* and *M. fascicularis*. *International Journal of Primatology* **29**(5): 1295–1311.

Koenig, A., Larney, E., Kreetiyutanont, K., and Borries, C. 2003. The primate community of Phu Khieo Wildlife Sanctuary, northeast Thailand. *American Journal of Primatology* **60**(Suppl. 1): 64.

Malaivijitnond, S, Hamada, Y., Varavuddhi, P., and Takenaka, O. 2005. The current distribution and status of macaques in Thailand. *Natural History Journal of Chulalongkorn University* Supplement 1: 35–45.

Malaivijitnond, S., Takenaka, O., Kawamoto, S., Urasopon, N., Hadi, I., and Hamada. Y. 2007. Anthropogenic macaque hybridization and genetic pollution of a threatened population. *Natural History Journal of Chulalongkorn University* **7**(1): 11–23.

Malaivijitnond, S. and Hamada, Y. 2008. Current situation and status of long-tailed macaques (*Macaca fascicularis*) in Thailand. *Natural History Journal of Chulalongkorn University* **8**: 185–204.

Malaivijitnond, S., Sae-Low, W., and Hamada, Y. 2008. The human-ABO blood groups of free-ranging long-tailed macaques (*Macaca fascicularis*) and parapatric rhesus macaques (*M. mulatta*) in Thailand. *Journal of Medical Primatology* **37**(1): 31–37.

Meijaard, E. and Groves, C. P. 2006. The geography of mammals and rivers in ainland Southeast Asia. In *Primate Biogeography*. S. M. Lehman and J. G. Fleagle (eds.). New York: Springer.

Nhan, N. T. 2004. The status of primates at Pu Mat National Park and suggestions for sustainable conservation approaches. In *Conservation of Primates in Vietnam*. T. Nadler, U. Streicher, and H. T. Long (eds.). Frankfurt Zoological Society. pp. 85–89.

Polet, G, Murphy, D. J., Becker, I., and Thuc, P. D. 2004. Notes on the primates of Cat Tien National Park. In *Conservation of Primates in Vietnam*. T. Nadler, U. Streicher, and H. T. Long (eds.). Frankfurt Zoological Society. pp. 78–84.

Richard, A. F., Goldstein, S. J., and Dewar, R. E. 1989. Weed macaques: The evolutionary implications of macaque feeding ecology. *International Journal of Primatology* **10**: 569–594.

Roos, C., Nadler, T., and Walter, L. 2008. Mitochondrial phylogeny, taxonomy and biogeography of the silvered langur species group (*Trachypithecus cristatus*). *Molecular Phylogenetics and Evolution* **47**: 629 – 636.

Satkoski, J., George, D., Smith, D. G., and Kanthaswamy, S. 2008. Genetic characterization of wild and captive rhesus macaques in China. *Journal of Medical Primatology* **37**: 67–80.

Smith, D. G. and McDonough, J. W. 2005. Mitochondrial DNA variation in Chinese- and Indian-rhesus macaques (*Macaca mulatta*). *American Journal of Primatology* **65**: 1–25.

Smith, D. G., McDonough, J. W., and George, D. A. 2007. Mitochondrial DNA variation within and among regional populations of long-tail macaques (*Macaca fascicularis*) in relation to other species of the *fascicularis* group of macaques. *American Journal of Primatology* **69**: 182–198.

Street, S. L., Kyes, R. C., Grant, R., and Ferguson, B. 2007. Single nucleotide polymorphisms (SNPs) are highly conserved in rhesus (*Macaca mulatta*) and cynomolgus (*Macaca fascicularis*) macaques. *BMC Genomics* **8**: 480–488.

Tosi, A. J., Morales, J. C., and Melnick, D. J. 2002. Y-chromosome and mitochondrial markers in *Macaca fascicularis* indicate introgression with Inodochinese *M. mulatta* and a biogeographic barrier in the isthmus of Kra. *International Journal of Primatology* **23**(1): 161–178.

2003. Paternal, maternal, and biparental molecular markers provide unique windows onto the evolutionary history of macaque monkeys. *Evolution* **57**(6): 1419–1435.

Wolfheim J. H, 1983. *Primates of the World: Distribution, Abundance and Conservation.* Seattle and London: University of Washington Press.

Part II

The human–macaque interface

4 Campus monkeys of Universiti Kebangsaan Malaysia: Nuisance problems and students' perceptions

BADRUL MUNIR MD-ZAIN,
MOHAMED REZA TARMIZI AND MASTURA
MOHD-ZAKI

Introduction

Malaysia, in Southeast Asia, has a total landmass of 329, 845 km², separated by the South China Sea into two regions, Peninsular Malaysia and Borneo (i.e., Sabah and Sarawak). Peninsular Malaysia is located south of Thailand, north of Singapore and east of the Indonesian island of Sumatra. The country comprises tropical rain forest with dipterocarp forests, peat swamp forests, and mangrove forests (Mohd-Azlan, 2006). The human population is about 28 million and consists of multi-racial ethnic groups with a majority of Malays, followed by Chinese, Indians and other minorities (Lim et al., 2004). The country is the home for 229 species of mammals (Lim, 2008) including 18–20 species of primate species (Brandon-Jones et al., 2004; Md-Zain et al., 2009).

In Malaysia, there are three species of macaques, pig-tailed macaques (*Macaca nemestrina*), stump-tailed macaques (*Macaca arctoides*) and long-tailed macaques (*Macaca fascicularis*) (Brandon-Jones et al., 2004). *M. nemestrina* is widely distributed within Peninsular Malaysia but mostly in the undisturbed forests and is hardly seen near to the coastal forests. *M. arctoides* can only be found in northwestern part of Malay Peninsula (Medway, 1969). Meanwhile, *M. fascicularis* can easily be found near rivers and low ground secondary forests and near to human settlements (Marsh and Wilson, 1981). Macaque populations in Malaysia have not received much attention as many researchers focus on ecology, behavior and genetics of their sister taxon, the leaf monkeys (e.g., Md-Zain et al., 2008, 2010a, 2010b, 2011; Matsuda et al., 2009).

Monkeys on the Edge: Ecology and Management of Long-Tailed Macaques and their Interface with Humans, eds. Michael D. Gumert, Agustín Fuentes and Lisa Jones-Engel. Published by Cambridge University Press. © Cambridge University Press 2011.

Long-tailed macaques are highly opportunistic omnivores and are known to inhabit a wide range of many tropical habitats (Poirier and Smith, 1974; Aldrich-Blake, 1980; Fooden, 1995). Past studies examined variation in population density between sites and forest types. This species has a high density in riverine and edge habitats (Chivers and Davies, 1978) and in lowland forests (Marsh and Wilson, 1981). The highest density of long-tailed macaques in Peninsular Malaysia and other areas is mainly freshwater swamp forest (Marsh and Wilson, 1981, Yanuar *et al.*, 2009).

Under the Malaysian Wildlife Protection Act 1974, *M. fascicularis* is listed as a protected animal that cannot be killed and captured without permission from Department of Wildlife and National Parks, Peninsular Malaysia (DWNP). Macaques have been used as experimental laboratory animal in Malaysia since the 1970s (Marsh and Wilson, 1981). Due to the trade ban against Indian rhesus macaques, the export of Malaysia's macaques increased in numbers to more than 10,000 animals per year from 1974–1982 (Muda, 1982). This situation caused an assessment to be conducted on the macaque trade in the late 1982 (DWNP, 1982). This assessment resulted in a ban on macaque exporting due to a serious population decline (DWNP, 1985). The government of Malaysia banned the export of long-tailed macaques in 1984, to prevent the abuse of the monkeys in bio-medical and military laboratories in the United State. Since 1985, no macaques have been exported from Malaysia, and the population of *M. fascicularis* has continuously increased and is now considered a pest around areas of human settlement.

Recent reports by DWNP indicate that *M. fascicularis* are often reported by the public as problem makers, and are perceived as pests. A total of 4,139 cases of disturbance by this species were recorded in Malaysia in 2005 (Saaban, 2006) and this figure had increased to 6,578 cases two years later (DWNP, 2007). The populations of these macaques can be found at parks, urban forests and farms (Furuya, 1965; Kurland, 1973; Osman, 1998). They cause damage and disturbance in towns (Southwick and Cadigan, 1972) and farms (Kavanagh and Vellayan, 1995). It appears that the ecological and behavioral flexibility of long-tailed macaques brings them into conflict with humans (Crockett and Wilson, 1980), because it allows macaques to adjust to the increase of human population and their agriculture activities (Else, 1991; Lee and Priston, 2005). Furthermore, historical food provisioning might be another important factor contributing to the overlap between humans and macaques in Malaysia, as this has been found in Singapore (Sha *et al.*, 2009a). Therefore, the reluctance of people to stop feeding the macaques and to take measures to secure garbage has also increased the conflict problem recently in many urban settings (Eudey, 2008).

There have not been many comprehensive studies carried out on the macaque nuisance problems in Malaysia. In Universiti Kebangsaan Malaysia (UKM) main campus, several surveys have been conducted on long-tailed macaque behavior. These include investigations of daily behavior and human-macaque conflict at student residential colleges (Zuraidah, 2003; Sia, 2004; Farhani, 2009), as well as studies of social behavior in the area of the Law faculty (Puyong, 2006). Outside UKM main campus, several studies have been conducted to understand conflict issues in human settlement areas. Tuan-Zubaidah (2003) carried out a study on disturbance behavior on long-tailed macaques at the human settlement in Bukit Lagi, Kangar, located in northern part of Malay Peninsula. A similar study by Sia (2005) focused on the long-tailed macaque's daily activities and disturbance behavior at other human settlements in Taman Tenaga, Puchong, Selangor. Suhailan (2004) focused a study on the behavior of macaques at the local residences in West Country, Bangi, which is located only 3 km away from the UKM main campus. Other publications are available on nuisance problems (Md-Zain *et al.*, 2004) and other aspects of social behavior of macaques in the context of human presence (Md-Zain *et al.*, 2003; Md-Zain and Siti Jamaliah, 2005).

This chapter presents findings on behavioral studies focusing on human-macaque conflict of *M. fascicularis* in the main campus of UKM. The study was initiated based on several complaints from students and staffs of UKM. It has been carried out with three main objectives. First, conduct a population survey of the long-tailed macaques living within the main campus of UKM. Second, identify common types of human-macaque conflict around student dormitories. Third, determine students' perceptions towards long-tailed macaques on campus.

Methods

Study site

This research was conducted at the main campus of Universiti Kebangsaan Malaysia (UKM), Kajang, Selangor, Malaysia. UKM is located 35km south of Kuala Lumpur (Figure 4.1). This area is composed of 1,100 hectares, which contains residential colleges (i.e., dormitories), staff quarters, administration buildings, faculties, club house, health centre, rest house, and stadium. The UKM main campus is surrounded by UKM Permanent Forest Reserve. This forest is a lowland secondary dipterocarp forest with a size of 100 hectares. It has been the original home range area of long-tailed macaques before they were dispersed to many locations in the campus during student dormitory and faculty building construction in the past 40 years. The UKM Permanent Forest Reserve although fragmented, still harbours interesting and important resources with high biomass constituents (Mat Salleh, 1999). There are a number of primate species available for our observation in the forest reserve,

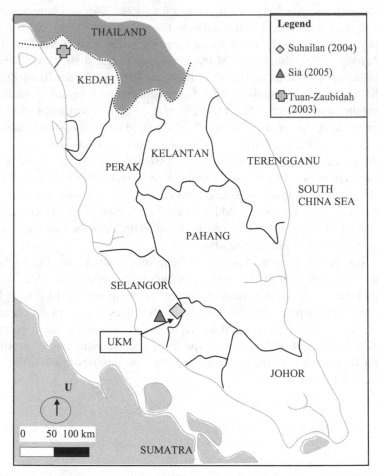

Figure 4.1. Map of our study site at the main campus of Universiti Kebangsaan Malaysia (UKM), and previous studies by Tuan-Zaubidah (2003), Suhailan (2004), and Sia (2005).

which include *M. fascicularis, M. nemestrina, Trachypithecus obscurus, Presbytis melalophos siamensis*, and *Nycticebus coucang*. The distribution of these species are sympatric as discussed by Bennett and Caldecott (1989) and Md-Zain and Abdullah (2008).

Population survey

The population of macaque on UKM's campus was studied and counted. Each macaque group was identified through the facial and physical features

Table 4.1. *Conflict-related behavior recorded during the study*

Behavior categories	Description
Take trash	Taking items from garbage cans.
Leave trash	Carrying and dispersing rubbish from garbage cans.
Enter residence	Enter into student's room or cafeteria at college residence.
Aggression	Behavior such as threatening or chasing students.
Take property	Taking food, clothes, or other belonging of students.
Damage property	Damaging clothes or other property (e.g., scratching cars)

of several individuals. Each individual was confirmed as belonging to a certain group by its movement in tandem with the movement of the group. This was to avoid observation errors associated with assigning individuals to more than one group. According to Chalmers (1979), there are four criteria in ascertaining whether an individual belongs to a particular group, which we used in this study. These are (1) observation through the tandem movements; (2) determining the occurrence of dispersal among individuals; (3) identifying whether spatial relationships influenced the individual to stay in a group; and (4) determining whether the formation of group is only for a short period. All groups interacted with humans and overlapped with the student dormitories.

Behavioral sampling method

From eight identified groups, four groups were chosen for observations. These groups were selected because they interfaced most closely with people. Observations of macaque behavior were conducted from 2003 until 2009. We began our observations with preliminary *ad libitium* observations to understand and describe both the subject and the behaviors that were intended for measurement purposes (Martin and Bateson, 1993). To obtain information on nuisance problems of *M. fascicularis*, full day observations from dawn to dusk were made, using focal sampling and *ad libitum* sampling methods (Altmann, 1974). Most of the surveys were conducted by Faculty of Science and Technology students in several student residential colleges where long-tailed macaques frequently range. In this survey, we only concentrated on conflict between humans and long-tailed macaques. The behavior categories (Table 4.1) were based on preliminary observation. Frequencies of nuisance behaviors were compared statistically using X^2 test.

Interviews

In order to get the feedback and perception from the college students, a questionnaire survey was conducted to find out whether the macaques had caused any problem to them. One hundred students of the Burhanuddin Helmi Residential College were interviewed and were given questionnaire regarding nuisance problems caused by long-tailed macaques. We interviewed 100 students in Burhanuddin Helmi Residential College (KBH) about the opinion, knowledge and attitudes towards long-tailed macaques. The majority of the respondents were Malays (75 percent), followed by Chinese (15 percent) and Indians (10 percent). Female students represented the highest percentage of the respondents as their residence is where the most nuisance problems from long-tailed macaque occurred.

Results

Population survey

In this study, eight groups of *M. fascicularis* were identified at eight different sites mostly in the area of student residential colleges (Table 4.2). The home ranges of some of the groups overlap with each another. Table 4.2 presents the total number of individuals in each group. Group eight, located at Perumahan Bukit Puteri (i.e, staff quarters) had the highest number of individuals, which was approximately 100. The second highest total number was recorded in group one at the Law faculty (55–65 individuals) and followed by group three (50–60 individuals) at Ibrahim Yaakob Residential College. Group six at Ibu Zain Residential College (45 individuals) had a unique characteristic as one Dusky leaf monkey (*Trachypithecus obscurus*) was seen to join the group. No agonistic interaction was recorded between them. Group two, located in Burhanuddin Helmi Residential College, was the main focus group for our human-macaque conflict behavior study and consisted of 36–42 individuals. During observation, one *M. nemestrina* was seen to search for food in the same area of group two. Some agonistic interactions occurred between members of *M. fascicularis* and the *M. nemestrina*. Meanwhile, in this survey, the smallest group was group four, at Aminuddin Baki Residential College, with only 18–27 individuals.

Table 4.2. *Individual counts of M. fascicularis on the main campus of Universiti Kebangsaan Malaysia based on our recent survey 2009 (OUR) and previous studies: Group 1: Law Faculty (LF), Group 2: Burhanuddin Helmi Residential College (KBH), Group 3: Ibrahim Yaakob Residential College (KIY), Group 4: Aminuddin Baki Residential College (KAB), Group 5: Rahim Kajai Residential College ((KRK), Group 6: Ibu Zain Residential College (KIZ), Group 7: Keris Mas Residential College (KKM), Group 8: Perumahan Bukit Puteri (PBP)*

Group	Site	Mastura (2008)							Other Survey
		Adult male	Adult female	Adolescent male	Adolescent female	Juvenile	Infant	Total	Individual Count
1	LF	6	8	10–15	10–13	15–17	6	55–65	Puyong (2006): 40
2	KBH	3	4	6–9	8–9	10–12	5	36–42	OUR (2009): 50–60
3	KIY	4	4	5–10	6–8	9–12	4	32–42	Zuraidah (2003): 20–30
4	KAB	3	4	4–5	5–6	6–8	1	18–27	Farhani (2009): 27
5	KRK	3	5	6–7	3–6	7–10	2	48–56	OUR (2009): 45
6	KIZ	4	6	9–12	9–11	15–18	5	26–33	Sia (2004): 20–30
7	KKM	5	6–8	8–11	7–12	14–17	7	47–52	OUR (2009): 40–50
8	PBP								OUR (2009): 100

Table 4.3. *Percentage and frequency (in parenthesis) of nuisance problems caused by Macaca fascicularis at the main campus of Universiti Kebangsaan Malaysia*

Survey	Farhani (2009)	Mastura (2008)	Sia (2004)	Zuraidah (2003)
Site	Aminuddin Baki Residential College	Burhanuddin Helmi Residential College	Ibu Zain Residential College	Ibrahim Yaakob Residential College
Take items from trash bins	21.88 (14)	47.61 (254)	47.9 (125)	25.0 (77)
Leave trash around	48.44 (31)	42.10 (259)	38.3 (100)	22.7 (70)
Enter a room	15.62 (10)	2.21 (12)	1.2 (3)	20.0 (62)
Take something	0	1.29 (7)	10.3 (27)	11.5 (35)
Disturb people	14.06 (9)	5.88 (32)	1.5 (4)	19.5 (60)
Damage facilities	0	0.90 (5)	0	0
Enter a cafeteria	0	0	0	1.3 (4)
Total	100 (64)	100 (245)	100 (259)	100 (308)
X^2 Value	(X^2 = 40.53) (df=5) P<0.05	(X^2 = 336.49) (df=6) P<0.05	(χ^2 = 60.137) (df=5) P<0.05	(X^2 = 19.48) (df=6) P<0.05

Human-macaque conflict

Conflict-related or nuisance behavior of *M. fascicularis* at the UKM main campus was classified into six categories as shown in Table 4.3. Taking items from trash bins (21.9–47.9 percent) and littering corridor areas (22.7–48.4 percent) were among the highest percentage of behavior caused by macaques. Mostly, rubbish was taken out of garbage cans by macaques, carried out and thrown along corridors in the student residences. Entering into rooms (1.2–20.0 percent), taking items (1.3–11.5 percent) and disturbing people (1.5–19.5 percent), were also among the nuisance problems caused by the long-tailed macaques in the Burhanuddin Helmi Residential College, Aminuddin Baki Residential College, Ibu Zain Residential College and Ibrahim Yaakob Residential College. Entering cafeteria (1.3 percent) and damaging facilities (0.9 percent) were among the lowest percentage of nuisance behavior observed in the UKM main campus.

There is a variation in the frequencies of conflict-related behavior across the study sites. This might be affected by factors such as the distance from the student residences to the forest fringe and college cafeteria and also how effective the garbage collecting schedule and residence cleaning process are. Table 4.3 also shows the comparison of X^2 values across the study sites indicating statistical differences occurred among the nuisance activities caused by the long-tailed macaques.

Student perceptions and attitudes

One-hundred percent of the students said that they can identify what a long-tailed macaques is based on morphological characters. Most respondents said that the species has a long tail with fur on their body, move in a large group and are smaller than pig-tailed macaque. Only 24.1 percent said they could differentiate between male and female macaques. Most of them gave the same answers saying that male is larger than the female, and some differentiated the sexes through observation of the genitals.

In this study, 88.6 percent of the respondents reported having a negative attitude and dislike for the presence of *M. fascicularis*. The reasons included that long-tailed macaques generated messes in their residence corridor and many reported being afraid of the macaques. 95.2 percent of respondents agreed that long-tailed macaques had made their residence halls dirty. Furthermore, 63.4 percent agreed that these macaques are dangerous to the safety and health of the college's residents. However, 11.4 percent of respondents reported having no problem with the macaques presence. Meanwhile, 60.8 percent of respondents reported that at some time they had been disturbed and/or chased by the long-tailed macaques. Most students (75.2 percent) agreed garbage cans in the hostel are inefficiently allocated and a proper place to dump rubbish is needed. Only 14.5 percent of respondents admitted to directly feeding the macaques. Bread and left-over food waste were examples of food given to macaques.

In this survey, most respondents had seen or experienced long-tailed macaques entering into rooms. In addition, 96.7 percent of the respondents said the macaques entered rooms through windows, while others (3.3 percent) said they entered through the door. Nearly half reported that that their rooms were messed up by macaques (51.5 percent) while 16.4 percent of respondents claimed the monkeys had taken items, particularly food, from their rooms. In addition, students claimed that macaques had taken down their hung-up laundry and that their vehicles had been scratched by macaques. Despite the numerous nuisances faced by the residents from macaques, only 10.1 percent of them

claimed to have ever reported the problem to the authorities at Burhanuddin
Helmi Residential College office. However, 87.8 percent of respondents recommended that the Department of Wildlife and National Parks (DWNP) set traps
to capture the macaques, while most of students (85.7 percent) did not agree of
poisoning or killing macaques to resolve human–macaque conflict. In addition,
25.4 percent recommended a more systematic and efficient schedule to collect
the rubbish from dumping areas in order to reduce nuisance problems.

Discussion

In this study, we found approximately 435 individuals in eight groups of
long-tailed macaques living on the UKM main campus. The high number of
macaques in UKM main campus is due to its location, which is adjacent to the
forest edge of UKM Permanent Forest Reserve. The macaques are drawn from
this edge into the campus because of the attractive food resources available.

One of the most common nuisances from the macaques on campus was taking items from trash (Table 4.3). Some of the factors that contribute to this high
frequency of the taking items from trash are the existence of many food sources,
the numerous trash bins, and the lack of appropriate lids and containment, which
all make it easy for long-tailed macaques to successful forage around the college residences. This finding is similar to those Omarrudin (1995) and Osman
(1998), who reported food resources in the garbage cans and dumping area to
be the main factor for the presence of long-tailed macaque at another university,
Universiti Malaya, in Kuala Lumpur. In addition, Eley and Else (1984) found
that messed trash bins was the main problem faced by certain hotels which were
built near primate populations. Suhailan (2004) who studied nuisance behavior of long-tailed macaques in human settlement of West Country, Kajang,
Malaysia, reported the strong smell produced from trash bins had influenced
the macaques to search for food. Numerous studies also indicated the majority
of macaque–human interactions are related to locating and obtaining food (Pirta
et al., 1997; Fuentes and Gamerl, 2005; Fuentes *et al.*, 2005).

Littering activity (i.e., macaques messing an area by carrying, moving, and
leaving human trash) was the second most frequent behavior observed in the
residential colleges. Littering activity normally occurred after taking items
from trash bins, and then dropping them elsewhere. Items such as plastic bags,
cans, papers, and other various rubbish were often thrown all over the place,
especially along corridors. Sia (2005) and Suhailan (2004) reported that the
long-tailed macaques were found carrying and leaving rubbish and plastic
bags on the main road and playground in human settlements. Medway (1969)
found that Penang Botanical Garden used to face the same problem when the

macaques moved rubbish, especially plastics, into the garden after foraging in the garbage cans. In this study, if the long-tailed macaques happened to enter any rooms, they were seen dropping food and plastic bags from the windows. Juveniles were seen carrying plastic bags into the nearby forest.

Macaques entered into students' rooms through windows that were left open when students were out. After entering rooms, they foraged for food and drinks, leaving the rooms in a messed condition because food, water, and personal belongings were scattered over the floor after their activity. In comparison with other studies in local residential areas, there is variation in the extent to which macaques enter homes. Suhailan (2004) observed long-tailed macaques entering into houses through the front doors left open by residents whereas in Taman Tenaga Puchong, Selangor, Sia (2005) reported that the long-tailed macaques only entered the yards of houses as there were many fruit trees within the house compound. In Amboseli National Park, vervet monkeys entered into cottages, kitchens and cars and caused a nuisance by occasionally attacking and biting the tourists and Park staffs (Lee *et al.*, 1986).

Frequency of taking items was low compared to other conflict-related activities. Macaques would sometimes take clothing out of student's rooms. In addition, hangers and clothes were used by macaques while playing with each other. In a local residential area of West Country, Suhailan (2004) had also reported the taking clothes and shoes. However, Sia (2005) reported more often seeing long-tailed macaques taking fruit from trees in people's private yards than clothes. In Sumatra, Crockett and Wilson (1980) reported *M. fascicularis* taking spiny-skinned durian fruit (*Durio zibethinus*), papaya (*Carica papaya*), and young corn plant (*Zea mays*), from orchards which were quickly picked and carried back to the forest fringe for undisturbed consumption.

The long-tailed macaques usually tended to chase students who were carrying food or food cues, and the student's reactions (e.g., resist, run, etc) to these chases sometimes triggered aggression from the macaques. However, no bites or scratches were observed to occur in this study, and this is the same as reports from Singapore (Fuentes *et al.*, 2008; Sha *et al.*, 2009b). Fuentes *et al.*, (2008) observed noncontact interactions such as approach towards and threats to humans in Singapore were related to the humans carrying food items. Meanwhile, Sia (2004) has reported that children in staff housing area of UKM main campus (i.e., Perumahan Bukit Puteri) threw stones at macaques. In response the long-tailed macaques were observed to chase and threatened these children in defense. Similar interactions frequently occur in Bali where the interactions between macaques and humans often involve physical contact and human-directed aggression by the macaques, of which most is related to the presence of food (Wheatley and Harya-Putra, 1994; Fuentes, 2006). Fuentes and Gamerl (2005) had reported 48 bites of macaques to human in the temple monkey of forest of Padentegal, Bali during a six-week study.

Damaging facilities was the least frequent behavior recorded at the UKM main campus. Long-tailed macaques were seen swinging on clothes lines and electric cables, which can be damaging to these structures. Sometimes, long-tailed macaques were seen to sit and rest on cars. Some of the car owners claimed that their cars were scratched by long-tailed macaques. According to Suhailan (2004), *M. fascicularis* at the West Country macaque were seen to destroy a few flower pots and vases when around people's homes. Sha *et al.*, (2009b) also found that taking food items or belongings, receiving aggression and experiencing property damage to fruits trees and ornamental gardens by long-tailed macaques were the most common types of complaints in Singapore. Suhailan (2004) found that the long-tailed macaques used to climb on house roofs, breaking shingles and damaging them. Many respondents claimed that their roof had to be changed at least once every three months as a result of damage caused by macaques. Tuan-Zaubidah (2003) observed macaques had broken a few roofs on local houses of Bukit Lagi, Perlis.

Based on the questionnaire, more than half of the respondents said the level of human-macaque conflict occurring in the residential area of the college was moderate compared to previous years. The macaques have no place to go after their natural habitats have been destroyed for campus development. However, most of the respondents agreed that the existence of long-tailed macaques in their residence creates a serious nuisance problem. The same situation observed by Suhailan (2004) in which 72.3 percent of his respondents agreed that the existence of long-tailed macaques in their residence was a serious pest problem. In questionnaire surveys, students also agreed that the trash bins were inefficient and a proper dumping area is needed to alleviate human-macaque conflict.

Our results suggest that there is a significant amount of nuisance problems from monkeys on the campus of UKM. We suggest that DWNP should take urgent actions such as scheduling the collections of rubbish from dumping areas more frequently so that trash does not lie around for macaques to obtain. In addition, UKM should provide proper macaque-proof trash bins with lids to cover the rubbish and restrict the macaque's access to food sources on campus. It may also be necessary to capture and move some nuisance macaques to other suitable areas. We hope that with the results of this study, the university authority and DWNP will take more immediate management actions in managing the human-macaque conflict at UKM, while also taking into account the importance to sustain this unique species in the area. Human and long-tailed macaques interface in many locations in Malay Peninsula, and thus human-macaque sympatry commonly occurs in Malaysia and is typical for this species. More population surveys should be conducted in interface zones to get a clearer picture on people's complaints and broader context of the human-macaque relationship in Malaysia.

Acknowledgements

We specifically thank Zuraidah Bukhari, Sia Wee Hock, Farhani Ruslin, Aminah Puyong, Mohd Hashim Abu, Ahmad Ampeng, Wan Mohd Suhailan for sharing their time during the surveys. We are indebted to Wan Mohd Razi Idris, Mohamed Afiq Karim, Hazwan Awang Kechik, Nor Iddiana Idris, Noor Ain Saari, Salsela Saidin, Aainaa Syazwani Hamzah and Kamarul Ariffin Hambali, Muhammad Khairul' Anwar and Syed Mohamad Fahmi for their assistance in preparing this manuscript. We are also extremely grateful to the student residential colleges especially students and staffs of Burhanuddin Helmi Residential College and the College Master, Assoc. Prof. Dr Kadderi Md Desa. We wish to thank Prof. Emeritus Dato' Dr Abdul Latiff Mohammad, Farhana Shukor and two anonymous reviewers for their comments on the manuscript. We also thank the Faculty of Science and Technology, Universiti Kebangsaan Malaysia and Department of Wildlife and National Park (DWNP) for granting us permission to conduct this research. The project was supported by Grant ST-019–2002, IRPA 0802020019- EA301, UKM-GUP-KRIB-16/2008.

References

Aldrich-Blake, F. P. G. 1980. Long-tailed macaques. In *Malayan Forest Primates*. D. J. Chivers (ed.). New York: Plenum Press. pp. 147–165.

Altmann, J. 1974. Observational study of behavior: Sampling methods. *Behavior* **49**: 227–267.

Bennett, E. L. and Caldecott, J. O. 1989. Primates of Peninsular Malaysia. In *Ecosystems of the World, 14B – Tropical Rain Forest Ecosystems: Biogeographical and Ecological Studies* H. Leith and M. J. A. Werger (eds.). Amsterdam: Elsevier. pp. 355–363.

Brandon-Jones, D., Eudey, A. A., Geissmann, T., *et al.* 2004. Asian primate classification. *International Journal of Primatology* **25**(1): 97–164.

Chalmers, N. 1979. *Social Behavior in Primates*. London: Edward Arnold.

Chivers, D. J. and Davies, A. G. 1978. Abundance of primates in the Krau Game Reserve, Peninsular Malaysia. In *The Abundance of Animals in Malesian Rain Forest*, G. Marshall (ed.). Misc PT. Series no 22, Department of Geography, University of Hull (Aberdeen-Hull Symposium on Malesia Ecology). pp. 9–32.

Crockett, C. M. and Wilson W.L. 1980. The ecological separation of *Macaca nemestrina* and *Macaca fascicularis* in Sumatera. In *The Macaques: Studies in Ecology,*

Behavior and Evolution, D. G. Lindburg (ed.). New York: Van Nostrand Reinhold. pp. 148–152.

DWNP (Department of Wildlife and National Parks,Peninsular Malaysia). 1982. Annual Report.

1985. Annual Report.

2007. Annual Report.

Eley, D. and Else, J. G. 1984. Primate pest problems in Kenya hotels and lodges. *International Journal of Primatology* **5(4)**: 334–337.

Else, J. G. 1991. Nonhuman primates as pests. In *Primate Responses to Environmental Change*, H.O. Box (ed.). London: Chapman and Hall. pp. 155–166.

Eudey, A. A. 2008. The crab-eating macaque (*Macaca fascicularis*): widespread and rapidly declining. *Primate Conservation* **23**: 129–132.

Farhani, R. 2009. Kajian aktiviti harian dan kelakuan gangguan kera (*Macaca fascicularis*) di Kolej Aminuddin Baki. BSc thesis, Universiti Kebangsaan Malaysia, Bangi, Selangor, Malaysia (unpublished).

Fooden, J. 1995. Systematic review of Southeast Asian long-tail macaques, *Macaca fascicularis* (Raffles, 1821). *Fieldiana Zool* **81**: 1–206.

Fuentes, A. 2006. Human culture and monkey behavior: assessing the contexts of potential pathogen transmission between macaques and humans. *American Journal of Primatology* **68**: 880–896.

Fuentes, A. and Gamerl, S. 2005. Disproportionate participation by age/sex class in aggressive interactions between long-tailed macaques (*Macaca fascicularis*) and human tourists at Padangtegal Monkey Forest, Bali, Indonesia. *American Journal of Primatology* **66**: 197–204.

Fuentes, A., Southern, M. and Suaryana, K. D. 2005. Monkey forests and human landscapes: Is extensive sympatry sustainable for *Homo sapiens* and *Macaca fascicularis* on Bali? In *Commensalism and Conflict: The Primate–Human Interface*, J. D. Paterson and J. Wallis (eds.). American Society of Primatology Publications. pp. 168–195.

Fuentes, A., Kalchik, S., Gettler, L., Kwiatt, A., Koneck, M., and Jones-Engel, L. 2008. Characterizing human-human-macaque interactions in Singapore. *American Journal of Primatology* **70**: 879–883.

Furuya, Y. 1965. Social organization of the crab-eating monkey. *Primates* **6**: 285–336.

Kavanagh, M. and Vellayan, S. 1995. Pest monkey problems in IADP mango projects in Perlis. Project report, WWF Malaysia.

Kurland, J. A. 1973. A natural history of Kra macaques (*Macaca fascicularis* Raffles, 1821) at the Kutai Reserve, Kalimantan Timur, Indonesia. *Primates* **14**: 245–262.

Lee, P. C., Brennan, E. J., Else, J. G., and Altmann, J. 1986. Ecology and behavior of vervet monkeys in a tourist lodge habitat. In *Primate Ecology and Conservation*, J. G. Else and P. C. Lee (eds.). Cambridge University Press. pp. 229–235.

Lee, P. C. and Priston, N. E. C. 2005. Human attitudes to primates: perceptions of pests, conflict and consequences for conservation. In *Commensalism and Conflict: The Primate–Human Interface*, J. D. Paterson and J. Wallis (ed.). American Society of Primatology Publications. pp. 168–195.

Lim, B. L. 2008. Critical habitats for the survival of Malayan mammals in Peninsular Malaysia. *Journal of Science and Technology in the Tropics* 4: 27–37.

Lim, L. S., Md-Zain, B. M., Mahani, M. C., and Shahrom, A. W. 2004. HVS I as a tool in phylogenetic analysis among human populations in Peninsular Malaysia. *Malaysian Journal of Biochemistry and Molecular Biology* 10: 1–8.

Marsh, C. W. and Wilson, W. L. 1981. A survey of primates in Peninsular Malaysian forest. Final report for the Malaysian Primates Research Programme Universiti Kebangsaan Malaysia and University of Cambridge, United Kingdom.

Martin, P. and Bateson, P. 1993. *Measuring Behavior: An Introductory Guide*, 2nd edn. Cambridge University Press.

Mat Salleh, K. 1999. The role and function of Universiti Kebangsaan Malaysia Permanent Forest Reserve in research and education. *Pertanika Journal of Tropical Agricultural Science* 22: 185–198.

Matsuda, I., Tuuga, A., and Higashi, S. 2009. The feeding ecology and activity budget of proboscis monkeys. *American Journal of Primatology* 71: 478–492.

Md-Zain, B. M., Norhasimah, M. D., and Idris, A. G. 2003. Long-tailed macaque of the Taman Tasik Taiping: Its social behavior. In *Prosiding Simposium Biologi Gunaan Ke-7*, Bangi, Selangor, Malaysia. pp. 468–470.

Md-Zain, B. M., Tuan-Zaubidah, T. H., and Zuraida, B. 2004. Kajian kelakuan gangguan kera (*Macaca fascicularis*) di kawasan kediaman. In *Prosiding Seminar Bersama FST, UKM-FMIPA, UNRI KE 3*, Bangi, Selangor, Malaysia. pp. 149–155.

Md-Zain, B. M. and Siti-Jamaliah, A. 2005. Activity budgets of long-tailed and pig-tailed macaques in Taiping Lake Garden, Perak. In *Proceedings of 2nd Regional Symposium on Environment and Natural Resources*, vol. 2, pp. 32–34. Bangi, Selangor, Malaysia.

Md-Zain, B. M. and Abdullah, M. 2008. Primat Bukit Belata. In *Bukit Belata, Selangor: Pengurusan Persekitaran Fizikal, Kepelbagaian Biologi dan Sosio-Ekonomi*, A. Muda, H. L. Koh, N. M. S. Nik Mustafa, S. A. Nawi and A. Latiff (eds.). Kuala Lumpur: Jabatan Perhutanan Semenanjung Malaysia. pp. 270–273.

Md-Zain, B. M., Morales, J. C., Hassan, M. N., *et al.* 2008. Is *Presbytis* a distinct monophyletic genus: Inferences from mitochondrial DNA sequences. *Asian Primates Journal* 1: 26–36.

Md-Zain, B. M., Vun, V. F., Ampeng, A., Rosli, M. K. A. and Mahani, M. C. 2009. Molecular systematics of the Malaysian leaf monkeys genus *Presbytis* and *Trachypithecus*. Paper presented in the 3rd International Congress on the Future of Animal Research (ICFAR): Biomedical and Field Research with Non-human PrimatesThailand,19–22 November 2009.

Md-Zain, B. M., Mohamad, M., Ernie-Muneerah, M. A., *et al.* 2010a. Phylogenetic relationships of Malaysian monkeys, Cercopithecidae, based on mitochondrial cytochrome c sequences. *Genetics and Molecular Research* 9(4): 1987–1996.

Md-Zain, B. M., Sha`ari, N. A., Mohd-Zaki, M., Ruslin, F., Idris, N. I., Kadderi, M. D., and Idris, W. M. R. 2010b. A comprehensive population survey and daily activity budget on Long Tailed Macaques of Universiti Kebangsaan Malaysia. *Journal of Biological Sciences* 10(7): 608–615.

Md-Zain, B. M. and Ch'ng, C. E. 2011. The activity patterns of a group of cantor dusky leaf monkeys (*Trachypithecus obscurus halonifer*). *International Journal of Zoological Research* **7**: 59–67.

Medway, L. 1969. *The Wild Mammals of Malaya and Offshore Islands Including Singapore*. Kuala Lumpur: Oxford University Press.

Mohd-Azlan, J. A. G. A. 2006. Mammal diversity and conservation in a secondary forest in Peninsular Malaysia. *Biodiversity and Conservation* **15**: 1013–1025.

Muda, H. 1982. *Perdagangan Primat di Malaysia Barat*. Kuala Lumpur: Jabatan Perlindungan Hidupan Liar Semenanjung Malaysia.

Omarrudin, B. I. 1995. Kajian ekologi dan pengurusan kera (*Macaca fascicularis*) di kampus Universiti Malaya: Keberkesanan tempat pembuangan sisa makanan alternatif. BSc thesis, Universiti Malaya, Kuala Lumpur, Malaysia (unpublished).

Osman, N. A. W. 1998. A study composition and density of long-tailed macaque (*Macaca fascicularis*) in Kuala Lumpur. *Journal of Wildlife and Parks* **16**: 80–84.

Pirta, R. S., Gadgil, M., and Kharshikar, A. V. 1997. Management of the rhesus monkey *Macaca mulatta* and hanuman langur *Presbytis entellus* in Himachal Pradesh, India. *Biological Conservation* **79**: 97–106.

Poirier, F. E. and Smith, E. O. 1974. The crab-eating macaque (*Macaca fascicularis*) of Angaur Island, Palau, Micronesia. *Folia Primatologica* **22**: 258–306.

Puyong, A. 2006. Study of social playing behavior of *Macaca fascicularis* at two different sites (Law faculty, UKM and West Country, Kajang). MSc thesis, Universiti Kebangsaan Malaysia, Bangi, Selangor, Malaysia (unpublished).

Saaban, S. 2006. *Pelan Pengurusan Kera Bermasalah di Semenanjung Malaysia*. Kuala Lumpur: Jabatan Perlindungan Hidupan Liar Semenanjung Malaysia.

Sha, C. M., Gumert, M., Lee, P. Y-H., Fuentes, A., Rajathurai, S., Chan, K. L., and Jones-Engel, L. 2009a. Status of the long-tailed macaque in Singapore and implications for management. *Biodiversity and Conservation* **18**: 2909–2926.

Sha, C. M., Gumert, M., Lee, P. Y-H., Jones-Engel, L., Chan. S., and Fuentes, A. 2009b. Macaque–human interactions and the societal perceptions of macaques in Singapore. *American Journal of Primatology* **71**: 825–839.

Sia, W. H. 2004. Kajian kelakuan *Macaca fascicularis* di kawasan sekitar kampus UKM. BSc thesis, Universiti Kebangsaan Malaysia, Bangi, Selangor, Malaysia (unpublished).

———. 2005. A study of pest behavior on long-tailed macaque (*Macaca fascicularis*) at Taman Tenaga, Puchong, Selangor. MSc thesis, Universiti Kebangsaan Malaysia, Bangi, Selangor, Malaysia (unpublished).

Southwick, C. H. and Cadigan, Jr., F. C. 1972. Population studies of Malaysian primates. *Primates* **13**: 1–18.

Suhailan, W. M. 2004. A behavioral study on the long-tailed macaque (*Macaca fascicularis*) in the residence of West Country, Bangi. MSc thesis, Universiti Kebangsaan Malaysia, Bangi, Selangor, Malaysia (unpublished).

Tuan-Zaubidah, B. T. H. 2003. Kajian terhadap kelakuan gangguan *Macaca fascisularis* di persekitaran Bukit Lagi, Kangar, Perlis. BSc thesis, Universiti Kebangsaan Malaysia, Bangi, Selangor, Malaysia (unpublished).

Wheatley, B. P. and Harya Putra, D. K. 1994. Biting the hand that feeds you: Monkeys and tourists in Balinese monkey forests. *Tropical Biodiversity* 2: 317–327.

Yanuar, A., Chivers, D. J., Sugardjito, J., Martyr, D. J., and Holden, J. T. 2009. The population distribution of pig-tailed macaque (*Macaca nemestrina*) and long-tailed macaque (*Macaca fascicularis*) in West Central Sumatra, Indonesia. *Asian Primates Journal* 1: 2–11.

Zuraidah, B. 2003. Kajian kelakuan harian dan kelakuan gangguan kera (*Macaca fascicularis*) di Kolej Ibrahim Yaakub, Universiti Kebangsaan Malaysia. BSc thesis, Universiti Kebangsaan Malaysia, Bangi, Selangor, Malaysia (unpublished).

5 Human impact on long-tailed macaques in Thailand

SUCHINDA MALAIVIJITNOND,
YOLANDA VAZQUEZ AND
YUZURU HAMADA

Introduction

Long-tailed macaques (*Macaca fascicularis*) are the most frequently seen species among the thirteen species of primates in Thailand (Lekagul and McNeely, 1988, Malaivijitnond *et al.*, 2005) and they were recently reported at 91 locations (Malaivijitnond *et al.*, 2009). In comparison, only nineteen, twelve, eleven, and nine locations were observed for rhesus (*M. mulatta*), pig-tailed (*M. nemestrina*), stump-tailed (*M. arctoides*) and Assamese macaques (*M. assamensis*) respectively (Malaivijitnond *et al.*, 2009). Long-tailed macaques inhabit a wide variety of habitats, including primary lowland rain-forests, disturbed and secondary rainforests, riverine forests, and coastal forests of nipa palm and mangrove. They have been frequently seen on the forest periphery, and at recreation parks, tourist attraction sites, temples, and other areas nearby human settlements (Aggimarangsee, 1992; Fooden, 1995; Malaivijitnond and Hamada, 2008). This reflects their wide adaptability to various ecological conditions. Humans are continually invading and disturb-ing the natural habitats of long-tailed macaques, by means of deforestation, conversion to agricultural land, development of infrastructure and widespread encroachment. As a result, natural habitats are being increasingly fragmented, degraded, and changed by human activity. Long-tailed macaque populations have been isolated from each other and the risk of inbreeding and/or outbreed-ing depression has increased (Malaivijitnond and Hamada, 2008). Additionally, anthropogenic habitat alteration has increased the overlap between long-tailed macaque and humans because many areas near human settlements, where macaques are now found, were not areas of interface in the past.

Monkeys on the Edge: Ecology and Management of Long-Tailed Macaques and their Interface with Humans, eds. Michael D. Gumert, Agustín Fuentes and Lisa Jones-Engel. Published by Cambridge University Press. © Cambridge University Press 2011.

There is some information on the status of long-tailed macaques in Thailand (Fooden, 1971; Aggimarnagsee, 1992; Malaivijitnond *et al.*, 2005; Malaivijitnond and Hamada, 2008). The most up-to-date information available regarding semi-tame troops of macaques living near human settlements was published in 1992 by Aggimarangsee, and the distribution and status of long-tailed macaques across Thailand published in 2008 by Malaivijitnond and Hamada. Although Eudey (2008) reported that long-tailed macaques in Southeast Asia were rapidly declining due to habitat loss, trapping, and trading for use in pharmaceutical research, many troops of long-tailed macaques in Thailand are locally overcrowded (i.e., a high number of macaques in a limited habitat area) and in conflict with humans (Malaivijitnond and Hamada, 2008). Another threat to long-tailed macaques in Thailand is hybridization with other macaque species as a consequence of pet release (Malaivijitnond *et al.*, 2005; Malaivijitnond and Hamada, 2008). This pollutes their gene pool and may cause a problem, especially as long-tailed macaques are used as animal models for drug developments in biomedical research.

This chapter presents an overview of information gathered to date regarding the distribution and status of long-tailed macaques in Thailand, addresses issues associated with the close human and long-tailed macaque sympatry, commensal and conflicting relationships, as well as the potential impacts of such relationships. The data presented were gathered using the methods subsequently outlined which include: questionnaire surveys, field surveys, capture-and-release techniques, literature and media reviews, and preliminary data from an intensive investigation of the human–long-tailed macaque interface at Lopburi, a renowned monkey city in Thailand, from interviews and direct field observations.

Materials and methods

Questionnaire survey

A questionnaire about thirteen species of non-human primates in Thailand was sent with a brochure and stamped envelope to the leader (or Kamnan) of each sub-district (or Tambon). Generally, most of Kamnans in Thailand are locals and live in the areas for a long time before they are elected, therefore we supposed that they knew the area well. In the year 2003, a website about each Tambon in Thailand was established and launched for the promotion of the program of "One Tambon One Product (OTOP)", and the address of each Tambon for our questionnaire survey was accessed from there. The questionnaires (see Appendix 5.1) were sent throughout Thailand, except the metropolitan Bangkok, to a total of 7,410 Tambons. After the completed questionnaires were returned, we classified locations that had a reported presence of primates in the area (i.e., positive reports) and mapped the distribution of five species of macaques in Thailand. We traveled to locations that had a reported presence of one of the five species of macaques, particularly for long-tailed macaques.

Field survey

We visited the Tambons where macaques were reported between December 2002 and September 2009. On these surveys we recorded the location names, geographical coordinates, macaque sub-species or species present (see Fooden, 1995), presence of hybrids, the number of macaques observed, evidence for the release of heterospecific or conspecific macaques to the troop, history of the troop, impact on and conflict with humans, habitat types (i.e., natural forest or anthropogenic), morphological characters and behavior (see Appendix 5.2). Photographs were also taken for further analysis and archival reference. The duration of observation in each location varied between a half-day and a week. Some locations were visited more than once, when it was an interesting site for the further study or it was a feasible site for monkey capturing. At each location, we interviewed a Kamnan, senior local people, or forest rangers.

During the survey, the morphological characters of the macaques, e.g., patterns of crest, crown, direction of cheek hair, pelage color, body size and tail length, swelling and reddening of sex skin in female monkeys (Engelhardt *et al.*, 2005; Malaivijitnond *et al.*, 2007a), were observed. The macaque sub-species or species was identified based on their morphological characters (Figure 5.1 and based on Fooden, 1975; 1990; 1995; 2000; Malaivijitnond *et al.*, 2005; Malaivijitnond and Hamada, 2008; Hamada *et al.*, 2005a; 2006; 2008). Once monkeys were observed with mixed morphological characters between at least two macaque species, they were identified as hybrid.

Animal capture

Following the field survey, eighteen troops of long-tailed macaques from selected feasible localities (*ca.* 16° 15' – 6° 30' N), (i.e., feasible to set a trap on the ground, monkeys were habituated and acquired food provisioned on the ground, and having more than 50 monkeys in the group), were temporarily caught with a net trap (6m x 6m x 1.5m or 6m x 10m x 1.5m). A total of

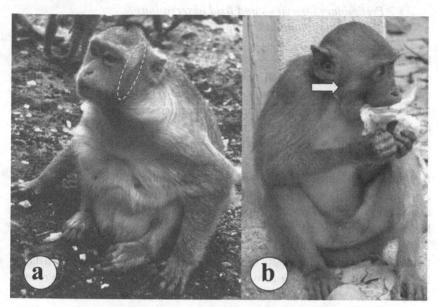

Figure 5.1. The cheek hair pattern variation found on the two subspecies of *Macaca fascicularis* living in Thailand. a) *Macaca fascicularis fascicularis* with a transzygomatic pattern of cheek hair living at Wat Thammasala, Nakhon Pathom Province (13° 48' 42"N, 100° 06' 52"E) and b) *M. f. aurea* with an infrazygomatic pattern of cheek hair living at Wat Paknam Pracharangsarith, Ranong Province (9° 57' 11"N, 98° 35' 41"E) (photographs by S. Malaivijitnond).

878 long-tailed macaques were caught between 2003 and 2007. All of them were wild or semi-wild macaques inhabiting the hills near temples (or Wat in Thai), recreation parks, or mangrove forests. Monkeys were anesthetized by intramuscular injection with 10mg/kg body weight of ketamine hydrochloride. While the monkeys were immobilized, body mass, tail length, crown-rump length, and pelage color were recorded (Hamada *et al.*, 2006; 2008), blood was collected by femoral venipuncture (Malaivijitnond *et al.*, 2007b), and feces were collected directly from the anus (Malaivijitnond *et al.*, 2006). Blood samples were preserved in heparin or EDTA buffer, and fecal samples were mixed vigorously and homogenously with 2–4 ml of 10 percent formalin buffer. The DNA was extracted from the white blood cells and kept as a DNA bank at the Primate Research Unit of Chulalongkorn University. The monkeys were subjected to various clinical inspections including external parasites before being released back into the troop after their complete recovery from anesthesia. The ages of monkeys were estimated from the dental eruption according to Smith *et al.*, (1994).

Preliminary interviews and field observations
of the human–long-tailed macaque interface at Lopburi

Lopburi city is home to intense human-macaque overlap because there are high populations of both species and close interactions between them. Some previous limited investigations have recorded evidence of both mutualism, i.e., beneficial relationships (e.g., provisioning of macaques and tourism revenues for the local community) and conflict, i.e., detrimental relationships (e.g., damage to households and physical injuries to the macaques) (Aggimarangsee, 1992; Malaivijitnond and Hamada, 2008). However, no in-depth study of this interface, its potential threats to either species, or of human attitudes exists. Threats associated with intense primate interrelationships include bidirectional infectious agent transmission, physical injury, conflicts over resources, economic losses and exploitation or condemning of the non-human primate species (Hill, 2005; Lee and Priston, 2005; Srivastava and Begum, 2005; Zhao, 2005). These threats not only affect the health of the primate species present but the sustainability of such co-existence (Srivastava and Begum, 2005; Lee and Priston, 2005). Therefore, it is necessary to assess the human-macaque interconnections in Lopburi in order to highlight any threats to its participants and to evaluate the need for any management incentives which encourage the sustainability of positive relationships. This preliminary investigation focused on using an ethnoprimatological approach to investigate the human-macaque interface within "Muang Lawo" or the Old City area surrounding two historical shrines – Prang Sam Yot and Sarn Pra Karn – in Lopburi, Thailand (Figure 5.2). Lopburi is one of the five sites in Thailand which have been highlighted as suffering from long-tailed macaque "overpopulation" (Malaivijitnond and Hamada, 2008), it makes a valuable study site into the human-macaque interface as it incorporates three high human traffic components: an urban residential area, a functioning religious complex and a popular tourist destination.

As part of the investigation of this interface, preliminary data were collected by means of verbal interviews and general observations.

Interviews

Interviews were conducted between May–July 2009 and included the opinions and observations of key persons and/or organization representatives involved in the human-macaque interface, i.e., local veterinarians, shrine staff, monkey hospital staff, monkey festival organizers, local people, NGO representatives, and the local monkey management committee (the "Monkey Foundation") representative. People chosen for interviews were sought out using snowball sampling in order to get information from people with high involvement in the

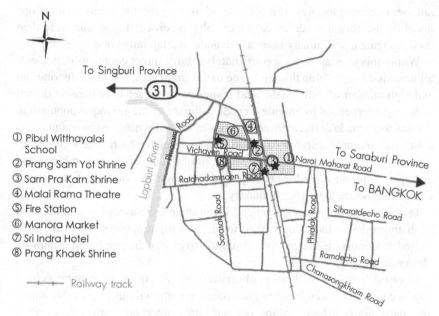

① Pibul Witthayalai
   School
② Prang Sam Yot Shrine
③ Sarn Pra Karn Shrine
④ Malai Rama Theatre
⑤ Fire Station
⑥ Manora Market
⑦ Sri Indra Hotel
⑧ Prang Khaek Shrine

—┼— Railway track

Figure 5.2. Home ranges of the five groups of long-tailed macaques (shading areas) and the provisioning sites (*) in the Old City area, Lopburi Province.

interface and who could provide a general overview of the current situation and attitudes, as well as insight into their specific role within the interface. Topics discussed during the interviews included; locations of macaques, interaction types, participants and stakeholder groups involved in interactions, persons and organizations involved in management concerning this interface, management initiatives relating to the interface, and general human attitudes towards the macaques. An outline of the pilot interview questions asked is described (see Appendix 5.3). However, the respondents were encouraged to add more detail, particularly where they had in-depth knowledge in any topic or to give any additional information they felt would be useful to this study. A total of nine people were interviewed.

Visual observations

Visual surveys were conducted between July and September 2009 and December 2009 and February 2010. Using the map generated from interview responses as a guide, demarking all the areas in the Old City area where respondents claimed to have seen macaques, four full days (dawn–dusk), selected at random, were dedicated to surveying the area for macaque presence. Macaque sightings were recorded using a global positioning system (GPS) and the number of macaques present recorded. *Ad libitum* sightings outside locations noted on the allocated survey days were also recorded. These data were then used to create a preliminary human-macaque overlap range map.

Within this overlap zone (approximately 1 km^2) direct counts were made of all macaques seen within this area. Due to the small area, clustered distribution and high number of individuals, total counts were chosen over transect or belt sampling alternatives to provide a crude estimate of the macaque population. Counts were made at three different times of day (morning, midday and afternoon) on randomly selected days and two samples for each time were taken. Each group count was carried out twice, minimum counts were combined and averaged to produce the lower range and the same was done with maximum counts to produce the higher boundary.

In order to identify groups: general location and site use, plus identification of distinguishable individuals were used. Two night surveys were also conducted to identify sleeping sites and attempt to clarify the number of groups by observing sleeping clusters.

Over 60 hours of *ad libitum* observations were made of general behaviors occurring between humans and macaques throughout the overlap range and particularly where groups of macaques were concentrated. General patterns of interactions were noted, participant groups involved in interactions and degree of contact involved. However, these were not quantified only described. The overlap zone was also surveyed for management measures linked to macaque–human interaction e.g., recording presence of any informative signs or warnings regarding the monkeys, and for any evidence of resource sharing between the two, a shared resource was anything which was observed as being used by both humans and monkeys, e.g., water taps, pavements, and homes.

Results

Information from the questionnaire survey

The questionnaires were sent to 7,410 Tambons in 75 provinces, which is separated into five regions: North, Central, Northeast, East and South (Malaivijitnond and Hamada, 2008). The largest numbers of questionnaires were sent to the northeastern region (2,795 Tambons) and the smallest to the eastern region (479 Tambons). A total of 1,417 questionnaires (19.1 percent) were returned. The largest number of questionnaires that were returned was from the northeastern region (642, 22.97 percent) and the smallest was from the eastern region (96, 20.04 percent). A total of 705 questionnaires were positive

Table 5.1. *The numbers of questionnaires (Qs) sent to and returned from the five regions of Thailand.*

Region	No. of provinces	No. of Qs sent	No. of Qs returned (% of Qs sent)	No. of Qs with no primates (% of Qs returned)	No. of Qs with primates (% of Qs returned)
North	16	1,488	226 (15.19%)	108 (47.79%)	118 (52.21%)
Central	19	1,520	278 (18.29%)	150 (53.96%)	128 (46.04%)
Northeast	19	2,795	642 (22.97%)	371 (57.79%)	271 (42.21%)
East	7	479	96 (20.04%)	38 (39.58%)	58 (60.42%)
South	14	1,128	175 (15.51%)	45 (25.71%)	130 (74.29%)
Total	75	7,410	1,417 (19.12%)	712 (50.25%)	705 (49.75%)

reports of the presence of primates, and 583 of the 705 questionnaires reported a presence of macaques in the areas. The highest frequency of positive replies was from the southern region (130, 74.3 percent). The highest percentage of negative reports (no primates) came from the northeastern region (371, 57.8) (Table 5.1).

When we carefully assessed the questionnaires reporting the presence of macaques in the areas, we found that many Tambons have neither long-tailed macaques nor other macaque species in their own areas, but they gave us the information about macaques living in other areas. For example, five questionnaires (from five Tambons) returned from Mahasarakham Province, northeastern Thailand together reported a single location of long-tailed macaques at Kosumphi Forest Park (16° 15'N, 103° 04'E), Tambon Huakhwang, that is four of the five Tambons had no primates in their own areas. Of 583 questionnaires reported a presence of macaques, only 386 Tambons were their habitats.

Based on information that each Kamnan added to the questionnaire as well as that acquired by the interview, we found that most of Kamnans in the northeastern region are long-term residents at their respective Tambons and have been able to observe environmental changes and the loss of wildlife over time. They described that local macaque populations went extinct or were exterminated about thirty years ago because of the rapid and extensive deforestation for agricultural development.

Information from the field survey

Of 386 Tambons, we went to 340 Tambons (88.1 percent) that had a reported presence of one of the five species of macaques, especially long-tailed macaques, and where access was possible. Most surveys were conducted in the northeastern and central regions (134 and 78 Tambons). Via the ground checking at 340 Tambons, we found that local people misidentified long-tailed macaques as other macaque species. The frequency and nature of misidentification varied between regions. In all five regions, mainly in the northeastern and central regions, long-tailed macaques were particularly misidentified as rhesus macaques and occasionally as pig-tailed macaques. Only one location in the northeast, at Wat Pa Sila Wiwek, Mukdahan Province, long-tailed macaques were misidentified as Assamese macaques.

Of the 340 Tambons surveyed, we found long-tailed macaques at 100 locations (the location is used instead of Tambon hereafter, because in some Tambons long-tailed macaques were seen in more than one location) which ranged from the lower northern and northeastern (*ca.* 16° 30' N) to the southernmost part (*ca.* 6° 20' N) of Thailand (Malaivijitnond and Hamada, 2008; Malaivijitnond *et al.*, 2009) (Table 5.2). In many locations, the population had several troops (two to five troops). The most frequently encountered subspecies was *M. f. fascicularis*, at 97 of the 100 locations visited. *M. f. aurea* was found at only three locations of Ranong Province 9° 35–57'N: Koh Piak Nam Yai and Koh Tao; Ngoa Mangrove Research Center; Wat Paknam Pracharangsarith (Malaivijitnond and Hamada, 2008). The habitat types at those three locations are mangrove forests or islands located on the Andaman sea coast.

Based on our field survey, the distribution maps of long-tailed macaques and other four species of macaques in Thailand were drawn in comparison with those of the questionnaire survey and the previous report of Lekagul and McNeely (1988) (Figure 5.3). One hundred, nineteen, twelve, eleven and nine locations were observed for long-tailed, rhesus, pig-tailed, stump-tailed and Assamese macaques, respectively. Comparing their distributions between three sources of information, they are similar, except for rhesus macaques where the distribution based on questionnaire survey is throughout Thailand. The distribution pattern of long-tailed macaques was similar to that described 30 – 40 years ago (Lekagul and McNeely, 1988), however, their habitats had changed greatly from natural forests to the recreational parks or temples close to human settlements. They can adapt well to disturbed habitats and some refer to them as a "weed species" (Richard *et al.*, 1989). At 100 locations visited, 57 locations were temples. Most temples visited were near forests or limestone hills. This suggests that the majority of the population of Thai long-tailed macaques is now at least partly dependent on the limited sanctuary provided by Buddhist temples.

Living in the temple complexes, vegetation for foraging is often sparse and they are utterly dependent on provisioning. Through provisioning, lack of

Table 5.2. *Names and geographical coordinates obtained using the Global Positioning System (GPS) for long-tailed macaques found in Thailand. Wat, Ban, Khao and Koh stand for temple, village, mountain or hill and island, respectively, in Thai.*

Region		Name of Location	GPS (N, E)	Number counted	Duration of observation	Estimated number of troops
Northeast (7 locations)	1	Kumpawaphi City Park, Udon Thani Province**	17° 06', 103° 01'	130	29 August 2003	2
	2	Wat Pa Sila Wiwek, Mukdahan Province	16° 32', 104° 43'	46	11 July 2004	1
				30	11 October 2004	1
	3c	Kosumphi Forest Park, Maha Sarakham Province	16° 15', 103° 04'	287	27–30 March 2003	3
	4	Don Poo Taa Monkey Park, Yasothon Province	15° 48', 104° 24'	(>1,000)*	18 February 2009	N/A
	5	Don Chao Poo Forest Park, Umnajiaroen Province	15° 36', 104° 50'	378 (600)*	11 October 2004	3
	6	Wat Pa Ban Paeu, Ubon Ratchathani Province	15° 33', 104° 37'	(300)*	25 March 2008	2
	7	Wat Kupra Kona, Roi Et Province	15° 30', 103° 49'	87	11 July 2004	3
	8	Wat Pa Tung Boon, Ubon Ratchathani Province	15° 30', 105° 16'	N/A	25 March 2008	1
	9	Muangling Ban Whan, Si Sa Ket Province	15° 22', 104° 10'	119	12 October 2004	>1
	10	Wat Pa Nam Boon, Ubon Ratchathani Province	15° 21', 105° 26'	16	25 March 2008	1
	11	Wat Ban Muangkhaen Potharam, Si Sa Ket Province	15° 21', 104° 13'	88	12 October 2004	2
	12	Wat Pa Pothiyan, Ubon Ratchathani Province	15° 11', 105° 25'	58	25 March 2008	1
	13	Chong Chom Boder Pass, Surin Province	14° 26', 103° 41'	4	23 March 2008	1
North (8 locations)	14	Wat Haad Moon Bang Kra Beau, Phichit Province	16° 30', 100° 16'	60	20 January 2007	1
	15	Wat Khao Nor, Nakhon Sawan Province	15° 57', 99° 52'	(>100)*	23–26 April 2007	2
				N/A (>200)*	18–21 September 2002	4
				932 (>1,000)*	5 September 2003	4
	16	Wat Tasang Tai, Nakhon Sawan Province	15° 56', 99° 57'	345	23 July 2004	3
	17	Wat Paa Khao Pha, Petchabun Province	15° 47', 101° 13'	24	16 May 2003	1
	18	Wat Tham Thepbandan, Petchabun Province	15° 44', 101° 02'	117	16 May 2003	2
				177	1–4 March 2004	2

Table 5.2. (cont.)

Region	Name of Location	GPS (N, E)	Number counted	Duration of observation	Estimated number of troops
	19 Wat Kriang Krai Klang, Nakhon Sawan Province	15° 44', 100° 11'	>200	15 May 2003	2
	20 Wat Khao Thong, Nakhon Sawan Province	15° 12', 100° 24'	100*	20 January 2007	1
	21 Wat Khao Wong, Nakhon Sawan Province	15° 10', 100° 24'	73	20 January 2007	1
Central (21 locations)	22 Wat Khao Patthawee, Uthai Thani Province	15° 28', 99° 45'	234 (>1,000)*	July 2004	2
	23 Khao Takhon, Lopburi Province	15° 20', 100° 51'	N/A	29 March 2008	N/A
	24 Khao Ling, Lopburi Province	15° 19', 100° 59'	N/A	29 March 2008	N/A
	25 Wat Phikun-ngam, Chai Nat Province	15° 16', 100° 03'	47	22 July 2004	1
	26 Khao Hin Kling, Lopburi Province	15° 10', 101° 06'	N/A	29 March 2008	N/A
	27 Wat Khao Sompod, Lopburi Province	15° 09', 101° 16'	258 (>1000)*	4 November 2007	>3
	28 Wat Khao Wongkot, Lopburi Province	15° 01', 100° 32'	(>140)*	29 March 2008	2
	29 Wat Khao Pra Ngam, Lopburi Province	14° 53', 100° 37'	84 (200)*	5 November 2007	2
	30 Sarn Pra Karn, Lopburi Province	14° 48', 100° 36'	850–1,100	24–25 November 2007	5
	31 Route to Jed Kod-Pongkonsao, Saraburi Province	14° 33', 101° 05'	30	16 July 2006	1
	32 Wat Kai, Ayutthaya Province	14° 30', 100° 31'	82	6 September 2003	2
			81	7 August 2004	1
	33 Wat Hansang, Ayutthaya Province	14° 30', 100° 31'	N/A	3 June 2004	N/A
	34 Wat Praputthachai, Saraburi Province	14° 28', 100° 56'	156 (>200)*	17 May 2003	>1
	35 Mahidol University of Kanchanaburi Campus Kanchanaburi Province	14° 07', 99° 09'	75	19 June 2004	1
			N/A	27–29 September 2005	1
	36 Wat Thammasala, Nakhon Pathom Province	13° 48', 100° 06'	49	23 August 2003	2
			64	20–22 February 2004	1
	37 Khao Ngu Rock Garden, Ratchaburi Province	13° 34', 99° 46'	(>1,000)*	22–25 September 2002	5
				10 December 2002	
				7 April 2003	
	38 Bangtaboon Witthaya School, Petchaburi Province	13° 15', 99° 57'	36	2 January 2003	1

No.	Location	Coordinates	Count	Date	
39	Ban Laem Mangrove Forest, Petchaburi Province	13° 14', 99° 58'	N/A	22 February 2009	N/A
40	Wat Tham Khao Yoi, Petchaburi Province	13° 14', 99° 49'	45	23 July 2005	1
41	Wat Kut Tha Raeng, Petchaburi Province	13° 09', 99° 56'	60 (100)*	22 February 2009	1
42	Pranakhonkhiri or Khao Wang, Petchaburi Province	13° 06', 99° 56'	N/A	22 July 2005	>1
43	Wat Khao Luang, Petchaburi Province	13° 07', 99° 55'	N/A (>500)*	15 April 2005 22 July 2005	5
44	Wat Khao Thamon, Petchaburi Province	13° 02', 99° 57'	(800)*	23 August 2003 24–27 February 2004	4 4
45	Wat Khao Krachiw or Wat Banpatawad, Petchaburi Province	12° 57', 99° 54'	50	24 August 2003	1
46	Sam Roi Yot National Park, Prachuap Khirikhan Province	12° 07', 99° 57'	111	16 July 2005	2
47	Wat Khao Takieb, Prachuap Khirikhan Province	12° 30', 99° 59'	(600)* (600)* N/A	25 August 2003 25–27 April 2006 26–29 April 2003	>1 >1 3
48	Wat Thammikaram Worawiharn or Wat Khao Chong Krachok, Prachuap Khirikhan Province	11° 48', 99° 48'	343	16 July 2005	
49	Wat Khao Chagun, Srakaew Province	13° 39', 102° 05'	(>1,000)*	23 September 2008	N/A
50	Khao Samsip Monastry, Chachoengsao Province	13° 35', 101° 44'	N/A	24 September 2008	N/A
51	Khao Ang Luea Nai Wildlife Reserve Center, Chachoengsao Province	13° 24', 101° 52'	N/A	20 December 2002	1
52	Khao Sam Muk, Chon Buri Province	13° 18', 100° 54'	(200)*	18 March 2005	2
53	Wat Santipuk or Wat Khao Phu, Chon Buri	13° 14', 100° 59'	(200–300)*	6 September 2009	2
54	Khao Khieow Open Zoo, Chon Buri Province***	13° 12', 101° 03'	(>100)*	15 April–20 November 2005 29 July–1 August 2006 8 December 2007	2
55	Khao Kaset of Kasetsart University, Chon Buri Province	13° 07', 100° 55'	N/A	7 September 2009	2

East
(7 locations)

Table 5.2. (cont.)

Region	Name of Location	GPS (N, E)	Number counted	Duration of observation	Estimated number of troops
	56	Wat Khao Cha-ang-on, Chon Buri Province	13° 12', 101° 39'	24	17 March 2005	1
	57	Wat Khao Cha-ang-on Nok, Chon Buri Province	13° 06', 101° 34'	25	17 March 2005	1
	58	Small hill close to Sudthangruk Restaurant, Chon Buri Province	13° 04', 100° 52'	27	18 March 2005	1
	59	Wat Tham Khao Bot, Rayong Province	13° 02', 101° 38'	(>200)*	21 September 2008	2
	60	Wat Tham Khao Pratun, Rayong Province	13° 00', 101° 36'	N/A	21 September 2008	N/A
	61	Wat Tham Suwanphupha, Rayong Province	12° 59', 101° 39'	35	21 September 2008	N/A
	62	Wat Tham Neramitr, Rayong Province	12° 58', 101° 39'	(100–200)*	21 September 2008	1
	63	Wat Tham Khao Wongkot, Chantha Buri Province	12° 52', 101° 49'	(100–200)*	21 September 2008	1
	64	Lan Hinkhong or Khao Plutaluang Monastery, Chon Buri Province	12° 42', 100° 58'	50*	18 March 2005	1
	65	Wat Aow Yai, Trat Province	12° 05', 102° 33'	3 (solitary males)	22 September 2008	N/A
South (31 locations)	66	Nong Yai Water Reservoir, Chumphon Province	10° 32', 99° 12'	(800)*	25 March 2005	>1
	67	Wat Paknam Pracharangsarith, Ranong Province	9° 57', 98° 35'	16	24 March 2005	1
	68	Suan Somdet Prasrinakharin Chumphon, Chumphon Province	9° 56', 99° 02'	20	7–9 May 2006	1
	69	Ban Koh Lhao, Ranong Province	9° 54', 98° 34'	9	25 March 2005	1
	70	Ngao Mangrove Research Center, Ranong Province	9° 52', 98° 36'	(200)*	2–4 May 2006	2
	71	Koh Payam, Ranong Province	9° 42', 98° 23'	(100)*	24 March 2005	1
	72	Koh Piak Nam Yai and Koh Thao, Ranong Province	9° 35', 98° 28'	N/A	22 July 2005	1
	73	Water Reservoir of Ban Na, Ranong Province	9° 32', 98° 42'	32	8–9 May 2006	1
	74	Wat Tham Silatieb, Surat Thani Province	9° 31', 99° 11'	N/A	11 December 2009	N/A
				70	23–24 March 2005	4
				65	1–6 December 2007	5
				N/A	22 July 2005	N/A
				29	18 July 2005	1

No.	Location	Coordinates	Count	Date	No.
75	Eastern Marine Research Center, Ranong Province	9° 22', 98° 24'	N/A	22 July 2005	N/A
76	Pak Nam Laempho, Surat Thani Province	9° 22', 99° 15'	13	17 July 2005	1
77	Ban Pak Nam 1, Surat Thani Province	9° 05', 99° 13'	13	18 July 2005	1
78	Decha Tukhan Monastery, Nakhon Si Thammarat Province	9° 04', 99° 50'	N/A	29 April–1 May 2006	3
			(300)*	19 July 2005	2*
79	Koh Klang, Ranong Province	8° 52', 98° 21'	20*	21 July 2005	1
80	Wat Nasarn, Surat Thani Province	8° 48', 99° 22'	N/A	20 July 2005	>1
81	Wat Ban Rhiang, Phang-nga Province	8° 35', 98° 40'	N/A	21 July 2005	1
82	Wat Tham Tapan, Phang-nga Province	8° 27', 98° 31'	N/A	21 July 2005	1
83	Wat Suwankhuha, Phang-nga Province	8° 25', 98° 28'	137	10–13 May 2003	2
			N/A	24 October 2005	N/A
84	Srinakharinthara Princess Garden, Phang-nga Province	8° 25', 98° 31'	20*	10 May 2003	1
85	Ban Chong Mai Kaew, Krabi Province	8° 24', 98° 44'	N/A	21 July 2005	1
86	Moo 4, Ban Kong Khong, Nakhon Si Thammarat Province	8° 23', 100° 11'	(>1,000)*	19 July 2005	>1
87	Wat Chai Khao, Nakhon Si Thammarat Province	8° 21', 99° 47'	32	20 July 2005	1
88	Wat Khuhasantayaram or Wat Tham Khao Daeng, Nakhon Si Thammarat Province	8° 14', 99° 52'	46	30 April 2003	1
				13–14 January 2007	
89	Wat Khao Kaew Wichian, Nakhon Si Thammarat Province	8° 12', 100° 05'	280	19 July 2005	3
90	Ban Kho-en, Phuket Province	8° 09', 98° 20'	N/A	24 January 2005	1
91	Monkey Bay, Phuket Province	N/A	11*	23 October 2005	1
92	Khao Chumthong Monastery, Nakhon Si Thammarat Province	8° 08', 99° 52'	N/A	1 May 2003	
93	Wat Thamsue, Krabi Province	8° 07', 98° 55'	6*	9 May 2003	N/A

Table 5.2. (cont.)

Region	Name of Location	GPS (N, E)	Number counted	Duration of observation	Estimated number of troops
94	Por Da Island, Krabi Province	7° 43', 98° 36'	6	28 December 2009	1
95	Ao Nang, Krabi Province	7° 42', 98° 45'	>20*	28 December 2009	1
96	Wat Khuha Sawan, Phatthalung Province	7° 37', 100° 5'	N/A	5 May 2003	
97	Khao Chaison, Phatthalung Province	7° 30', 100° 10'	N/A	10 October 2005	N/A
98	Khao Noi/Khao Tangkuan, Songkhla Province	7° 12', 100° 35'	(>300)*	1–5 May 2003	3
99	Wat Khuha Phimuk, Yala Province	6° 31', 101° 13'	55	5–10 May 2003	2
100	Tarutao National Park, Satun Province	6° 19', 99° 19'	N/A	20 October 2008	N/A

* Number of monkeys that we could observe on the day of survey, while other individuals fled away or foraged in other areas. The number in parenthesis is the estimated number of the population.

** Long-tailed macaques, rhesus macaques and monkeys showed morphological characters between long-tailed, rhesus and pig-tailed macaques were found.

*** Long-tailed macaques, rhesus macaques and monkeys showed morphological characters between long-tailed and rhesus macaques were found.

N/A = numbers of population or troop are not available. For some troops we could not see monkeys, but we confirm the species by scrutinizing photos taken by the local people. For some troops we did not have enough time to count all monkeys.

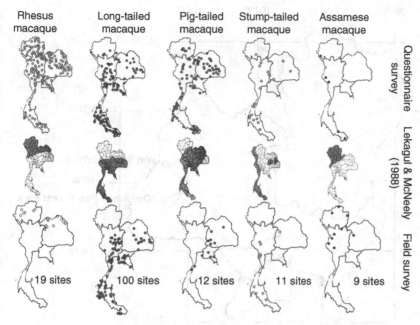

Figure 5.3. The distribution of five species of macaques in Thailand based on our questionnaire survey, field survey, and a report by Lekagul and McNeely (1988)

predators and the constraints of limited viable natural habitat; Thai long-tailed macaque populations in and around temple areas are increasing, while in contrast, worldwide populations of long-tailed macaques are in general decreasing (Eudey, 2008). The average population size at identified long-tailed macaque locations is 200 monkeys, but five locations had more than 500 individuals and eight locations had more than 1,000 individuals. Compared with the populations reported in each locations by Aggimarnagsee (1992), 20 years ago, the population has increased two to eight times. At these locations overpopulation is clearly an issue and human residents have had to take precautions in order to protect their homes against macaque damage.

Translocation of long-tailed macaques

Translocations have influenced the distribution of long-tailed macaques in many locations in Thailand. Monkeys at Wat Pa Sila Wiwek, Mukda Han Province (16° 32' 63" N, 104° 43' 43" E), which were regarded as the northern-most population of typical long-tailed macaques, were reported to have been translocated from Don Chao Poo Forest Park, Umnajjaroen Province (15° 36' 36" N, 104° 50' 53" E), approximately 100 km away, several years

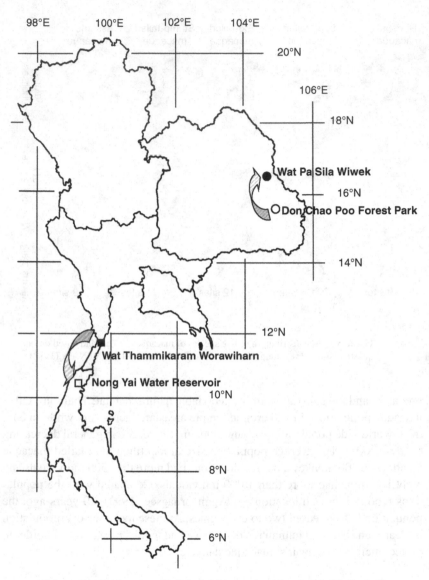

Figure 5.4. The location of long-tailed macaque translocations in Thailand. The arrows indicate the direction of translocation.

ago (Figure 5.4). They were translocated by the chief monk of Wat Pa Sila Wiwek in an effort to attract visitors to the temple, according to an interview with a deacon of Wat Pa Sila Wiwek in the year 2004. Another population at Nong Yai Water Reservoir, Chumphon Province (10° 32' 59" N, 99° 12' 26" E) was translocated from Wat Thammikaram Worawiharn, Prachuap Khirikhan Province (11° 48' 41" N, 99° 48' 05" E). Wat Thammikaram Worawiharn is a recreational park on a hill, which is surrounded by human settlements, the sea and the temple. Since there is limited habitat and heavy provisioning by many tourists, the population at Wat Thammikaram Worawiharn became locally overcrowded. To alleviate this overpopulation issue, some of the monkeys were trapped and translocated by the local government. Nong Yai Water Reservoir also had groups of long-tailed macaques. Subsequently, a native troop of conspecifics at the Nong Yai Water Reservoir faced competition for resources from the translocated troop.

The release of heterospecific pet macaques

Another potential threat to long-tailed macaques in Thailand is the release of pet macaques. Although keeping macaques as pets is illegal in Thailand, many pet macaques, especially infants, can be found. Frequently, when the monkeys reach sexual maturity and become aggressive, the pet is released. Thai people rearing pet macaques usually release them into the vicinity of wild or semi-wild macaque troops. Because long-tailed macaques can easily be seen, the pet owners release their pet macaques into those troops. We found released female pig-tailed macaques in Wat Tasang Tai, Nakhon Sawan Province (15° 56' 53" N, 99° 57' 15" E), Sarn Pra Karn, Lopburi Province (14° 48' 08" N, 100° 36' 53" E), and Khao Sam Muk, Chon Buri Province (13° 18' 45" N, 100° 54' 14" E). A released female stump-tailed macaque was also found in Sarn Pra Karn. In addition to the release of heterospecific macaques into a group of long-tailed macaques, a released long-tailed macaque was also found in a troop of stump-tailed macaques at the Khao Krapuk-Khao Tao Mo Non-hunting area, Petchaburi Province (12° 47' 46" N, 99° 44' 42.4" E) (Figure 5.5).

Sympatry between long-tailed and heterospecific macaques

The monkeys in Kumpawaphi City Park, Udon Thani Province (17° 06' 37" N, 103° 01' 14" E) showed mixed morphological characteristics of long-tailed, rhesus, and pig-tailed macaques (Figure 5.6). Based on reports from the local residents, the pig-tailed macaques in this population were released in the recent past. However, the presence of the other two species in the past was only anecdotal. A gazette printed by the local governor for the opening ceremony of the city hall in the year 2000, stated that macaque monkeys (without specifying the species) had lived in the area for hundreds of years. Museum records suggest that this area is on the edge of the range of rhesus and long-tailed macaques (Fooden 1995; 2000). In fact, long-tailed and rhesus macaques are segregated

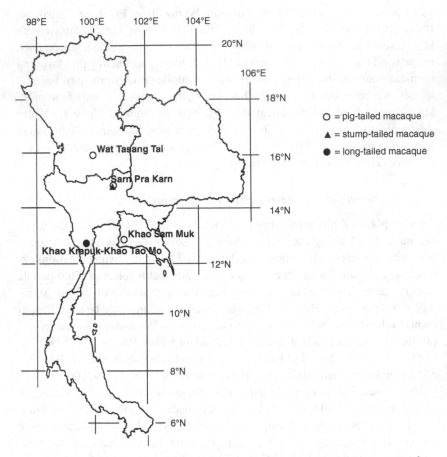

Figure 5.5. The released pig-tailed (o) and stump-tailed macaques (▲) to troops of long-tailed macaques, and a released long-tailed macaque to a group of stump-tailed macaques (•).

by different ecological niches, long-tailed macaques live in a warmer climate (south of 18° N) and rhesus macaques live in a colder climate (north of 18° N) (Fooden, 1982). During our survey, the relative tail length of monkeys in this troop was noted and varied between 60–100 percent. The pelage color pattern in some monkeys was a bipartite pattern of rhesus macaques (Hamada *et al.*, 2006; Malaivijitnond *et al.*, 2007b). Some monkeys had crests at the crown as long-tailed macaques or posteriorly directed crown hairs like rhesus macaques or a dark brown crown patch like pig-tailed macaques.

We also observed a free-ranging troop of macaques at the Khao Khieow Open Zoo, Chon Buri Province (13° 12' 56" N, 101° 03' 19" E) with morphological

Figure 5.6. Monkeys in Kumpawaphi City Park. Male monkeys with morphological characters of rhesus macaque (a), long-tailed macaque (b), and long-tailed macaque with a dark-brown crown patch of pig-tailed macaque (c) (photographs by Y. Hamada).

Figure 5.7. An adult male long-tailed macaque (a) and lactating female rhesus macaques (b) in a free-ranging group around Khao Khieow Open Zoo. The male shows a pelage color of long-tailed macaque, a whorl of hair at cheek of rhesus macaque and intermediate relative tail length between those of long-tailed and rhesus macaques (c) (photographs by J. Jadejaroen).

characters of long-tailed macaques, rhesus macaques or mixed morphological characters between these two species (Malaivijitnond and Hamada, 2008) (Figure 5.7). We previously reported one adult male pig-tailed macaque approached the group, successfully mated a female long-tailed macaque, but no hybrid offspring was observed (Malaivijitnond and Hamada, 2008). However, during our visit in September 2009, we found the juvenile male showed mixed morphological characters between the relative tail length (tail length/crown-rump length × 100) of 120 percent of the long-tailed macaque (Fooden, 1995) and a dark brown crown patch of the pig-tailed macaques (Fooden, 1975).

At Wat Tham Khao Daeng, Nakhon Si Thammarat Province (8° 14' 29" N, 99° 52' 11" E), long-tailed macaques were sympatric with stump-tailed macaques. The Wat Tham Khao Daeng is located near a limestone

Figure 5.8. Sympatry between long-tailed and stump-tailed macaques at the feeding area of Wat Tham Khao Daeng. White arrows indicate stump-tailed macaques (photographs by C. Richter and L. Mevis).

mountain with mining for limestone occurring on one side of it. During observations made in May 2003, long-tailed macaques were seen avoiding and always keeping a distance from stump-tailed macaques (Malaivijitnond and Hamada, 2005). However, observations made in February 2007 at the same site, revealed the two species of macaques came closer to each other and were seen grooming and playing (Malaivijitnond and Hamada, 2008) (Figure 5.8).

At Wat Tham Kham, Sakhon Nakhon Province (17° 13' 07.2" N, 103° 54' 02.2" E), a group of pig-tailed macaques with 86 individuals were counted in October 2004. Their relative tail lengths varied (35–75 percent) and were rather higher than that reported by Fooden (1975). Three female long-tailed macaques and one juvenile male with hybrid morphological characters between the two species (crest at the crown and pelage color of long-tailed macaques, but with a slender, thinly covered in fur and short tail of pig-tailed macaques) were observed. However, when we first visited the site in January 2002, only four long-tailed macaques were found at the time.

Contraception on long-tailed macaques

As previously stated, many populations of long-tailed macaques in Thailand are overpopulated and human-macaque conflict is taking place. In January 2008, the Ministry of Natural Resources and Environment announced the launch of a population control program, by means of contraception, targeting 40 groups of long-tailed macaques throughout Thailand (Kom Chad Luek Newspaper, 2008). At Khao Wang (13° 06' 45" N, 99° 56' 16" E), the Petchaburi Provincial Livestock Office started conducting a contraception project for long-tailed macaques, by orchidectomy in males and levonorgestrel hormone (a synthetic progesterone hormone) implantation in female. The program has been underway since October 2005. However, after being orchidectomized, adult male monkeys lost their social rank and were chased out of the troop, and had to live peripherally to the troop, resulting in more problems for people living nearby. Since January 2007, vasectomys were used as an alternative form of male sterilization. This same group of veterinarians also helped to perform sterilization procedures on long-tailed macaques at Khao Toh Phayawang Recreation Park, Satun Province (*c.a.* 6° 30' N, 100° 01' E) in September 2009. A well-reported contraception project for long-tailed macaques was recently launched at Sarn Pra Karn, Lopburi Province (Thaiwildlife, 2009; Reuters, 2009). The veterinarian of the project, Juthamas Supanam said that there were four groups of long-tailed macaques at Lopburi; San Pra Karn, Manora Market, Building area and Malai Rama Theater (Figure 5.2). Her unverified estimations were that there were approximately 2,900 monkeys that gave birth to 500 infants per year. The project committee planned to control the birth rate to be not more than 300 infants per year. Thus, at least half of the 1,500 male macaques were expected to be sterilized as part of the project (more detail about sterilization program for San Pra Karn long-tailed macaques is in the section below on management).

Many locations, such as at the Khao Khieow Open Zoo; Wat Khao Chong Krachok, Prachuap Khirikhan Province (11° 48' 41" N, 99° 48' 05" E); and Khao Noi/Khao Tangkuan, Songkhla Province (7° 12' 33" N, 100° 35' 49" E), were planning to follow with contraceptive projects like those at Khao Wang and Sarn Pra Karn. However, local people in these areas had varying attitudes to the enforcement of long-tailed macaque population control. From newspaper interview statements it was revealed that a German tourist at Wat Khao Chong Krachok and the veterinarian at Khao Noi/Khao Tang Kuan, viewed the contraception as a negative approach; while others, tourist organization at Khao Noi/Khao Tang Kuan and those that live near the sites and are disturbed by monkeys at Khao Wang, had a positive attitude to the population control (National News, 2009; Thai News, 2009). The students and staff from Ban

Figure 5.9. A feeding party for long-tailed macaques at Prang Sam Yot shrine, Sarn Pra Karn, Lopburi Province (a), and Kosumphi Forest Park, Maha Sarakham Province (b) (photographs by S. Malaivijitnond).

Laad Industrial and Community Education College who were bitten by long-tailed macaques living at Khao Wang expressed a positive attitude to the population control (Cablephet, 2009). Monkey-food vendors at Wat Khao Chong Krachok had inconsistent opinions regarding population control. Nonetheless, it seems that Thai people can accept contraceptive methods of population control, whilst in Nepal, the local residents perceived contraceptive methods of population control as unacceptable due to their religious beliefs, whilst translocation was generally accepted (Chalise and Johnson, 2005).

Conflict and commensalism between long-tailed macaques and humans

In many locations, the local people or local organizations hold a feeding party for monkeys once a year, each with different names, such as a Monkey Buffet at Sarn Pra Karn, Lopburi in November (Figure 5.9a); a Boon Pa Khaow Ling (Boon Pa Khaow Ling = bringing food for monkeys, northeastern Thai dialect) at Kosumphi Forest Park, Maha Sarakham Province (16° 15' 13" N, 103° 04' 18" E) in April (Figure 5.9b); a Monkey Palaeng (Palaeng = party, northeastern Thai dialect) at Don Chao Poo Forest Park, Umnajjaroen Province (15° 36' 36" N, 104° 50' 53" E) in April; a Monkey Buffet at Muangling Ban Whan, Si Sa Ket Province (15° 22' 44" N, 104° 10' 59" E) in March; and a Monkey Buffet at Khao Wang, Petchaburi Province in January. The feeding party for monkeys at Lopburi has become an internationally renowned event. The main group of monkeys attending the party lives at the Prang Sam Yot shrine (Figure 5.2). Local people believe that Lopburi's monkeys are the descendants of Hanuman, the monkey-hero of the mythical Hindu figure Rama, and respect and tolerate them. In addition to the positive attitude to the monkeys, there also were negative attitudes towards the macaques. The monkeys were chased off when they ventured inside buildings. At Kosumphi Forest Park where the golden long-tailed macaques appeared (5.2 percent of the total population, Hamada *et al.*, 2005b), the local residents used this to advertise for Monkey Buffet Festival (Figure 5.9b). Most of people here have a positive attitude to monkeys and have established a plan for conservation.

At some locations, interview reports indicate that monkeys damaged crops and houses during the dry season when the natural foods are scarce and thus they are viewed as pests. Residents living around these populations protected their houses with metal fencing and television antennae with steel guards that blocked the macaques from climbing them (Malaivijitnond and Hamada, 2008).

Preliminary interviews and field observations at Lopburi

Human and macaque groups

Our work at Lopburi is very preliminary and largely descriptive at this point. In our preliminary observations at Lopburi we estimated that there were five troops. The largest group used two shrines and adjacent school grounds (400–450 individuals). Another group was based at an abandoned Malai Rama Theater behind Prang Sam Yot (110–150 individuals). A third group was concentrated around the fire station including Manora Market (140–190 individuals). A fourth large troop (180–240 individuals) resided directly south of the Prang Sam Yot shrine on residential buildings and Sri Indra Hotel. The fifth and smallest troop was usually found near the Prang Khaek shrine (70–80 individuals). Preliminary counts suggested that there were approximately 900–1150 individuals in the Old City area, and this is much lower than the amount estimated by the sterilization program. The macaques are mostly concentrated within a 1 km² area in the center part of town. However, interview answers regarding the macaque population show that people estimate a larger number than there actually are. People reported personal estimations of macaques in Lopburi to range between 1,500–4,000 and the estimated number of groups ranged between three and eight.

Human groups coming into contact and interacting with macaques could be split into two broad categories: local and non-local. Locals consisted of people living and/or working in Lopburi, who were mostly Thai nationals. Non-locals included domestic and international tourists or visitors who tended to spend just a few hours or overnight within the Old City area, most of whom came to see the main attractions in town and the monkeys.

The main groups earning revenues from macaque-centered tourism were hotel owners, the Sarn Pra Karn shrine and the associated Monkey Foundation, Prang Sam Yot and the Fine Arts department, monkey-food and souvenir vendors, and to a lesser extent some of the local restaurants and taxi drivers. Interestingly, although bus-loads of tourists arrived on a daily basis to see the monkeys at the shrines, there were no tour companies in Lopburi which were responsible for these tours; they were all organized and paid for outside of Lopburi. Therefore, much of the tourism revenue was not made available to local people.

Resource sharing

All areas used by the macaques were used by humans, with the exception of several buildings immediately surrounding the Prang Sam Yot shrine which have been reputedly abandoned by previous residents and owners due to the intolerable damage and nuisance from the monkeys. The overlap areas included religious sites used for worship and tourism, people's homes, businesses (e.g., several restaurants and hotels), banks, a fire station, main roads, train tracks, a school, and a market area, electricity and telecommunications cables, etc. All these areas have suffered from some form of damage, ranging from direct sabotage and destructive behavior by the monkeys (e.g., bending and ripping of tin roofs, bringing down television antenna, and dislodging bricks from historical ruins), to daily deposits of feces, urine, food remains, and rubbish. The macaques were often seen drinking water from sources also used by humans and they turned on and drank from taps around the town, which were shared by street market vendors for washing food and utensils. They also drank from leaky air-conditioner units and household water pipes. Additionally, they took drinks from passers-by. A large proportion of the food the macaques ate was intentionally provisioned by humans, but macaques also frequently snatched food or bags from people and took food from unguarded food stalls and vehicles.

Provisioning

There were four main provisioning sites within the Old City. The first at Sarn Pra Karn shrine, the second behind Prang Sam Yot shrine, the third behind the fire station, and the fourth next to the rail tracks opposite the Hotel Sri Indra. The majority of provisioned food was given at the Sarn Pra Karn shrine (Figure 5.2). Here the monkeys got more food than could be consumed in a single day. The food came from visitors who bought food to give directly to the monkeys, from food donations given to the temple at Sarn Pra Karn to make merit, and from food purchased with monetary donations made by shrine visitors. At the provisioning site opposite the Hotel Sri Indra, food was provided

Figure 5.10. Long-tailed macaques jump on shoulders and backs, and cling on hair of the non-local Thai (a) and foreign visitors (b) while they gave foods to the monkeys at Prang Sam Yot shrine, Lopburi (photographs by Y. Vazquez and S. Malavijitnond).

twice a day and consisted of food bought with donation money collected at Sarn Pra Karn. Foods provisioned behind Prang Sam Yot and behind the fire station appeared to be mostly donated by local people at various times through-out the day. However, the monkeys were also provisioned on a lesser scale throughout the main 1 km² overlap zone. Some local people fed the monkeys daily on their rooftops, outside their windows or on their doorsteps, or just dropped food on the pavements as they passed by, or threw food out of their car windows. Tourists tended to feed the macaques mostly in and around the grounds of Prang Sam Yot and Sarn Pra Karn shrines.

Interactions

Observed human-monkey interactions were almost always associated with pro-visioning, whether "intentional provisioning" or "unintentional provisioning." Although provisioning was offered both by locals and non-locals, Thais and foreigners alike, the interactions associated with such provisioning appeared to be quite different. Provisioning initiated by locals rarely involved contact with the macaques, as food was dropped on the ground and the provisioner moved away as the monkeys approached to eat. In contrast, non-locals, par-ticularly foreigners, often handed over food morsels by hand and encouraged the monkeys to approach, frequently allowing one or several monkeys to jump on them and cling to their clothing, head, limbs, and hair (Figure 5.10). As non-locals were concentrated mostly within the grounds of the two shrines, it is likely that most close contact interactions took place in these areas with the largest macaque group.

Other observed human-macaque interactions were those related to discouraging the monkeys from entering or loitering around people's properties. These generally involved using some form of apparatus to chase or scare the monkeys away. Some of the items used to deter macaques were sling-shots, sticks, toy machine guns, and buckets of water. Additionally, some people used stuffed crocodile dolls to keep the monkeys away, as it is believed in local folklore that the monkeys are scared of crocodiles. These behaviors were almost solely observed around the entrances of retail shops and street stalls located in and around the residential areas rather than within the shrine areas.

Monkeys were also seen jumping in and out of slow-moving pick-up trucks looking for food, taking food and shopping bags out of moped baskets, pulling jewellery off tourists, and snatching water bottles and drinks from people. People were seen teasing monkeys, pulling their tails and slowing down their cars to allow monkeys to cross the road.

Attitudes

According to preliminary interviews, there were varying attitudes towards the macaques. In the most part people liked the macaques and perceived them as a symbol of Lopburi city. In addition, as is the case at various other sites around Asia, the traditional Buddhist beliefs afford the monkeys some protection and people believe they have to be kind to the monkeys (Zhao, 2005; Fuentes *et al.*, 2005; Aggimarangsee, 1992). There were, of course, contrasting opinions, as some people disliked them, found them intolerable, and wanted to get rid of them.

Local people also expressed conflicting attitudes towards the sterilization program underway at the time of this study. Some reported it as being a great initiative that will solve the overpopulation problem and reduce human–macaque animosity. Others felt that it was good, but all the monkeys should be included to humanely exterminate the population. Others felt that by reducing the population there may not be enough monkeys to attract tourists and therefore visitor revenues could decrease.

Management

In 1992, Aggimarangsee reported that the shrines were under the care of the Fine Arts Department, but maintained by the private sector, which earns revenue from visitors and street vendors. More recently, interviews revealed that it is only Prang Sam Yot which is under the management of the Fine Arts Department, and Sarn Pra Karn shrine is managed by the local committee which is also responsible for managing monkeys within Lopburi. This group is known locally as the "Monkey Foundation." There are no entrance fees to the Sarn Pra Karn shrine, but revenues from voluntary donations are in part awarded to the Monkey Foundation to employ staff who are responsible for feeding the monkeys and managing human-monkey interactions at the Sarn Pra Karn site. As they are separate organizations, each shrine is responsible for the management of the macaques and visitors in its own area.

As for the residents of Lopburi, it is considered their personal responsibility to protect their homes and personal belongings from the macaques. Protective barriers (e.g., metal fences) around some people's homes and steel guards on television antennae, as reported by Malaivijitnond and Hamada (2008) are still in use today, and some additional barriers have been added, including barbed wire and electric fences. The efficacy of these obstructions is debatable as monkeys were frequently observed agilely climbing over barbed wire, over steel guards, shaking antennae, and touching, climbing, and even chewing on the electric fences.

The main organization responsible for the management of macaques and human-monkey interactions in Lopburi is the Monkey Foundation. The foundation has been established for over ten years and had a large committee of thirty-one members representing local people, several government bodies, an NGO (WARF- Wildlife Aid and Rescue Foundation of Thailand), infrastructure associations, local veterinarians, the headmaster of a local school, and several other stakeholder groups which are affected by or had knowledge concerning this interface. Although this committee contained a broad representation of most groups affected by or directly involved in the human-monkey interface, it lacked one vital representative – a primatologist. We suggest this needs rectifying, as management planning would benefit from having primate experts, experienced in studying human-macaque interfaces, on the board.

As part of a government plan to control populations of these commensal macaques in Lopburi, a sterilization program organized by the Monkey Foundation was initiated in May 2009. A local veterinarian and her team of three to seven assistants were observed regularly seizing adult and subadult males mostly (but not exclusively) from the groups living outside the shrine areas – mainly those living around the Fire Station and those on the residential area south of Prang Sam Yot. By August 2009 the local veterinarian had castrated over 140 males. The veterinarian told the investigator that the monkeys were selected according to scrotum size and, in particular, monkeys which were showing evidence of hernias. The reason given for the choice of castration as the sterilization method was to reduce aggression with the aim of reducing human–monkey conflict. Whilst the monkeys were anaesthetized for castration, the veterinarian also inspected the monkey for general health and injuries and any open wounds. The wounds were cleaned and stitched, and gun pellets were frequently removed from their bodies. Injured macaques were given antibiotics and treated for mange. The monkeys were subsequently photographed, their ears were pierced with an identification number, and they were marked with temporary purple spray across their chest to distinguish them from untreated males once they were released. As part of this management initiative, there was supposed to be a monitoring program to count the population before and after the sterilization program. However, full details were not outlined in the Monkey Foundation meeting notes and the investigator was not aware of any accurate count having been undertaken either before or during the investigation period.

Besides the sterilization program, the government and the Monkey Foundation had another management initiative which has been the creation of three main provisioning sites: at Sarn Pra Karn, behind Prang Sam Yot and opposite the Sri Indra Hotel. In addition, signs warning of 2,000 baht fines for feeding the monkeys along the streets and outside the designated provisioning sites were dotted around the Old City areas. The initiative to restrict feeding to specialized sites was aimed at luring the macaques away from residential buildings and towards the shrine areas. Our observations suggest these signs were largely ineffective, as there was frequently evidence of provisioning all around the areas where there were monkeys including directly in front of these warning signs. It is unlikely this fine is enforced, as one of the signs was directly in front of a police booth and provisioning was frequently observed in this area.

Other than these signs warning of fines, there were few other informative signs in relation to the monkeys: there were several "Beware Monkey Zone" signs directly surrounding Sarn Pra Karn shrine, and some information boards listing the monkeys' favorite foods and drinks within the Sarn Pra Karn grounds. No information was displayed anywhere on how to behave near the monkeys, what attire to avoid, or what to do if a monkey became aggressive towards you.

Discussion

In the most recent worldwide review of the status of non-human primates, experts stated that in Asia alone, over 70 percent of non-human primates are listed by the World Conservation Union (IUCN) as "vulnerable," "endangered" or "critically endangered" (IUCN, 2009). However, the two subspecies of long-tailed macaques in Thailand, *M. f. fascicularis* and *M. f. aurea*, are classified as of least concern and data deficient, respectively on the IUCN Red

List 2009. IUCN has also listed the long-tailed macaque as one of the "100 Worst Alien Invasive Species." Although long-tailed macaques are the most frequently observed species among five species of macaques in Thailand and the populations are locally overcrowded (Malaivijitnond and Hamada, 2008; Malaivijitnond *et al.*, 2009), their current numbers may not be comparable with populations that previously occupied natural forest habitats and were never assessed. Based on the summary of IUCN 2009, their overall population has a declining trend, but our data suggests that maybe the macaques are increasing in number in human-landscaped environments. A recent article by Eudey (2008) calls for a reassessment of the conservation status of this species due to its rapid decline, particularly as a result of trapping and trade for use in the pharmaceutical industry. There are suspicions that individuals are being illegally sourced for the trade from wild populations. Furthermore, it has been suggested that trafficking of "temple monkeys" to Cambodia from Thailand for further export may have existed for years (Eudey, 2008). Similar concerns and suspicions over illegal sourcing of wild-caught long-tailed macaques have also been expressed by other studies and organizations (Yiming and Dianmo, 1998; Malone *et al.*, 2003; Hamada *et al.*, Chapter 3; see Box 1.2 and 3.1)

Elements of Asian culture have granted some primate species relative protection when living within temple complexes (Fuentes *et al.*, 2005). Thailand's main religion is Buddhism, with 85–94 percent of the total population practicing Theravada Buddhism (International Religious Freedom Report, 2005). Buddhism frowns upon harming and/or consuming primates (Sponsel *et al.*, 2002; Wolfe, 2002; Zhao, 2005). They often encourage the giving of "offerings" in order to gain spiritual merit. Offerings frequently take the form of food items and can become provisioning for primates, either directly (people feeding the monkeys by actively handing or throwing food at them) or indirectly (providing edible offerings at the shrine and the monkeys later taking them) (Aggimarangsee, 1992; Fuentes *et al.*, 2005; Zhao, 2005). The result of this human support has led to the population of long-tailed macaques becoming overcrowded in many anthropogenic environments in Thailand.

Information from the questionnaire survey

From the questionnaire survey, the largest number of negative replies (i.e., no monkey observed) is in the northeastern region. Most of Kamnans in the northeastern region have lived long in respective sub-districts to observe the history of environmental change and the loss of wildlife. They described that local macaque populations became extinct or were exterminated about thirty years ago because of the rapid and extensive deforestation for agricultural development. As the climate in the northeastern region is seasonal and dry, the forest could not recover well (WorldClimate.com). On the other hand, most of the replies were positive from the southern region.

Hybridization between long-tailed and rhesus macaques

In general, local Thai residents do not always precisely identify the species of macaque monkeys, especially the distinctions between rhesus and long-tailed macaques. One reason is that rhesus macaques are called "Wok" in Thai, originating from the ancient Northern Thai or Lan Na language and it means "monkey" (Center for the Promotion of Arts and Culture, 2009). Thus, irrespective of the species, Thai people identified any macaque monkeys, especially long-tailed macaques that were the most frequently seen, as "Wok" monkeys. In addition, in the northern, northeastern and central regions at the latitude range of 15–20° N, it has been a postulated hybrid zone between long-tailed and rhesus macaques (Fooden, 1995; Hamada *et al.*, 2006; Hamada *et al.*, Chapter 3). Fooden (1964) and Hamada *et al.*, (2006) proposed that hybridization is occurring between rhesus and long-tailed macaques at the boundary areas, and this is based on tail-length variation and body size. The work of Malaivijitnond *et al.*, (2008) has supported this hybridization hypothesis between Thai long-tailed and Thai rhesus macaques by showing evidence of mixing during human ABO-blood group analysis of wild-caught macaques. The morphological characters of the two species in Thailand are difficult to distinguish. The morphological characters of Thai rhesus macaques (i.e., tail length and body mass), are different from those of typical (i.e., Indian and Chinese) conspecifics. The relative tail length of Thai rhesus macaques was much longer (52.30 – 66.20 percent) than the average values from China (35.30 percent) and India (42.50 percent), whereas the body size was much smaller (Hamada *et al.*, 2005a; 2006). The bipartite pelage color pattern of Thai rhesus macaques is also not different enough for the local people to notice (Hamada *et al.*, 2006). In addition, the northern long-tailed macaques have shorter tails (<120 percent) and lighter, and yellowish pelage close to that of rhesus macaques (Hamada *et al.*, 2006; 2008).

In comparison, two other macaque species that are sympatric with long-tailed macaques in the southern region, that is, stump-tailed and pig-tailed macaques showed extensively different morphological characters (Fooden 1975; 1990; 1995), thus misidentification by the southern people was only 17.7 percent of the questionnaires surveyed. Moreover, the southern people are very familiar with pig-tailed macaques because they are used for coconut-picking. Among the five species of macaques in Thailand, stump-tailed macaques have distinct morphological characters and only their newborn infants have a whitish pelage color (Malaivijitnond and Hamada, 2005; Koyabu *et al.*, 2008). They are, therefore, the only macaque species not misidentified by local Thai people.

Preliminary interviews and field observations at Lopburi

Lopburi is situated in Central Thailand, 154 km north of Bangkok, and is now renowned as Thailand's "monkey-city" (WARF, 2007) due to a population of long-tailed macaques that extensively interfaces with the people there. Lopburi has been a site of human settlement since prehistoric times but reached its first peak of prosperity in the Khmer period, between the ninth and eleventh centuries C.E. (Ooi, 2004). It was during this period that the Prang Sam Yot, Sarn Pra Karn, and other Khmer-style shrines in the town were constructed. These religious shrines are still held sacred in Thai society, and it is around the shrines that Lopburi's monkey population is concentrated (Aggimarangsee, 1992; Watanabe *et al.*, 2007; Malaivijitnond and Hamada, 2008).

The exact history of this monkey occupancy is not officially documented, but Fooden (1971) reported Lopburi human residents stating that the monkeys were thought to have descended to the shrine area from adjacent forests at least 50–60 years before his observations in 1967. If so, they have been in the city for over 100 years. Today, the nearest "wild" long-tailed macaque population is reportedly found at another temple site – Khao Phra Ngam – approximately 20 km away (Malaivijitnond and Hamada, 2008). The estimated population size and number of groups found during visual surveys in this investigation are similar to observations made by Watanabe *et al.* (2007) who recorded five troops located in the Old City area in and around Prang Sam Yot, comprising 850–1,100 individuals within a 300 x 400 m^2 area. Interestingly, interviewees all overestimated the population size, which implies that without specific counts people tend to overestimate the number of macaques present. The variation in answers regarding number of troops can be explained by the proximity of groups to one another and the overlap in site use observed in some groups, which make it difficult to distinguish between groups.

The difference in the two main human participant groups, locals and non-locals, is similar to that found at monkey tourism sites around the world (Fuentes, 2006; Zhao, 2005; Sha *et al.*, 2009). Preliminary observations suggest interactions appear to differ in distinct areas and between various participants. The interactions between tourists and macaques are intense, close in contact, and brief, while the interactions between macaques and locals appear to exhibit little direct contact. These observations are similar to those made by Fuentes (2006) when he observed differences between types of people and the interactions between humans and macaques at Bali and Gibraltar. In addition, work in Singapore has shown that residents and visitors to nature parks with macaques have very different attitudes and very different experiences with macaques (Sha *et al.*, 2009)

Due to the urban nature of Lopburi, there is very little vegetation within the macaque-inhabited areas. The macaques are almost entirely dependent on human provisioning for their survival, and all of the macaques in Lopburi receive provisioning from people. As a result, there is a very close interface between people and macaques, perhaps one of the most intense interfaces on earth. People are highly affected by macaques in Lopburi and so measures are needed to properly manage this population. People use a variety of short-term tactics to deter macaques and the preferred tools for monkey deterrence appear to be sling-shots, toy machine guns and sticks. A sterilization program is underway as well, and there was evidence of shotgun injuries in approximately half of the male monkeys captured, with some individuals having several pellets embedded over their bodies. This suggests more harmful deterrent methods may be employed than were directly observed, and that the human-macaque interface does not only have negative consequences to the humans involved. Overall, management programs, involving primatologists, will need to be developed to manage and control the population and nuisances of the macaques at Lopburi, while also working to sustain a healthy population of macaques at this very unique interface.

Conclusion

In Thailand, there are pros and cons to having a troop of long-tailed macaques living close to humans. Many people benefit from the monkeys, such as banana vendors, hotel owners and shop keepers, but there are also people who live near monkey populations who do not gain any economic benefits but receive damage. Up to now, there are neither concrete management plans nor primatologist participation to overcome the problem of local overcrowding and conflict with humans. The local overcrowding of long-tailed macaques is a delicate matter, and to solve the problem we need a mutual understanding among people. In many macaque localities the local governors have made short-term plans such as catching and translocating monkeys or performing contraceptive operations, without considering their long-term consequences. The translocation may destroy the natural genetic diversity of the species, and makes it difficult to understand the evolutionary history of macaques. Moreover, inexperienced and unplanned contraceptive operations, such as castration of adult male monkeys, may actually cause more problems in the future than they solve in the present.

Management and conservation plans will require cooperation from various groups, including primatologists, veterinarians, local residents, conservationists, governmental agencies, and NGO's. Educational programs are needed to raise awareness in the public, government, and conservation sectors. Education will need to focus on population management and controlling human interaction with long-tailed macaques. Particular emphasis will need to be placed on lobbying to the Thai government for initiating macaque management programs that will decrease macaque overpopulation and alleviate human-macaque conflict, while also protecting and maintaining a healthy long-tailed macaque population. Specific strategies initially should include 1 controlling provisioning, 2. preventing the translocation of macaques, 3. developing well-managed contraception programs, 4. curbing pet release, and 5. seeking government-level protection for populations with unique characteristics. Therefore, the conservation and management plan for Thai long-tailed macaques will require a multi-faceted solution that is unique to the needs of Thailand's people and macaques.

Acknowledgements

This work was supported by the Thailand Research Fund (grant numbers RSA/02/2545 and RMU4880019), the Biodiversity Research and Training Program (BRT), Chulalongkorn University (Grant for the Primate Research Unit), the Commission on Higher Education, Ministry of Education, Thailand (grant no. CHE-RG-01), and the Japanese Society for the Promotion of Science (grant numbers 1645017 and 2025506).

References

Aggimarangsee, N. 1992. Survey for semi-tame colonies of macaques in Thailand. *Natural History Bulletin of the Siam Society* **40**: 103–166.

Cablephet. 2009. A student of Ban Laad Industrial and Community Education College bitten by monkeys. www.cablephet.com/board/n-view.php?nc_id=1505 (downloaded 13 September 2009).

Center for the Promotion of Arts and Culture, Chiang Mai University. 2009. Wok. *Matichon Weekly* **29**(1483):72.

Chalise, M. K. and Johnson, R. L. 2005. Farmer attitudes toward the conservation of "pest" monkeys: the view from Nepal. In *Commensalism and Conflict: The*

Human-Primate Interface, J. D. Paterson and J. Wallis (eds.). Norman, OK: American Society of Primatologists Publications. pp. 223–239.

Engelhardt, A., Hodges, J. K., Niemitz, C., and Heistermann, M. 2005. Female sexual behavior, but not sex skin swelling, reliably indicates the timing of the fertile phase in wild long-tailed macaques (*Macaca fascicularis*). *Hormones and Behavior* **47**: 195–204.

Eudey, A. A. 2008. The crab-eating macaque (*Macaca fascicularis*): widespread and rapidly declining. *Primate Conservation* **23**: 129–132.

Fooden, J. 1964. Rhesus and crab-eating macaques: intergradation in Thailand. *Science* **143**: 363–365.

    1971. Report on primates collected in western Thailand January–April, 1967. *Fieldiana Zoology* **59**: 1–62.

    1975. Taxonomy and evolution of liontail and pigtail macaques (Primates: Cercopithecidae). *Fieldiana Zoology* **67**: 1–169.

    1982. Ecogeographic segregation of macaque species. *Primates* **23**: 574–579.

    1990. The bear macaque, *Macaca arctoides*: a systematic review. *Journal of Human Evolution* **19**: 607–686.

    1995. Systematic review of Southeast Asia longtail macaques, *Macaca fascicularis* (Raffles, 1821). *Fieldiana Zoology* **81**: 1–206.

    2000. Systematic review of rhesus macaque, *Macaca mulatta* (Zimmermann, 1780). *Fieldiana Zoology* **96**: 1–180.

Fuentes, A. 2006. Human culture and monkey behaviour: assessing the contexts of potential pathogen transmission between macaques and humans. *Amercian Journal of Primatology* **68**: 880–896

Fuentes, A., Southern, M., and Suaryana, K. 2005. Monkey forests and human land-scapes: is extensive sympatry sustainable for *Homo sapiens* and *Macaca fascicularis* on Bali? In *Commensalism and Conflict: the Human-Primate Interface*. J. D. Patterson and J. Wallis (eds.). Norman, OK: American Society of Primatologists Publications. pp. 168–195

Hamada, Y., Watanabe, T., Chatani, K., Hayasawa, S., and Iwamoto, M. 2005a. Morphometrical comparison between Indian- and Chinese-derived rhesus macaques (*Macaca mulatta*). *Anthropological Science* **113**: 183–188.

Hamada, Y., Hadi, I., Urasopon, N., and Malaivijitnond, S. 2005b. Preliminary report on golden long-tailed macaques (*Macaca fascicularis*) at Kosumpee Forest Park, Thailand. *Primates* **46**: 269–273.

Hamada, Y., Urasopon, N., Hadi, I., and Malaivijitnond, S. 2006. Body size and proportions and pelage color of free-ranging *Macaca mulatta* from a zone of hybridization in northern Thailand. *International Journal of Primatology* **27**: 497–513.

Hamada, Y., Suryobroto, B., Goto, S., and Malaivijitnond, S. 2008. Morphological and body color variation in Thai *Macaca fascicularis fascicularis* north and south of the Isthmus of Kra. *International Journal of Primatology* **29**: 1271–1294.

Hill, C. M. 2005. People, crops, and primates: a conflict of interests. In *Commensalism and Conflict: the Human-Primate Interface* J. D. Patterson and J. Wallis (eds.). Norman, OK: American Society of Primatologists Publications. pp. 41–59

International Religious Freedom Report. 2005. www.state.gov/g/drl/rls/irf/2005/51531. htm (downloaded 13 June 2009).

IUCN. 2009. IUCN Red List of Threatened Species. Version 2009.2. www.iucnredlist. org (accessed 22 February 2010).

Kom Chad Luek Newspaper. 2008. http://vivaldi.cpe.ku.ac.th/~note/event/?id=367086 (downloaded 3 September 2009).

Koyabu, D. B., Malaivijitnond, S., and Hamada, Y. 2008. Pelage color variation of *Macaca arctoides* and its evolutionary implications. *International Journal of Primatology* **29**: 531–541.

Lee, P. C. and Priston, N. E. C. 2005. Human attitudes to primates: perceptions of pests, conflicts and consequences for conservation. In *Commensalism and Conflict: the Human-Primate Interface*. J. D. Patterson and J. Wallis (eds.). Norman, OK: American Society of Primatologists Publications. pp. 1–23

Lekagul, B., and McNeely, J. A. 1988. *Mammals of Thailand*. Darnsutha Press, Bangkok, Thailand.

Malaivijitnond, S., Chaiyabutr, N., Urasopon, N., and Hamada, Y. 2006. Intestinal parasites of long-tailed macaques (*Macaca fascicularis*) inhabiting some tourist attraction sites in Thailand. Proceedings of the 32nd Thai Veterinary Medical Association Meeting. 1–3 November 2006, Thailand.

Malaivijitnond, S. and Hamada, Y. 2005. A new record of stump-tailed macaques in Thailand and the sympatry with long-tailed macaques. *Natural History Journal of Chulalongkorn University* **5**: 93–96.

Malaivijitnond, S., Hamada, Y., Varavudhi, P., and Takenaka, O. 2005. The current distribution and status of macaques in Thailand. *Natural History Journal of Chulalongkorn University* **Suppl 1**: 35–45.

Malaivijitnond, S., Hamada, Y., Suryobroto, B., and Takenaka, O. 2007a. Female long-tailed macaques with scrotum-like structure. *American Journal of Primatology* **69**: 721–735.

Malaivijitnond, S., Takenaka, O., Kawamoto, Y., Urasopon, N., Hadi, I., and Hamada, Y. 2007b. Anthropogenic macaque hybridization and genetic pollution of a threatened population. *Natural History Journal of Chulalongkorn University* **7**: 11–23.

Malaivijitnond, S., Sae-low, W., and Hamada, Y. 2008. The Human-ABO blood group of free-ranging long-tailed macaques (*Macaca fascicularis*) and its parapatric rhesus macaques (*M. mulatta*) in Thailand. *Journal of Medical Primatology* **27**: 31–37.

Malaivijitnond, S. and Hamada, Y. 2008. Current situation and status of long-tailed macaques (*Macaca fascicularis*) in Thailand. *Natural History Journal of Chulalongkorn University* **8**: 185–204.

Malaivijitnond, S., Urasopon, N., Goto, S., and Hamada, Y. 2009. Diversity study of primates in Thailand. The 2nd International Symposium on Southeast Asian Primate Research, Biodiversity Study of Primates in Laos. 22–23 January 2009, National University of Laos.

Malone, N. M., Fuentes, A., Purnama, A. R., and Adi Putra, I. M. W. 2003. Displaced hylobatids: biological, cultural, and economic aspects of the primate trade in Jawa and Bali, Indonesia. *Tropical Biodiversity* **8**: 41–49.

National News. 2009. http://breakingnews.nationchannel.com/read.php?newsid= 361628 (downloaded 13 September 2009).

Ooi, K. G. 2004. *Southest Asia: A Historical Encyclopedia from Angkor Wat to East Timor.* vol. 1. California: ABC-CLIO. pp. 793–794.

Reuters. 2009. No monkey business: Thailand launches primate birth control. *Reuters Life*, August 21, 2009.

Richard, A. F., Goldstein, S. J., and Dewar, R. E. 1989. Weed macaques: The evolutionary implications of macaque feeding ecology. *International Journal of Primatology* 10: 569–594.

Sha, J. C. M., Gumert, M. D., Lee, B. P. Y.-H., Jones-Engel, L., Chan, S., and Fuentes, A. 2009. Macaque-human interactions and the societal perceptions of macaques in Singapore. *American Journal of Primatology* 71: 825–839.

Smith, B. H., Crummett, T. L., and Brandt, K. L. 1994. Ages of eruption of primate teeth: A compendium for aging individuals and comparing life histories. *Yearbook of Physical Anthropology* 37: 177–231.

Sponsel, L. E., Ruttanadakul, N., and Natadecha-Sponsel, P. 2002. Monkey business? The conservation implications of macaque ethnoprimatology in southern Thailand. In *Primates Face to Face: Conservation Implications of Human-Nonhuman Primate Interconnections.* A. Fuentes and L. Wolfe (ed.). Cambridge University Press. pp. 288–309

Srivastava, A. and Begum, F. 2005. City monkeys (*Macaca mulatta*): a study of human attitudes. In *Commensalism and Conflict: the Human-Primate Interface.* J. D. Paterson and J. Wallis (ed.). Norman, OK. American Society of Primatologists Publications. pp. 259–269.

Thaiwildlife. 2009. Thai Wildlife: Wildlife Conservation Network in Thailand. www. thaiwildlife.org/main/forums/wildlife-meeting-room/topic-93. (downloaded 13 September 2009).

Thai News. 2009. http://travel.sanook.com/news/news_05322.php (downloaded 13 September 2009).

WARF. 2007. The Wild Animal Rescue Centre Lopburi. www.warthai.org/product. php?id=4 (downloaded 6 June 2009).

Watanabe, K., Urasopon, N., and Malaivijitnond, S. 2007. Long-tailed macaques use human hair as dental floss. *American Journal of Primatology* 69: 940–944.

Wolfe, L. D. 2002. Rhesus macaques: a comparative study of two sites, Jaipur, India and Silver Springs, Florida. In *Primates Face to Face: Conservation Implications of Human-Nonhuman Primate Interconnections.* A. Fuentes and L. Wolfe (eds.). Cambridge University Press. pp. 310–330.

Yiming, L. and Dianmo, L. 1998. The dynamics of trade in live wildlife across the Guangxi border between China and Vietnam during 1993–1996 and its control strategies. *Biodiversity and Conservation* 7: 895–914.

Zhao, Q. K. 2005. Tibetan macaques, visitors, and local people at Mt. Emei: problems and countermeasures. In *Commensalism and Conflict: the Human-Primate Interface.* J. D. Paterson and J. Wallis J (eds.). Norman, OK: American Society of Primatologists Publications. pp 376–399.

Appendix 5.1.   The questionnaire sent to the heads of sub-districts as translated from Thai to English

1. First name...... Family name..... Head of sub-district......................
   Address...
   Telephone   no..............Fax   no............Cell   phone
   no......................
2. Have you ever seen monkeys in your area or vicinity? (please mark one)
   □ Ever □ Never □ Other ....
3. What species of monkeys did you see? (Please refer to the attached brochure)
   Answer..
4. When was the last time that you saw monkeys? (Approximate date, month, year)
   Answer...
5. How many groups of monkey did you see?
   Answer...
6. How many times did you see those monkeys?
   Answer...
7. If you could see monkeys more than one time, was the number of monkeys that you saw for the last time increased or decreased compared to the first time?
   Answer...
8. How many monkeys were in each group?
   Answer...
9. In one group of monkeys, how many species of monkeys did you see?
   Answer...
10. Please give us the name of the location where you saw monkeys.
    Answer...
11. Please give us detailed information about the location where you saw monkeys, e.g., close to the village, in the temple, close to the agricultural area or in the forest.
    Answer...
12. How well have monkeys been habituated to humans? (Please mark at least one)
    □ When monkeys found humans, they fled away and we could not see them
    □ When monkeys found humans, they escaped or climbed up trees, but we could see them.
    □ Monkeys came to get food from humans' hands when they were provisioned

13. Where did monkeys forage for foods?
    Answer..

14. If we want to get more information or to perform survey, whom should we contact and how should we contact them?
    Answer..

15. Are there any researchers/persons who studied those monkeys? If yes, please give us the information, e.g., names of researchers and when they performed their study?
    Answer..

16. Have you ever seen monkeys with a name printed in the brochure that showed a different morphology from the picture in the brochure? Please tell us the details, such as body size, pelage color, hair characters, group or solitary monkeys, color of baby.
    Answer..

17. Have you ever seen other monkeys in your area or vicinity that do not appear on the brochure? If yes, what is the morphological character of those monkeys?
    Answer..

Appendix 5.2.   The form of data recorded during the field survey

Observer..Date............Time............
Address ....
Local name of the place..
GPS....................N..................E UTM:.............., H.............m
Name of person interviewed..
Habitat: Forest (primary/secondary/disturbed)/Temple/Park/Other..............
Species observed.......................Local name of species......................
Number of troop: 1. Observed.............. 2. Interviewed....................
Remark: Morphological, Physiological, Ecological characters

Sex	Estimated body weight (kg)	Infant/lactating mother	Juvenile	Sub-adult/Adult	Total
F					
M					
		Infant/lactating mother	Juvenile	Sub-adult/Adult	Total
F					
M					

	Infant/lactating mother	Juvenile	Sub-adult/Adult	Total
F				
M				

....

....

Foraging...

Habituation to human..

Present status (Neighboring troop: Y/N; Continuity of forest or corridor forest :
 Y/N; Other................)

Damage on agricultural crop?. Y/N..

....

Comments..

....

....

....

Location Map:

Appendix 5.3. Outline of preliminary interview

1. **Locating macaque range**
   1.1. Please indicate on the map of Lopburi any locations where you think there are monkeys.
2. **Identifying participant groups**
   2.1. Who interacts with the monkeys? Please give a list of people or groups that you know interact with the monkeys.
3. **Identifying attitudes towards the monkeys**
   3.1. How do you feel about the monkeys?
   3.2. What feelings do you think people in Lopburi have towards the monkeys?
4. **Identifying parties responsible for managing monkeys and/or human-monkey interactions**.
   4.1. Please give the names of any people/organizations that you know of that take responsibility for any of the following:
      4.1.1. Feeding monkeys
      4.1.2. Controlling the monkey population
      4.1.3. Managing Sarn Pra Karn and Prang Sam Yot shrines

      4.1.4.   Protecting people and their homes from monkeys

      4.1.5.   Managing tourist visitors

      4.1.6.   Monitoring monkey health/welfare

      4.1.7.   Providing medical care/first aid for people who have become injured as a result of interacting with monkeys

   4.2.   Can you think of anyone you have not mentioned already that would be useful for us to talk to regarding the human-monkey interaction in Lopburi?

   4.3.   Do you know of any management plans in place concerning the monkeys and/or human-monkey interactions? If so, please give details.

5. **Human-monkey interactions**

   5.1.   Do you know if any monkeys are:

      5.1.1.   Captured by people

      5.1.2.   Kept as pets

      5.1.3.   Killed by people

      5.1.4.   Injured by people/cars

      5.1.5.   Fed by people

   5.2.   Do you know if any people in Lopburi:

      5.2.1.   Are attacked/injured by the monkeys

      5.2.2.   Have things taken from them by monkeys (food, bags, etc)

      5.2.3.   Suffer damage to their property due to monkeys

      5.2.4.   Suffer any agricultural damage due to the monkeys

6. **Additional information**

   6.1.   How many monkeys do you think there are in the Old City?

   6.2.   Is there any additional information you would like to give that you think might be useful to us in this study?

6   Macaque behavior at the human–monkey interface: The activity and demography of semi-free-ranging Macaca fascicularis at Padangtegal, Bali, Indonesia

AGUSTIN FUENTES, AIDA L. T. ROMPIS, I. G. A.
ARTA PUTRA, NI LUH WATINIASIH, I. NYOMAN
SUARTHA, I. G. SOMA, I. NYOMAN WANDIA,
I. D. K. HARYA PUTRA, REBECCA STEPHENSON
AND WAYAN SELAMET

Introduction

Macaca fascicularis is an excellent species to examine adaptation to a particularly wide array of habitats and environmental variables, especially where human impact is a core component of the landscape. Within the long-tailed macaque species (*Macaca fascicularis*) there are at least ten subspecies, dramatic variation in facial hair patterns, and body size varies from 2.5–7.0 kg for females and 4.7–14 kg for males (Gumert, Chapter 1; Fooden, 1995; Napier and Napier, 1967; Rowe, 1996). Despite their well-documented occurrence and utilization of primary tropical rainforest (up to 2,000m elevation), the long-tailed macaques appear to prefer riverine habitats, coastal forests, swamp or mixed forests and secondary forest habitats (Crockett and Wilson, 1980). The ability to thrive in a variety of environmental types probably played a role in this group's evolutionary success throughout Southeast Asia especially during the last 5–8,000 years of human-induced (agricultural) environmental change.

The island of Bali is approximately 5632 km² and has a rich history of volcanic activity and thus some of the most fertile soils in the world. There are

Monkeys on the Edge: Ecology and Management of Long-Tailed Macaques and their Interface with Humans, eds. Michael D. Gumert, Agustín Fuentes and Lisa Jones-Engel. Published by Cambridge University Press. © Cambridge University Press 2011.

approximately 247 rivers all cascading down slopes from the central volcanic range. During both the wet and dry seasons moisture accumulates above the volcanoes in the center of the islands providing a nearly year-round supply for the rivers that course rapidly down towards the sea, creating deep ravines and ready access to water for the south central portion of the island. Bali's landscape can be currently described as a highly human-modified environment, where nearly all land in south central Bali is human villages, rice agriculture and rivers with deep ravines. The fertile soils combined with nearly 1,000 years of highly successful wet rice agriculture has facilitated some of the highest human densities in the world for a primarily rural population (avg. 482/km^2, range 259–1104/km^2) (Mantra, 1995). Currently, three primate species are indigenous to the island: *Macaca fascicularis*, *Trachypithecus auratus* and *Homo sapiens*.

The large population of long-tailed macaques on the island of Bali has been the focus of a number of studies. Previous to this study the majority of information on the Balinese macaques comes from published studies of behavior at the Padangtegal Monkey Forest in Ubud by Bruce Wheatley and colleagues (Wheatley, 1999; 1991, 1988; Wheatley *et al.*,1996; Wheatley and Harya Putra, 1994). Aside from these only a few researchers conducted short-term behavioral, physiological or pathogen-related studies at monkey forests in Bali (Angst, 1975; Dolhinow *et al.*, 1995; Engel *et al.*, 2006, Jones-Engel *et al.*, 2008; Kawamoto, *et al.*,1984; Koyama *et al.*, 1981; Fuentes, 1992; Fuentes *et al.*, 2000; and Suaryana *et al.*, 2001). Human-macaque interactions at Padangtegal and on Bali more generally, have also been the focus of a few publications (Fuentes, 2006; Fuentes and Gamerl, 2005). Current research on Bali macaques includes work on behavior, human-macaque interactions, population genetics, pathogens and infectious diseases, mechanisms of obesity, reproductive and dietary endocrinology, and the role of macaques in the human social context of the Balinese (Fuentes, 2006; Fuentes *et al.*, 2005; Engel *et al.*, 2006; Lane *et al.*, 2010).

The Bali macaque is an integral part of the culture, economy, and everyday life of Bali. While some have argued for the sacredness of macaques, more recent research has demonstrated that this sacredness is context-dependent (Lane *et al.*,2010; Schillaci *et al.*, 2010; Loudon *et al.*, 2006; Fuentes *et al.*, 2005). In contexts outside temples, the Balinese view macaques as potential pets, crop-raiding pests, or sometimes even a source of food. Throughout the island, macaques are chased away from rice fields, kept as pets by people of all socio-economic levels, and allowed to thrive across the island in and around temples. The macaque's habituation and presence at temples is a great promoter for the substantial tourism sector in Bali, and some villages are able to exploit such benefits for their local economy. The multi-faceted and intimate nature of the human-macaque relationship on Bali has required researchers to take an interdisciplinary approach and consider religion, culture, and biology simultaneously (Lane *et al.*, 2010; Loudon *et al.*, 2006).

On Bali, at least 43 monkey forest populations exist (Southern, 2002; Fuentes *et al.*, 2005). Variation between the populations exists in local habitat, climate, food resources, group size, responsiveness to humans, arboreality, provisioning, and to some extent, morphology (Lane *et al.*, 2010). The largest macaque populations on the island, are found at Padangtegal-Ubud, Uluwatu, Sangeh, Pulaki, and Alas Kedaton, which are all established tourism sites with routine, large-scale provisioning of food. Smaller populations are located in areas where provisioning occurs only during temple ceremonies and where the macaques are often considered pests by local villagers. These smaller populations are often located in the geographic extremes of the Balinese landscape – to the north, east, and west – while the largest temple populations are located predominantly in the central core of the island, along with most other tourist destinations and the largest human populations. The macaques of Bali are particularly striking as they live in forest patches, coastal, dry scrub habitats and lush rainforests to isolated beaches and the summit of an active volcano (Whitten *et al.*, 1996).

Bali provides a valuable opportunity to examine an island population of *Macaca fascicularis* that has co-existed with large human populations and substantial habitat alteration for at least 1,000 years. Here we report on the findings of a five-year long behavioral study of the three groups of *Macaca fascicularis* at Padangtegal, Bali, Indonesia.

Methods

The study site of Padangtegal Wanara Wana is located in the villages of Padangtegal and Ubud in the Gianyar regency of South-Central Bali, Indonesia (Figure 6.1). The site consists of approximately 7 ha of mixed forest and a temple complex (i.e., three temples and various shrines/statues), and is bordered by two towns, rice fields, a road, and two rivers. One of the rivers marks the eastern edge of the forest and forms a ravine that extends southward out of the forest and is fringed by riparian growth for at least 2 kilometers away from the site. At the site 116 species of tree/shrub/liana are found with the top three genera being: *Artocarpus*, *Cocos* and *Ficus* (Kriswiyanti and Watiniasih, 1999). The forest is described as secondary forest with a broken canopy and has a maximum height of 35m. The ground level consists of tree litter, dirt, and stone paths. Humans regularly use the site. Local Balinese use the temples at the site for ceremonies and use the paths through the forest to move between village areas. Between 40,000 and 110,000 domestic and international tourists visited the site annually between 1998 and 2002.

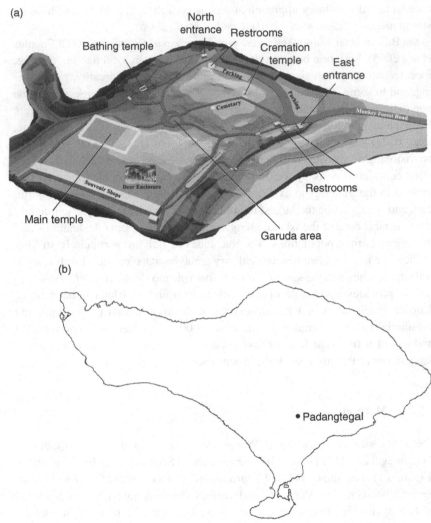

Figure 6.1. (a) Map layout of Padangtegal Temple Forest. (b) Location of Padangtegal in Bali, Indonesia.

From 1998–2002 we collected behavioral data on the Padangtegal population for between six to ten weeks per year during the months of June–August (i.e., the dry season for this part of Bali). Data on area use, activity budgets, and specific social behaviors were collected between 1999–2002, with the 1998 field season being devoted to initial behavioral observations, range use patterns and establishment of the methodology for the subsequent years. Nine to 23 observers annually collected group composition, diet, ranging, and general activity data across the 1999–2002 study periods resulting in 1,568 hours of behavioral activity data. Dominance ranks for adult males and females were calculated annually from approach-retreat interactions (i.e., displacements), the results of aggressive contests and priority of access to resources (i.e., provisioned food and favored resting sites). Over 500 hours of additional data from affiliated projects related to diet, copulatory behavior, object rubbing, temple licking, intestinal parasites, and wounding patterns were also collected and their results are reported elsewhere (Arta Putra *et al.*, 2001; Emel *et al.*, 2001; Fraver *et al.*, 2001; Kriswiyanti and Watiniasih, 1999; Loudon *et al.*, 2002; Suartha *et al.*, 2002; Suaryana *et al.*, 2000; Truce and Fuentes, 2002; Velucci *et al.*, 2000; Welch *et al.*, 2001). In addition to the behavioral data, Universitas Udayana Pusta Kajian Primata staff and members of the Padangtegal Wanara Wana management staff collected 46 months of general demographic data between 1998 and 2002.

Observers were trained in age/sex class recognition of the macaques and utilized a modified behavioral checklist. Individually known adult macaques were assigned sex and number indicators (M1 for male number 1, for example) which remained consistent across the data collection periods. Females were termed adult when they reached approximately 3.5 years old (85–90 percent adult body size). Males were initially added to the identified individuals list at age five (~80 percent adult body size) but termed adolescent until approximately age six to seven (i.e., time of full canine eruption and testicular size). General behaviors were recorded on customized record sheets during 20-minute focal follows, at one-minute intervals, using the naked eye and binoculars to facilitate observation. A scan sample method with a customized behavior repertoire was used (Altmann, 1974; Dolhinow, 1978). Only identified individuals who had more than three hours of follows per six-week observation period were included in behavioral data analysis. All data collectors were tested as exceeding 90 percent inter-observer reliability in data recording with project PIs before initiating sample collection. Video and still image records were made of adult animals and specific behaviors at the Padangtegal site annually between 1999 and 2002.

Results

Demographic patterns

Table 6.1 shows the population changes from 1986, 1990–1992 (from Wheatley, 1999) and from 1998–2002 (this study). The data since 1998 include changes in group size. Although Wheatley (1999) provides group sizes they are

Table 6.1. *Population size history at Padangtegal*

Year	Group 1	Group 2	Group 3	Total population	Deaths recorded (over 12-month period)	Population increase
1986*				69		
1990*				97		
1991*				122		20.5%
1992*				133		8.3%
1994			Streptococcus outbreak	ND	10–20% mortality estimated	
				1993–1997 no population data		
1998	20	45	57	122		
1999	22	49	54	125		2.4%
2000	24	62	53	139	5	10%
2001	28	75	58	161	15	13.7%
2002	31	100	63	204	7	21%

* 1986–1992 data from Wheatley 1999, ND=no data

not included in Table 6.1 due to the unclear relationship between Wheatley's groups 1, 2 and 3 and groups 1, 2 and 3 during this study. Because of methodological inconsistencies in reporting of deaths between August 1998 and June 1999 data on deaths are not given for either 1998 or 1999.

The population at Padangtegal underwent a 40 percent increase in size between 1998 and 2002, with 80 percent of this growth occurring since 2000 (Table 6.1). This increase in population size was not equally distributed across the three groups in the population. The smallest group, group 1, increased by 35.5 percent between 1998 and 2002, and underwent a transition from a one-adult male group to a multi-adult male group (one adult male to three adult males). Group 2, the largest group, increased by 65 percent, and group 3 (the largest group in 1998) only increased by 10.5 percent.

Table 6.2 shows the age and sex class distribution for the total population for 1986, 1990–1992 and 1998–2002. During the 1990–1992 period the number of adult males increased by 27 percent, adult females by 31 percent, and immatures by 25 percent. During the 1998–2002 period adult males increased by 34 percent, adult females by 46 percent, and immatures by 37 percent.

Table 6.2. *Demography: Age and sex make-up of population at Padangtegal*

Year	Adult males	Adult females	Immatures	Infants born during major birth peak (May–August)	Observed deaths (June–August only)
1986*	5	26	38	2 (11 total infants nursing)	
1990*	8	29	60	3 (15 total infants nursing)	
1991*	11	35	76	7 (19 total infants nursing)	
1992*	11	42	80	3 (19 total infants nursing)	
1998	14	43	65	7	1
1999	11	48	66	16	2
2000	10	55	74	15	3 (2 inf, one ad M7)
2001	10	60	91**	19	3 (3 inf)
2002	21	80	103	23	5 (3 inf, one ad male, one ad female)

* 1986–1992 data from Wheatley 1999
** includes 12 older subadult males

Table 6.2 includes the newborn, "black infants," and nursing "brown" infants for the period 1986, 1990–1992 as these are how the data were reported by Wheatley (1999). The Table 6.2 entries for this study (1998–2002) contain only those infants born during the May–August period as we have observed that some individuals nurse for more than one year and pelage changes can vary somewhat between individuals. From 1998–2002 we observed two birth/mating peaks annually (May–August and December–January), with a small number of births occurring outside of these peaks. Sixteen neonates were born during May–August of 2000, nineteen in 2001, and 23 during the same period in 2002 (Table 6.2). Females are observed copulating almost immediately after giving birth, suggesting a post-partum false estrus period. Interbirth intervals ranged between 16–36 months. A majority of the adult females in the population gave birth to infants during the 1998–2002 period, with some giving birth to as many as three offspring. One female successfully gave birth to a set of twins that have survived past the age of two years. Wheatley *et al.* (1996) reported an extremely high infant mortality rate (~90 percent) in 1990 at this site, however our observations suggest that the current rate averages less than 30 percent for the period 1998–2002.

Between 1998 and 2000 at least ten adolescent males disappeared from the site. We assume that they attempted emigration as neither our team nor the temple staff found their bodies. One adult male is known to have emigrated from the site during this time as well. Between 2000 and 2002 only three adolescent males were confirmed to have left the population. We observed three immigration attempts by males from outside of population, two of which were successful. We also observed three inter-group transfers by males and one by a female at the site.

Range use

As of 2002 the macaques at Padangtegal used approximately 24.3 ha of area, including the temple and temple forest proper (Figure 6.1). Each group used all areas within the ~7 ha temple and temple forest (i.e., 100 percent overlap in areas used over the five-year observation period). However, groups were rarely in the same area at the same time. Groups 2 and 3 consistently displaced group 1 (i.e., the smallest group) within the temple forest area. Groups 2 and 3 had variable dominance interactions across the five year study, with group 2 displacing group 3 in over 70 percent of observed inter-group displacements since 2000. Over 90 percent of the inter-group displacements occurred at the central area of the temple forest (i.e., main tourist location) or the plaza area on the west side of the main temple (i.e., secondary tourist area). Temple staff provided provisions for the macaques at both locations. Other inter-group displacements took place in the forest area just south of the main temple and in the graveyard in the eastern portion of the temple forest.

Outside of the temple forest area group 1 used the roadside forest and the riparian growth up to 200 m away from the eastern forest edge along the main road. Group 2 used the southern rice fields and the northwestern rice fields and riparian forest up to 300 m away from the temple forest proper. Group 3 used areas southwest, west and northwest of the forest proper. Although area usage varied across years and seasons, group 1 had a range averaging approximately 7.2 ha, group 2 had a range of approximately 11 ha, and group 3 ranged across 17 ha.

Activity patterns

Across the four study periods in which activity data are available (1999–2002) adult male and adult female macaques exhibited similar behavioral patterns. Figure 6.2 shows the per-scan frequency of thirteen activity categories as mean values across 1999–2002 for the combined categories of adult males and adult females. Females huddled (including sit in contact) with

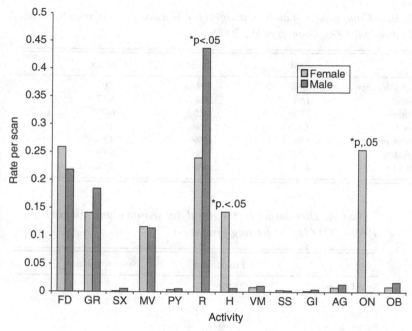

Figure 6.2. Adult male and adult female mean activity patterns 1999–2002. FD: Feed, GR: Groom, SX: Sex, MV, Move, PY: Play, R: Rest, H: Huddle, VM: Visual monitor SS: Sexual solicit, GI: Genital inspect, AG: Aggression, ON: On ventrum, OB: Object manipulate.

other individuals more frequently ($\chi^2=12.4$, df=1, p<.05) and had infants on their ventrum more frequently ($\chi^2=25.26$, df=1, p<.05) than did males. Males rested nearly twice as frequently as females ($\chi^2=5.88$, df=1, p<.05). When compared with other *Macaca fascicularis* sites in southeast Asia the general activity budgets of the Padangtegal macaques fit well within the species normative range (Table 6.3).

Dietary patterns

We examined a record of 11,761 feeding instances during the months of June–August (1999–2001) for feeding patterns. During these months the macaques at this site received nearly 70 percent of their diet from provisioning (Table 6.4). Monitoring during other times of the year indicated that the percentage of provisioned food drops to ~50 percent during February–May and October–November as the overall human traffic at the site is reduced. The temple management committee staff provisioned the three macaque groups with

Table 6.3. *Comparison of activity budgets for Macaca fascicularis (from this study, Fooden, 1995, Gumert et al., 2009a)*

	Feed	Travel	Other
Bali (Padangtegal)	23%	12%	58%
Sumatra	15%	26%	59%
Mauritius	32%	4%	64%
Malaysia	35%	20%	45%
Kalimantan	13%	45%	42%
Bangladesh	39%	9%	53%
Singapore	43%	14%	43%

Table 6.4. *Diet during Padangtegal dry season (June–August) 1999–2001 (11,761 feeding instances)*

	Provisioned	Non-provisioned
Sweet potato	28.9%	
Banana	21.9%	
Peanuts	12.4%	
Papaya leaves	5.7%	
Coconut	1%	2.8%
Invertebrates		7.5%
Forest vegetation		9%
Other		10.3%

sweet potatoes, papaya leaves, and various fruits and vegetables. Tourists visiting the site primarily provided bananas and peanuts. It is clear that ongoing study is needed to effectively document the annual dietary variation in these macaques. It is also clear from monitoring the appearance of the individuals that they were quite well fed.

The "other" category in Table 6.4 consists of temple offerings, chicken eggs provided by temple staff, a few types of pre-packaged food provided by tourists, and vertebrate animals (e.g., primarily infant squirrels, infant rats and adult frogs). All the macaques in this population had sporadic access to rice, a variety of flowers, cakes, chicken, and assorted fruits as these are frequently the contents of offerings placed at the temples by the local Balinese. However, amounts of offerings varied dramatically across ceremony and time and access to offerings is generally mediated by dominance rank of the macaques. Once a year the temple staff provided numerous eggs to the macaques as part of a festival honoring the monkeys. Throughout the year temple staff occasionally provided an egg or two to adult males as well. Although only seventeen instances of vertebrate consumption (e.g., lizards, frogs, infant squirrels) were observed, it was predominantly immature individuals (i.e., between two and four years of age) who consumed vertebrate prey (i.e., fifteen of seventeen instances). Therefore, distribution of the nutritional content of offerings and other items in the "other" category is not equal across age/sex classes and individuals.

Dominance

Top-ranked males' tenure ranged from 8 to >40 months, and males have been observed changing groups within the monkey forest populations in which they live (Loudon *et al.*, 2002). Rank and copulatory success were loosely correlated in group 2 but not in group 3 at the Padangtegal site (Welch *et al.*, 2001). Group 1 had only one copulating male for the majority of the observation periods. Preliminary data analyses suggest that adult males may practice both aggression and affiliative strategies to gain access to females (Loudon *et al* 2002). Although high intensity fighting occurs amongst adults, often with substantial wounding, the conflicts rarely result in death (Suartha *et al.*, 2002). During the period 1998–2002 only two adult males and two adult females died from wounds inflicted by other macaques or from injuries resulting from falls during arboreal chases. However, it is not clear how many of the immature deaths resulted from wounds inflicted by other macaques.

Infanticide

No instances of infanticide by adult males were observed at Padangtegal between 1998 and 2002. None were reported by Wheatley (1999) for the 1986, 1990–1992 period. We observed ten females taking infants and holding them away from their mother for between three minutes and four days. In the majority of these cases the female taking the infant held it, groomed it, allowed other females to touch and inspect it, and on occasion allowed the infant to suckle (i.e., only in females who were lactating at the time). In a few cases, females were observed to act aggressively towards infants they had taken, and we observed two cases of infanticide by females. In one case during July 2000, a female had taken an infant from a mother, and a third female bit the infant leaving a lethal head wound and killing the infant. The infant was subsequently retrieved by its mother and died within 24 hours. The other observed infant death resulted from a mother refusing to allow her newborn to nurse (i.e., a maternal infanticide). The infant died three days after birth. We also suspect that at least two additional infant deaths were related to females taking infants from their mothers, but cannot rule out infant illness or congenital disorders.

Object manipulation

Object manipulation is common in the population at Padangtegal and has been observed at other macaque sites in Bali (i.e., Alas Kedaton, Uluwatu, Pulaki, Sangeh, and Bedugal) where the macaques also have frequent interaction with humans (Fuentes *et al.*, 2005). Object manipulation observed includes stone play (i.e., stone stacking and general use in play), the use of stones to acquire foods, use of leaves to groom other individuals, multi-object rubbing, and food washing.

All age/sex classes were observed to manipulate objects, with adolescent males and immatures being the only class to manipulate objects more than expected. Non-food items were manipulated significantly more often than food items. There was no significant difference between manipulated food items being eaten or not eaten. Approximately 82 percent of all items manipulated were not eaten (Truce and Fuentes, 2002). There were significant differences in the frequency of manipulation styles between the age/sex classes and between food and non-food items. Non-food items were rubbed on the ground and pounded more often than food items, while food items were rubbed between the actor's hands more often than non-food items (Truce and Fuentes, 2002).

Discussion

These results can be compared to data reported by Wheatley (1999) from his research on this population between 1986 and 1992 to provide an overview of the longest study of semi-free-ranging *Macaca fascicularis* groups at a highly anthropogenic site.

Demographics

Ranging patterns reported by Wheatley (1999) for the three groups suggest that his groups 1–3 may be the same as groups 1–3 in this study, however given the lack of monitoring during 1993–1997 we cannot be sure of this. If these are the same groups as Wheatley's, then between 1992 and 1998 group 1 decreased in size by 41 percent, group 2 decreased by 4.5 percent, and group 3 increased in size by 173 percent. This may suggest that groups 1 and 2 were impacted most severely by the streptococcus outbreak in 1994 which caused high mortality in many macaque groups in central Bali (see Table 6.1). However, local temple officials and villagers have stated that group 3 suffered significant losses as well. The demographic patterns between 1992 and 1998 at this site remain unclear.

Wheatley (1999) also reports a relatively cohesive all male group from 1990–1992 at Padangtegal. We observed frequent play groups composed of juvenile and sub-adult males from either group 2 or group 3. Occasionally young males from both groups formed a play group in the space between groups. However, these play groups were temporary lasting from a few minutes to multiple hours. There was no evidence of any permanent or temporally cohesive male groups. It is possible that the increase in overall population has created substantial peer clusters within each group and led to the reduction in inter-group play by immatures.

In 1999, a new management program was initiated which included removal of plastic and other wastes from the forest, anti-erosion control, reclamation and re-forestation of fields bordering the forest, and occasional medical care for injured or sick individual macaques. It is clear that this change in management style combined with the increased diversity, amount, and regularity of provisioning, as well as the increased tourist presence and feeding from them, resulted in the population increasing in size at a dramatic rate.

The number of new infants in the May–August birth peak (Table 6.2) has increased annually. These births make up approximately half the annual births. Anecdotal information suggests that rate of increase have continued through 2010 with estimates of the current population nearing 500 individuals (see Box 6.1).

As the diet of this population is heavily provisioned, the death rate is fairly low and infant mortality appears low as well, population growth appears to be primarily curbed by epidemics and at least two have occurred between 1986 and 2002. This suggests that pathogen environment may be a significant selective force on these macaques. A significant role for epidemics/disease has also been proposed for the semi-free-ranging population of macaques (*Macaca sylvanus*) in Gibraltar (Fa and Lind, 1996; O'Leary and Fa, 1993) and a large population of chimpanzees in Tanzania (Nishida *et al.*, 2003).

Immigration/emigration

Koyama *et al.*, (1981) report the immigration of two adult males into this population. Wheatley did not observe any immigration and we observed three attempts of which two were successful. We also noted thirteen assumed or observed emigrations by adolescent males and one by an adult male. This suggests that there is potential gene flow between populations of macaques in south-central Bali. Elsewhere we propose a model for the appearance and maintenance of gene flow across many of the macaque groups despite high human density on the island of Bali (Fuentes *et al.*, 2005; Lane *et al.*, 2010).

Range use and inter-group dominance

Wheatley reports his group 1 dominant to group 2 and group 2 dominant to group 3 in 1986–1990, however by 1992 group 2 regularly displaced both groups 1 and 3 from preferred sites (Wheatley, 1999). This reflects the situation during 1999–2002, with group 2 frequently displacing the other groups. In 1998 groups 2 and 3 displaced each other equally. This further suggests the possibility that the Wheatley's and our groups may be the same.

The pattern of inter-group dominance suggests that group size is a key factor in achieving access to desired areas (see Table 6.2). As reported by Wheatley (1999) we also observed single males, or a few young males, from one group (usually group 2) displacing entire other groups. However, in inter-group conflicts where a number of individuals were involved adult females and adult males (usually high ranking) appeared at the forefront of the displacement activities. The locus of inter-group dominance activities remains similar across the two studies with both Wheatley (1999) and this study reported the majority of inter-group displacements occurred in areas where the majority of provisioning occurred.

Activity patterns and feeding/provisioning

General activity patterns in the Padangtegal population are by and large similar to other *M. fascicularis* groups in Asia (both free and semi-free-ranging). At Padangtegal, male and female activity patterns are similar for the behaviors recorded except that females engage in friendly affiliative contact with other individuals and hold infants more often than do males. Males, alternatively, rest significantly more frequently than females. Given the basic distribution of individuals in a macaque group and the relative adult sex ratios it is not surprising that females are more frequently in contact with other individuals than are males.

Interactions with humans are a daily part of the behavioral repertoire and human-made structures constitute a substantial segment of the environment of these macaques. This is not uncommon for populations of *M. fascicularis* or *M. mulatta* (Agaramirasee, 1992; Fuentes, 2006; Fuentes and Wolfe, 2002) and therefore documenting the basic activities of such populations is important for understanding the potential range of macaque behavioral patterns.

Wheatley (1999) reports that between 39 and 81 percent of this population's diet came from provisioning during 1990–1992. As provisioning was less consistent and less diverse during that period, it is probable that the mean percentage of provisioned food in the diet of all groups in this population is above 50 percent, and probably closer to the 70 percent reported here. There are a number of potential food items growing in the forest (Kriswiyanti and Watiniasih, 1999) and the macaques appear to exploit invertebrates more consistently (7.5 percent of diet) between 1998–2002 than in the previous study (1–9 percent for 1986,1990–1992) (Wheatley, 1999). Given the high rate of provisioning, the current relatively diverse composition of the provisioned food and the lack of mortality tied to obesity or other diet-based health issues, we propose that this population is not under nutritional stress. In fact, the abundance of foods and the relative protection offered by the temple forest most likely combined to produce the rapid population growth observed between 1998 and 2002.

Infanticide

Although male infanticide has been inferred in two groups of long-tailed macaques in Sumatra, Indonesia, (van Schaik, 2000) no infanticide by males has been observed or inferred at Padangtegal either by Wheatley (1999), this study, or in any other studies of this species. However, infant deaths did result from female infant taking. This has been reported for two Balinese macaque populations (Padangtegal and Sangeh, Wheatley, 1991, 1999, Chapter 10). Wheatley (1991, 1999) and Wheatley *et al.* (1996), report infant injuries resulting in deaths occurred during inter-group aggression at the Padangtegal site between 1986 and 1992. While we observed frequent inter-group aggression we did not record any consistent association between these conflicts and infant injury at Padangtegal between 1998 and 2002. Wheatley (1999) also reports four observations of infant taking resulting in infant death. He refers to this behavior as "kidnapping." Wheatley (1999) suggests that this is an example of reproductive competition between females, with higher ranking females taking and killing the infants of competing lower ranking females.

However, Wheatley clearly stresses that the infant deaths in these cases probably did not result from "intent to kill" but rather from higher-ranking females' attempts to harass lower ranking females (Wheatley, 1999). While infants are frequently held by females other than their mothers, we suggest it still remains unclear why in a few instances this results in the death of an infant. The two cases we observed are substantially different from each other precluding any single explanatory hypothesis. A possible byproduct of infant taking is the potential for successful adoption. Between 1998 and 2002 we observed the successful adoption of two infants younger than six months whose mothers died.

Object manipulation

While early reports suggested that the majority of manipulation behavior may be a form of food acquisition (Wheatley, 1988, 1999), recent research suggests that food preparation may account for only a relatively small fraction of the object manipulation observed at monkey forest sites (Fuentes, 1992; Truce and Fuentes, 2002). In fact, preliminary research suggests that variation in, and overall rates of, object manipulation may be tied to the levels of macaque-human interaction, provisioning, and specific structural factors associated with monkey forests and their proximity to human villages (Fuentes, 1992; Truce and Fuentes, 2002). That is, low-foraging pressure, no predator pressure, high degree of time spent on the ground, and exposure to a variety of diverse objects may lead to high frequencies of object manipulation. However, there are also reports of stone tool use on this species in wild populations (Gumert *et al.*, 2009b) suggesting that this species uses objects across a wide range of ecological and social circumstances.

Relevance of studying a macaque population in a highly anthropogenic environment

There is a great need for the study of long-term sympatry between human and non-human primates. Most socioecological investigations into primate groups do not incorporate their interactions (i.e., beyond predation or crop raiding), potential pathogen sharing, or the role of the anthropogenically impacted environment. This and other ongoing research projects, and a growing number of publications, demonstrate that long-term sympatry between human and non-human primates can create a complex web of behavioral, ecological, epidemiological, and economic relationships, suggesting a need for increased attention to this topic to uncover how all of these factors interact. As the overall range of many free-ranging primates continues to diminish, understanding the patterns of interaction between non-human primates and sympatric humans will facilitate future management issues, disease research, and behavioral investigations (Chapman and Peres, 2001; Fuentes and Wolfe, 2002; Fuentes, 2006,). Studying the macaques at Padangtegal can contribute to the overall attempt to understand the *Macaca mulatta-fascicularis* group's relative success in south and southeast Asia where they have overlapped with human settlements and use areas for many millennia.

Specifically documenting the pattern of demographic changes across time in the Padangtegal population might allow us to better understand how to facilitate management procedures that will maintain a healthy macaque population and minimize the potential negative impacts (e.g., pathogen transmission, crop raiding, or other elements) to the human populations. Also, by examining behavioral patterns such as object manipulation and infant taking we might be able to gain insight into the range of potential behavioral expression when macaques are relieved of specific selective stressors such as food and predation stress.

Conclusions

Given the results of research at the Padangtegal site a few general conclusions can be made:

1. Activity patterns of *Macaca fascicularis* groups in this population are generally similar to other *M. fascicularis* populations.
2. General activity patterns are similar for adult females and males in this population, with females engaging in more affiliation and holding of infants, and males resting more than females.
3. The diet of these macaques is heavily provisioned and thus this population may be free from nutritional stress and therefore exhibit behaviors that otherwise may be suppressed by increased foraging pressures (e.g., increased object manipulation).
4. The demographic history suggests that population growth is high and will continue to increase in rate with current management practices. It may be that the population growth was curbed by epidemics in the past and therefore the pathogen environment might be a significant selective force on commensal populations of macaques (or at least on this population).

5. The correlation between release of foraging and predation pressure and high rates of object manipulation may be important for management and assessment of similar semi-free-ranging populations.

Acknowledgements

In memoriam of Dr. Komang Gde Suaryana, a pioneer of Balinese Primatology.

This project was conducted under research permission from Lembaga Ilmu Pengatahuan Indonesia (LIPI), the Provincial Government of Bali and the Regency of Gianyar, and the Padangtegal Wanara Wana Management Committee. We wish to extend special thanks to the Rector of Universitas Udayana, the Governor of Bali, the Directorate of Social and Political Affairs, The Padangtegal Wanara Wana Management Committee and the Wanara Wana staff. We also thank The University of Notre Dame, Central Washington University, Dr. Hiro Kurashina, the University of Guam, Universitas Udayana Pusat Kajin Primata, and the student team members and faculty of the Balinese Macaque Project field school 1998–2002. Partial support for this study came from the Central Washington University Office of International Studies and Programs and Office of Graduate Studies.

References

Aggiramangsee, N. 1992. Survey of semi-free ranging macaques of Thailand. *Natural History Bulletin of the Siam Society* **40**:103–166.

Altmann, J. 1974. Observational study of behavior: Sampling methods. *Behaviour* **49**: 227–267.

Angst, W. 1975. Basic data and concepts on the social organization of *Macaca fascicularis*. In *Primate Behavior:Developments in Field and Laboratory Research*. L. A. Rosenblum (ed.). New York: Academic Press. pp. 325–388.

Arta Putra, I. G. A., Fuentes, A., Suaryana, K. G., and Rompis, A. L. T. 2001. Perilaku makan monyet ekor panjang (*Macaca fascicularis*) di Wanara Wana, Padangtegal, Ubud, Bali. in *Konservasi Satwa Primata: Tinjuan ekologi, sosial eknomi dan medis dalam pengembangan ilmu pengetahuan dan teknologi*. Universitas Gadjah Mada Press, Yogyakarta. pp. 132–140.

Chapman, C. A. and Peres, C. A. 2001. Primate conservation in the new millennium: the role of scientists. *Evolutionary Anthropology* **10**(1):16–33.

Crockett, C. M. and Wilson, W. L. 1980. The ecological separation of *Macaca nemestrina* and *Macaca fascicularis* in Sumatra. In *The Macaques: Studies in Ecology,*

Behavior and Evolution, D. G. Lindburg (ed.). New York: Van Nostrand Reinhold. pp.148–181.

Dolhinow, P. 1978. A Behaviour Reportoire for the Indian Langur Monkey (*Presbytis entellus*). *Primates* **19**(3): 449–472.

Dolhinow, P., Fuentes, A., and MacKinnon, K. 1995. The four faces of irus: behavioral variability in Balinese *Macaca fascicularis*. Abstract *American Journal of Physical Anththropolgy* Supp. **20**: 84.

Emel, G., Fuentes A., and Suaryana, K.G. 2001. Visual monitoring by *Macaca fascicularis* at Padangtegal, Bali, Indonesia. *American Journal of Physical Anthropology* Supp. **32**: 62.

Engel, G., Hungerford, L. L., Jones-Engel, L., Travis, D., Eberle, R., Fuentes, A., Grant, R., Kyes, R. and Schillaci, M. 2006. Risk assessment: a model for predicting cross-species transmission of simian foamy virus from macaques (m. fascicularis) to humans at a monkey temple in Bali, Indonesia. *American Journal of Primatology* **68**: 934–948.

Fa, J. E. and Lind, R. (1996) Population management and viability of the Gibraltar Barbary Macaques. In *Evolution and Ecology of Macaque Societies*, J. E. Fa and D. J. Lindburg (eds.). Cambridge University Press. pp. 235–262.

Fooden, J. 1995. Systematic review of Southeast Asian long-tailed macaques, *Macaca fascicularis* (Raffles, 1821). *Fieldiana Zoology* **81**.

Fraver, J. B., Emel, G., Fuentes, A., Suaryana, K. G., and Harya Putra I. D. K. 2001. An ongoing study of the female copulation call in long-tailed macaques (*Macaca fascicularis*). *American Journal of Physical Anthropology* Supp. **32**: 65.

Fuentes, A. 1992. Object rubbing in Balinese macaques (Macaca fascicularis). *Laboratory Primate Newsletter* **31**(2): 14–15.

2006. Human culture and monkey behavior: Assessing the contexts of potential pathogen transmission between macaques and humans. *American Journal of Primatology* **68**(9): 880–896.

Fuentes, A. and Gamerl, S. 2005. Disproportionate participation by age/sex classes in aggressive interactions between long-tailed macaques (*Macaca fascicularis*) and human tourists at Padangtegal monkey forest, Bali, Indonesia. *American Journal of Primatology* **66**(2): 197–204.

Fuentes, A., Southern, M., and Suaryana, K.G. 2005. Monkey forests and human landscapes: is extensive sympatry sustainable for *Homo sapiens* and *Macaca fascicularis* in Bali? In *Commensalism and Conflict: The Primate–Human Interface*, J. Patterson and J. Wallis (ed.). American Society of Primatology Publications. pp. 168–195.

Fuentes, A., Harya Putra, I. D. K., Suaryana, K. G., *et al.* 2000. The Balinese macaque project: Background and stage one field school report. *Jurnal Primatologi Indonesia* **3**(1): 29–34.

Fuentes, A. and Wolfe, L. D. 2002. *Primates Face to Face: The Conservation Implications of Human–Nonhuman Primate Interconnections*. Cambridge University Press.

Gumert, M. D., Sha, J. C. M., Lee, B. P. Y-H., and Chan, S. 2009a. Factors influencing the interface between humans and long-tailed macaques in Singapore. *American Journal of Primatology* **71**(Supp. 1): 56.

Gumert, M. D., Kluck, M., and Malaivijitnond, S. 2009b. The physical characteristics and usage patterns of stone axe and pounding hammers used by long-tailed macaques in the Andaman Sea region of Thailand. *American Journal of Primatology.* **71**(7): 594–608.

Jones-Engel, L., May, C. C., Engel, G. A., *et al.* 2008. Diverse contexts of zoonotic transmission of simian foamy viruses in Asia. *Emerging Infectious Diseases* **14**(8):1200–1208.

Kawamoto, Y., Ischak, T. B. M., and Supriatna, J. 1984. Genetic variation within and between troops of the crab-eating macaque (*Macaca fascicularis*) on Sumatra, Java, Bali, Lombok and Sumbawa, Indonesia. *Primates* **25**(2):131–159.

Koyama, N., Asnan, A., and Natsir, N. 1981. Socio-ecological study of the crab-eating macaques (*Macaca fascicularis*) in Indonesia. *Kyoto University Overseas Research Report of Studies on Indonesia Macaque* 1–10.

Kriswiyanti, E. and Watiniasih, N. L. 1999. *Inventarsi jenis dan manfaat tumbuhan di "Monkey forest" desa adapt Padangtegal, Ubud.* Publikasi Jurusan Biologi, Universitas Udayana Press.

Lane, K. E., Lute, M., Rompis A., *et al.* 2010. Pests, pestilence, and people: The ongtailed macaque and its role in the cultural complexities of Bali. In *Indonesian Primates, Developments in Primatology: Progress and Prospects*, S. Gursky-Doyen and J. Supriatna (eds.). New York/Heidelberg: Springer Science. pp. 235–248.

Loudon, J., Fuentes, A., and Welch, A. 2005. Agonism and affiliation: Adult male sexual strategies across one mating period in three groups of long-tailed macaques (*Macaca fascicularis*). *Laboratory Primate Newsletter* **44**(3): 12–15.

Loudon, J., Howell, M., and Fuentes, A. 2006. The importance of integrative anthropology: A preliminary investigation employing primatological and cultural anthropological data collection methods in assessing human-monkey co-existence in Bali, Indonesia. *Ecological and Environmental Anthropology* **2**(1): 2–13.

Mantra, I. B. 1995. Population of Bali. In *Bali: Balancing Environment, Economy and Culture,* S. Martopo and B. Mitchell (eds.). Department of Geography Publication series, University of Waterloo. pp. 45–55

Napier, J. R. and Napier, P. H. 1967 *A Handbook of Living Primates.* New York: Academic Press.

Nishida, T., Corp, N., Hamai, M., *et al.* 2003. Demography, female life history, and reproductive profiles among the chimpanzees of Mahale. *American Journal of Primatology* **59**(3): 99–121.

O'Leary, H. and Fa, J. E. (1993.) Effects of tourists on Barbary macaques at Gibraltar. *Folia Primatologica* **61**(2): 77–91.

Rowe, N. 1996. *A Pictorial Guide to the Living Primates.* Pogonias Press.

Sartono, S. 1973. On Pleistocene migration routes of vertebrate fauna in southeast Asia. *Geol. Soc. Malaysia Bulletin* **6**: 273–286.

Schilaci, M. A., Engel, G. A., Fuentes, A., *et al.* 2010. The not-so-sacred monkeys of Bali: A radiographic study of hHuman-primate commensalism. In *Indonesian Primates, Developments in Primatology: Progress and Prospects,*

S. Gursky-Doyen and J. Supriatna (eds.). New York/Heidelberg: Springer Science. pp. 249–256.

Southern, M. W. 2002. An assessment of potential habitat corridors and landscape ecology for long-tailed macaques (*Macaca fascicularis*) on Bali, Indonesia. Master's thesis, Central Washington University (unpublished).

Suartha, I. N., Watiniasih, N. L., and Fuentes, A. 2002. Kesembuhan luka monyet ekor panjang di obyek wisata Wanarawana Padangtegal. *Jurnal Veteriner* 3(2):50–54.

Suaryana, K. G., Arta Putra, I. G. A., and Fuentes, A. 2000. Parasite of the long-tailed macaque (*Macaca fascicularis*) and its potential risk to human health. *The International Seminar on Soil transmitted Helminthiasis*. Denpasar, Bali, Indonesia.

Suaryana, K. G., Fuentes, A., Arta Putra, I. G. A., Harya Putra, I. D. K., and Rompis, A. L. T. 2001. Ekologi dan distribusi monyet ekor panjang (*Macaca fascicularis*) di Bali. In *Konservasi Satwa Primata: Tinjuan ekologi, sosial eknomi dan medis dalam pengembangan ilmu pengetahuan dan teknologi*, pp. 120–131. Universitas Gadjah Mada Press, Yogyakarta.

Truce, W. and Fuentes, A. 2002. Object manipulation in a population of semi-free ranging *Macaca fascicularis* in Bali, Indonesia. *American Journal of Physical Anthropology* 117(Supp. 34):157.

van Schaik, C. P. 2000. Infanticide by male primates: The sexual selection hypothesis revisited. In *Infanticide by Males and its Implications*, C. P. van Schaik and C. H. Janson (eds.). Cambridge University Press. pp. 27–60

Vellucci, S. M., Fuentes, A., Suaryana, K. G., and Harya Putra I. D. K. 2000. Sex differences Balinese macaque (*Macaca fascicularis*) temple licking: Requirement of lactation or taste for salt? *American Journal of Physical Anthropology* 111(Supp. 30): 310.

Welch, A. E. , Loudon, J. E. and Fuentes, A. 2001. Male sexual strategies in a semi-free ranging group of long-tailed macaques (*Macaca fascicularis*). *American Journal of Physical Anthropology* 114(Supp. 32): 163.

Wheatley, B. P. 1988. Cultural behavior and extractive foraging in *Macaca fascicularis*. *Current Anthropology* 29: 516–519.

1991. The role of females in inter-troop encounters and infanticide among Balinese *Macaca fascicularis*. In *Primatology Today* Ehara *et al.* (ed.). Elsevier Science Publishers. pp. 169–172.

1999. *The Sacred Monkeys of Bali*. Waveland Press, Inc.

Wheatley, B. P. and Harya Putra, D. K. 1994. Biting the hand that feeds you: Monkeys and tourists in Balinese monkey forests. *Tropical Biodiversity* 2(2): 317–327.

Wheatley, B.P., Harya Putra, D. K., and Gondar, M. K. 1996. A comparison of wild and food enhanced long-tailed macaques (*Macaca fascicularis*). In *Evolution and Ecology of Macaque Societies*, J. E. Fa, and D. G Lindberg (eds). Cambridge University Press. pp. 182–206.

Whitten, A.J., Soeriaatmadja, R.E., and Affif, S. 1996. *The Ecology of Java and Bali*. Singapore: Periplus Editions.

Box 6.1   Recent demographic and behavioral data of *Macaca fascicularis* at Padangtegal, Bali, Indonesia

F. Brotcorne, I. Nyoman Wandia, Aida L. T. Rompis, I. G. Soma, I. Nyoman Suartha and M. C. Huynen

We conducted an assessment of the long-tailed macaque population at Padangtegal Monkey Forest from October to December 2009. In this study, we documented recent changes in the synanthropic population of macaques occupying this site, which has a growing tourist interest and is actively managed by a local village committee. We present the latest trends in demography, activity and dietary patterns, to compare with previous data collected by Fuentes *et al.* (this chapter) between 1998 and 2002. We will then discuss the effects of management strategies on the population since 2002.

During this study, we collected 181 hours of behavioural data using 20-minute focal sampling of identified individuals belonging to various age/sex classes. In total, we also performed four population censuses and 20 group counts using two census methods. First, to census the whole population, observers (i.e., researchers and students from the Universitas Udayana and volunteer members of the village committee), trained to identify age/sex classes, were spread in 24 sectors of the Monkey Forest. Each observer had to count the number of individuals in each age/sex class present in his respective sector. Second, throughout the study, one observer counted the groups during collective travels as soon as they crossed over an open area.

Demography: The recent census at Padangtegal revealed a marked increase in the population size, with 498 individuals including 34 adult males, 39 subadult males, 155 adult females and 270 immatures. This means the population increased by 59 percent between 2002 and 2009. If we consider changes in population size since 1986 (Figure 6.3), we note that the population growth was not stable over time. The annual population growth rate* was 11 percent between 1986 and 1992, 12 percent between 1998 and 2003, and rose to 20 percent between 2007 and 2009. The growth of this population has thus been continuing at a dramatic rate, especially since 2007, apparently without any population crashes to curb this positive trend. This growth was related to a change in demographic structure where the population increased from three groups in 2002 to five groups in 2009. Based on the groups' ranging patterns reported by Fuentes *et al.* we assume group 1 and group 2 each split in two groups at some time between 2006 and 2009. The increase in abundance and diversity of food provisioning and

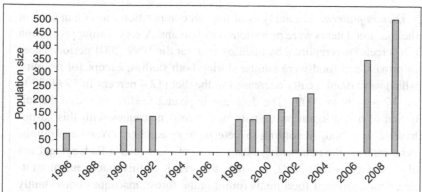

Figure 6.3. Population size history at Padangtegal. 1986–1992 data from Wheatley (1999); 1998–2002 data from Fuentes *et al.* (this chapter), 2003 data from Loudon *et al.* (2006), 2007 unpublished data from management staff.

the management style at Padangtegal can both explain the reasons behind this growth.

Ranging: The range of the population slightly increased over time. Our assessment in 2009 revealed that macaques used about 30 hectares of area while Fuentes *et al.* reported 24.3 hectares were used in 1998–2002. The local committee under the direction of Pak Wayan Selamet (i.e., the director) actively carried out a program of re-planting trees in order to increase the space and food resources available to macaques. However, this practice was limited by the presence of fields and human infrastructure surrounding the Monkey Forest. The high demographic growth combined with the small range increase led to a population density that was twice as high in 2009 (16.6 macaques per hectare) than in 2002 (8.4 macaques per hectare). During this study, we observed that macaque groups were raiding crops in surrounding fields at an average frequency of 0.7 raids per day. Although Fuentes *et al.* did not supply data on crop-raiding frequency for the 1998–2002 period, the management committee (pers. comm.) reported that the introduction of guards at the edges of the Monkey Forest has been effective in decreasing the frequency of crop raiding episodes.

Activity Pattern: The time budget of the macaque population appeared similar over time. The frequency and the diversity of the macaques' interactions with humans remained as reported for the 1998–2002 period. Macaques at Padangtegal spent 1.5 percent of their time budget manipulating objects and the "stone play" behavior was the most common of these activities. As already reported by Fuentes *et al.* immature males were more frequently involved in object manipulation than other age/sex classes.

Dietary pattern: The analysis of the diet composition shows that 68% of the consumed items were provisioned by humans. A very similar proportion (70 percent) was reported by Fuentes *et al.* for the 1999–2001 period. Types of provisioned food were similar during both studies, except for peanuts, which were dramatically decreased in the diet (12.4 percent in 1999–2001 *vs.* 3.2 percent in 2009). The decrease in peanut feeding followed the initiation of a restriction on tourists not to feed macaques with this carbohydrate-rich food. According to veterinary researchers (Wandia, Rompis, pers. comm.), the number of obesity cases of macaques at Padangtegal has also decreased since 2002, possibly as a result of this management strategy. Among natural food items found in the forest, macaques consistently exploited forest vegetation (12.6 percent of diet), fruits (9 percent) and invertebrates (8.2 percent).

In conclusion, the long-tailed macaques' activity and dietary patterns at Padangtegal show some stability over time despite considerable changes in demography. The population has grown at a high rate since 1998, and this allows us to expect the same positive trend in the future. Current management practices at Padangtegal appear to be beneficial to macaques, as shown by the continuous demographic growth and the decreased obesity trend. This management could be maintained in the future, but consideration needs to be given to the large increase in population size over the last decade. Moreover, we have to keep in mind that this population is threatened by the risks of inbreeding depression and increased disease transmission due to the close interface with humans.

*According to the formula from Kurita *et al.* (2008): $\{(N_2/N_1)^{1/(\text{year } 2 - \text{ year } 1)} -1\} * 100$ wherein N_1, N_2=population size in year 1, year 2; year 2 – year 1= the number of years between year 1 and year 2.

References
Kurita, H., Sugiyama, Y., Ohsawa, H., Hamada, Y., and Watanabe, T. 2008. Changes in demographic parameters of *Macaca fuscata* at Takasakiyama in relation to decrease of provisioned foods. *International Journal of Primatology* **29**: 1189–1202.

Wheatley, B. P. 1999. *The Sacred Monkeys of Bali*. Waveland Press, Inc.

Loudon J. E., H. M. E., and Fuentes A. 2006. The Importance of integrative anthropology: A preliminary investigation employing primatological and cultural anthropological data collection methods in assessing human-monkey co-existence in Bali, Indonesia. *Ecological and Environmental Anthropology* **2**(1): 1–13.

7 The role of Macaca fascicularis in infectious agent transmission

GREGORY ENGEL AND LISA JONES-ENGEL

Introduction

In Chapter 1, Gumert discusses the various ways in which humans interact with long-tailed macaques, pointing out how human and macaque behaviors shape these encounters. Understanding interspecies interactions provides an indispensable backdrop for considering how infectious agents are transmitted both from humans to primates and from primates to humans (Daszak *et al.*, 2001). It should be appreciated that, while we focus our discussion here on cross-species transmission between humans and primates, this dyadic interaction takes place within a much larger and more complex pathogen landscape that includes many other species present in the environment, as well as the environment itself, which constitutes an additional reservoir of infectious agents.

The likelihood of interspecies infectious agent transmission depends on numerous factors, including prevalence of infectious agents in the reservoir population, capacity of the potential "recipient" to sustain infection, and the manner in which species interact (Jones *et al.*, 2008). Consider, for example, a pet monkey and its owner. Pet owners may have close contact with a pet for years. They may share food and water with their pets, allow the pets to climb on their shoulders or head, even sleep near their pet. In some cases, the monkey is considered a part of the family, interacting with other family members, neighbors, and other animals in and around the home. Intimate contact over an extended period of time provides the opportunity for the bidirectional transmission of multiple infectious agents via multiple routes. Face-to-face contact facilitates the transmission of respiratory pathogens – likely from human to pet – such as a wide range of respiratory viruses, as well as tuberculosis. Physical contact, especially with the monkey climbing about on the owner's head and shoulders, may provide access for respiratory viruses present on the monkey's hands and feet to the owners' mucous membranes: nose, mouth and eyes. Infections with gastrointestinal pathogens are transmitted through

Monkeys on the Edge: Ecology and Management of Long-Tailed Macaques and their Interface with Humans, eds. Michael D. Gumert, Agustín Fuentes and Lisa Jones-Engel. Published by Cambridge University Press. © Cambridge University Press 2011.

the "oral/fecal" route – facilitated by prolonged close physical contact. Most pet owners report some kind of injury inflicted by their pet, a bite or scratch, that breaks the skin, allowing pathogens to enter the owner's blood directly ("parenterally"). In addition, the owner and pet may share exposure to vector-borne pathogens, such as malaria, originating in humans, non-human primates or other animals reservoirs. Whether or not transmission occurs depends to a large degree on the prevalence of infectious agents in the community. This in turn is influenced by factors such as demography (e.g., children are commonly affected by a range of respiratory and gastrointestinal pathogens), infrastructure (e.g., sewage disposal has an important bearing on prevalence of diarrheal illness) and immune status (e.g., immunization and nutritional status).

The above example illustrates the complexities of cross-species transmission of disease in the "real world." Over the past decades, scientific approaches to studying human infectious diseases have been informed, to a great extent, by laboratory research on non-human primates, in particular rhesus (*M. mulatta*) and long-tailed (*M. fascicularis*) macaques. The genetic, physiological and behavioral similarities of humans and primates contribute to a shared susceptibility to infectious agents. In the laboratory, over 200 infectious agents, including viruses, bacteria, fungi and parasites, have been shown capable of infecting both humans and primates (Fiennes, 1967). However, a discussion of this entire range of infectious agents is beyond the scope of this chapter, which will focus on a small subset of infectious agents, of particular interest from the perspective of primate conservation and public health.

It is worth noting that, while laboratory-based research on cross-species transmission has gained momentum over the past decades, still relatively little is known about how infectious agents are transmitted to and from populations of free-ranging primates. With respect to our pet monkey example, laboratory data is helpful is defining the pathophysiology of a given infection, but is ill-suited to shedding light on the how cross-species transmission occurs, which populations are affected, and what management strategies are likely to be successful in controlling disease spread. There remains a great need for further field research to link laboratory data to real world infectious challenges that affect primate and human populations.

Transmission of infectious agents from long-tailed macaques to humans

Cercopithecine herpesvirus 1

Macaques (*Macaca sp.*) are the only known natural carriers of *Cercopithecine herpesvirus* 1 (commonly known as herpes B, *Herpesvirus simiae*, B virus).

This virus has gained notoriety over the past few decades, and is often greatly misunderstood by the general public as well as by veterinarians, physicians and animal managers (Abbott, 2008; CDR, 2000; Ritz *et al.*, 2009). B virus, which is enzootic in macaques, is a close relative of the Herpes Simplex Virus endemic in human populations. In macaques, B virus can cause intermittent eruption of mucosal ulcers but otherwise typically remains latent in dorsal root and trigeminal ganglia of infected animals (Eberle and Hilliard, 1995). Within macaque populations, B virus is thought to be transmitted horizontally through bites, sexual contact, and mucosal contact with infectious body fluids (Hilliard and Ward, 1999).

There are very few studies of B virus seroprevalence in macaques outside of a laboratory setting. Among pet macaques in Asia, the prevalence appears to be much lower than that seen in temple or free-ranging populations (Jones-Engel *et al.*, 2006c). The low seroprevalence rate we observed among the pet macaques is likely attributable to the young age at which pets are typically taken from the wild. These very young animals typically do not engage in aggressive and sexual behaviors that are likely to lead to infection with B virus. Weigler and colleagues detected antibodies to B virus in only 22 percent of laboratory macaques < 2.5 years of age, but in 97 percent of the macaques older than 2.5 years of age (Weigler *et al.*, 1993). This same pattern of increasing seroprevalence with age has been seen among populations of free-ranging macaques (Engel *et al.*, 2002; Jones-Engel *et al.*, 2006a; Kessler *et al.*, 1989). Importantly, although infection rates are high as determined by the presence of antibodies, rates of viral shedding are very low (Weigler *et al.*, 1993; Weir *et al.*, 1993). In laboratory settings it is estimated that 2–3 percent of animals in a group may be shedding the virus at any one time (Elmore and Eberle, 2008; Huff *et al.*, 2003). It is hypothesized that stress associated with captivity and/ or the breeding season (e.g., rhesus and Japanese macaques (*M. fuscata*) may increase the likelihood of shedding in an animal) (Elmore and Eberle, 2008; Mitsunaga *et al.*, 2007). We have performed viral culture on mucosal specimens from hundreds of long-tailed macaques from free ranging populations in Asia and have yet to find evidence of active viral shedding (Eberle *et al.*, unpublished data).

B virus' notoriety as a human pathogen can be traced to its identification as the etologic agent of more than 40 documented cases of severe meningoencephalitis, 70 percent of which resulted in death, among workers at primate laboratories and/or zoos (Huff and Barry, 2003). The most recent reported case of human Herpes B infection occurred in 1996 and involved a laboratory researcher who developed severe disease and died after accidental mucosal contact with body fluids of a rhesus macaque. Previous cases of human herpes B infection have been attributed to contact with long-tailed and rhesus macaques, though it should be noted that in all cases of human infection with B virus, the infected individual had contact with rhesus macaques and in only some of those cases did he or she also have contact with long-tailed macaques (Smith *et al.*, 1998). Therefore, it is not possible to clearly indentify whether long-tailed macaques have ever truly been involved in a human infection of B virus. In the 1990s, concern over B virus infection led primate laboratories to institute stricter guidelines to protect lab workers from agents transmissible in primate body fluids (Holmes *et al.*, 1995). They appear to be successful, since no cases of B virus have been reported since 1996.

Interestingly, no human B virus cases have ever been documented outside the laboratory context, in spite of the abundant human-primate contact that occurs in many countries in south and Southeast Asia (Engel *et al.*, 2002; see also Jones-Engel and Eberle, unpublished data in 2006a). A possible explanation for this observation is that only the strain of B virus circulating among laboratory macaques is capable of causing severe human disease. There are some data that suggest that this hypothesis should be further explored. Smith and colleagues showed the presence of four distinct genotypes of B virus in four species of macaques, *M. mulatta, fascicularis, nemestrina* and *fuscata* (Smith *et al.*, 1998). However, Smith and others have also noted that at least within a laboratory setting there was evidence of cross-species transmission of B virus between rhesus and long-tailed macaques (Hilliard and Weigler, 1999). The possibility that a pathogenic B virus resulted from the mixing of a variety of macaque species in laboratory settings also exists.

The above hypothesis may explain the lack of reported human herpes B infections in south and southeast Asia, where human/macaque contact is abundant. Alternatively, some have theorized that human infection with B virus actually does occur, but goes undiagnosed owing to poor disease reporting, or mild symptomatology, perhaps as a result of high titres of anti-Herpes simplex antibodies in these human populations (Elmore and Eberle, 2008; Hilliard and Weigler, 1999; Ohsawa *et al.*, 2002). However, this hypothesis fails to explain the lack of reported B virus disease among the hundreds of thousands of visitors to Asia who may be exposed to the virus through bites, scratches, and mucosal contact with macaque body fluids (Fuentes, 2006; Fuentes and Gamerl, 2005). A portion of these visitors return to their home country and seek medical evaluation/treatment, yet no cases of B virus have been reported (Gautret *et al.*, 2007; Ritz *et al.*, 2009).

Notwithstanding these scientific considerations, fear of B virus infection has led to the occasional and, we would argue, irrational culling of macaques. For example, in March 2000, after testing rhesus macaques at the Woburn Safari Park in the UK for herpes antibody and learning, not surprisingly, that herpes virus was present, managers of the park, acting on recommendations of the Health and Safety Executive and the zoo licensing authority authorized the shooting of an entire colony of more than 200 monkeys. Similarly, in August, 2008 a colony of Tonkean macaques (*M. tonkeana*) at the Louis Pasteur University in Strasbourg, France was culled out of concern that laboratory workers would be infected with herpes B (Abbott, 2008; CDR, 2000).

Clearly, more research is needed to understand the natural epizootiology of B virus in non-laboratory macaques and to compare the B virus strains that have caused significant mortality and morbidity in laboratory workers with the strains that circulate naturally in free-ranging populations of macaques.

Simian Retroviruses

Simian immunodeficiency virus (SIV)

Though simian immunodeficiency virus (SIV) has gained renown as the hypothesized progenitor of human immunodeficiency virus (HIV), it is not known to naturally infect macaques. (Hahn *et al.*, 2000; Wolfe *et al.*, 1998). During the early years of the HIV epidemic *M. fascicularis* was proposed as a model for studying the immune response to HIV infection (Gardner *et al.*, 1988). However, *M. fascicularis* turned out not to be an appropriate animal model for HIV, as inoculated animals failed to produce viremia or a neutralizing antibody response (Li *et al.*, 1995; Reimann *et al.*, 2005).

SIV has not been shown to naturally infect Asian macaques (Hayami *et al.*, 1994; Jones-Engel *et al.*, 2006c; Jones-Engel *et al.*, 2006a). Experimental infection of long-tailed macaques with SIV derived from African monkeys has been demonstrated. In a study of thirty-three *M. fascicularis* experimentally inoculated with SIVsm (SIV from sooty mangabeys), 29 of the long-tailed macaques died after infection with SIV. Infection in these animals caused splenic and lymph node enlargement, chronic diarrhea, weight loss and a very high incidence of malignant lymphomas (Putkonen *et al.*, 1992). Secondary infection has been observed in experimentally infected pig-tailed macaques (Kuller *et al.*, 1992; Morton *et al.*, 1989). Laboratory studies indicate that SIV is highly pathogenic in Asian macaques.

SIV is of potential concern in situations such as zoos, exotic animal trade and wet markets where African primates may come into contact with Asian primates including long-tailed macaques. It is possible that SIV could be introduced into free-ranging, long-tailed macaque populations with unknown but potentially grave results for the animals. Because these contexts of inter-species transmission of SIV exist, we advocate monitoring populations of long-tailed macaques that have potential contact with African primates for SIV. It should be noted that there is no reason to believe that long-tailed macaques pose any risk for the transmission of immunodeficiency viruses to humans.

Simian foamy virus (SFV)

Simian foamy viruses (SFV) are members of the taxonomic subfamily of retroviruses known as *Spumavirinae*. Foamy viruses are found in many mammals including cats, cows, rodents and sea lions as well as several species of non-human primates (Meiering and Linial, 2001). Molecular studies suggest that foamy viruses are ancient, having coevolved with their host species over tens of millions of years (Switzer *et al.*, 2005). Foamy viruses are unique among retroviruses in that, in spite of the virus's persistence in a variety of tissues, no pathogenic state has been associated with infection (Meiering and Linial, 2001). Research in non-human primates suggests that although host animals mount an antibody response to SFV, the virus continues to be detected in the host throughout its life (Falcone *et al.*, 2003). Seroprevalence of SFV reaches 100 percen in adult macaques in the laboratory and in free-ranging populations of long-tailed macaques (Jones-Engel *et al.*, 2007).

Although to date no endemic human foamy virus has been discovered, human infection with SFV has been detected. The first research documenting human infection with SFV described infection in individuals who had contact with non-human primates at zoos and primate laboratories (Brooks *et al.*, 2002; Heneine *et al.*, 1998). Since then, SFV infection has been shown in bushmeat hunters in Central Africa and in people who came into contact with a variety of Asian primates in multiple contexts in south and southeast Asia (Calattini *et al.*, 2007; Falcone *et al.*, 2003; Jones-Engel *et al.*, 2008; Wolfe *et al.*, 2004). SFV infection in humans is not well studied. Follow-up of a very small number (i.e., fewer than ten) infected individuals, infected in the context of a primate laboratory suggests that infection is persistent but there is no evidence that infected individuals developed disease or transmitted the virus to close contacts, including sexual partners (Boneva *et al.*, 2007). Data thus far indicates humans may be dead end hosts for SFV (Callahan *et al.*, 1999; Meiering and Linial, 2001). However, more data is certainly needed to assess whether this is actually the case. In particular, it will be important to follow the course of SFV infection in immunocompromised (e.g., through malnutrition or coinfection with other immune compromising pathogens) humans who may present a more "permissive" environment for this retrovirus.

Transmission of SFV has been shown to occur from long-tailed macaques to several humans who came into contact with macaques at monkey temples in Bali (Jones-Engel *et al.*, 2005; Jones-Engel *et al.*, 2008). Phylogenetic data showed that the virus in the infected humans were almost identical to that isolated from long-tailed macaques at the temple where they worked. These findings are particularly important in light of the brisk tourism industry in southeast Asia, which brings millions of visitors from all over the world every year to the region, many of whom take advantage of opportunities to come into contact with non-human primates (Fuentes, 2006). We used mathematical models to estimate the risk of SFV transmission from a macaque to a visitor to a monkey temple in Bali (Engel *et al.*, 2006). This model predicted that six out of 1000 visitors would become infected with SFV. Of course, given SFV's current status as a "virus without a disease," there is no evidence that SFV infection, in and of itself, is a risk to public health. However, the fact of primate-to-human transmission of a retrovirus in multiple contexts in southeast Asia should give us pause. SFV should be regarded as a "marker" for the potential transmission of other agents, including agents as yet unknown between humans and non-human primates. As such, this information should reinforce efforts to reduce human-primate contact, especially in high-risk contexts, such as monkey temples, pet markets, and in urban areas with primate populations.

Simian type D retrovirus (SRV)

Simian type D retrovirus (SRV), is a multi-serotype group of retroviruses of the taxonomic family *Retroviridae*. This virus is found almost exclusively in the genus *Macaca*. SRV serotypes 1 and 2 are associated with long-tailed macaques (Lerche *et al.*, 1997; Wilkinson *et al.*, 2003). The prevalence and characteristics of this virus are virtually unknown in most free-ranging populations of primates (Engel *et al.*, 2008; Jones-Engel *et al.*, 2006a; Schillaci *et al.*, 2005). Laboratory data suggests that the virus is transmitted among animals through bite wounds (Lerche *et al.*, 1987). SRV is known to cause epidemics of an AIDS-like syndrome in laboratory macaques (Guzman *et al.*, 1999; Marx *et al.*, 1985; Tsai *et al.*, 1990). There is some evidence that laboratory and zoo workers exposed to SRV + animals can become infected with the virus though to date no disease has been associated with infection in humans (Heneine *et al.*, 1993; Lerche *et al.*, 2001).

Rabies

We discuss the issue of rabies virus here, because it comes up frequently in the context of travel medicine. It is common practice in Southeast Asia for rabies prophylaxis to be given to people who present to a healthcare provider having been injured, usually bitten, by a monkey (Altmann *et al.*, 2009; Gautret *et al.*, 2007; Ramos and Ramos, 2002; Reynes *et al.*, 1999). This widespread practice may give the false impression that rabies is actually enzootic or common among free-ranging non-human primates. In fact, there is no evidence that this is the case and little reason to believe that bites from Asian non-human primates pose a risk for rabies infection. There is no documented evidence of transmission of rabies virus from *M. fascicularis* to humans. We can only hypothesize about why rabies does not appear to naturally infect macaques even in areas where rabies transmission from animals to humans is common. In general, macaques and dogs/rodents/bats (the latter which have been shown to carry and transmit the rabies virus) do not occupy the same ecological niche. In areas where dogs and monkeys do overlap, monkeys are typically scared of dogs because dogs are often used to chase monkeys from the fields. Additionally, unlike humans, monkeys can easily escape a rabid dog by fleeing up into the trees. There is, however, limited data suggesting that pet New World monkeys that have been exposed to other rabid animals may be capable of transmitting rabies to humans (Favoretto *et al.*, 2001).

Ebola Reston virus

Ebola Reston is one of five known strains of filovirus. Of the four other strains, the Zaire, Sudan and Bundibugyo strains are associated with severe human disease characterized by high mortality (Gonzalez *et al.*, 2007). Ebola Reston was discovered in 1989, linked to an epidemic of viral hemorrhagic fever among long-tailed macaques imported from the Philippines and quarantined at a facility in Reston, Virginia (Miranda *et al.*, 2002). Other epidemics among long-tailed macaques imported from the Philippines were reported during the 1990's (Jahrling *et al.*, 1996). Ebola Reston causes high mortality rates in infected macaques, but no disease has been observed in any of the two dozen humans with serological evidence of infection. Recently, epizootics (i.e., new outbreaks in animals) in pigs in the Philippines have been linked with Ebola

Reston virus infection (Normile, 2009). The reservoir of Ebola Reston virus is still unknown, but it appears that long-tailed macaques are not the natural reservoir for this pathogen.

Human-to-primate transmission

As pointed out in the above sections, although much is known about physiology and disease in laboratory macaques which have been used as biomedical models for human disease, relatively little is known about how human pathogens may impact free-ranging macaque populations, including *M. fascicularis*. Scattered anecdotal reports hint at die-offs that could be attributed to outbreaks of infectious disease in long-tailed macaque populations, however, there are virtually no studies that have verified that human infectious agents are the culprit. In general, obstacles to research on die-offs in free-ranging macaque populations include logistical difficulty of the research (i.e., population-level baseline serological data prior to infectious agent exposures) and the difficulty in funding these studies on long-tailed macaques, considered by some as a nuisance, or "weed" species (Richards, et al.,1989). Below we consider several endemic human pathogens that have been shown experimentally to infect and in some cases cause disease in long-tailed macaques.

Tuberculosis

Globally, tuberculosis is one of the most significant infectious threats to human populations. It is estimated that one-third of all humans living are infected with *Mycobacteria tuberculosis* (TB), the organism that causes most human disease (Jasmer *et al.*, 2002). Of these 2 billion individuals, most have inactive (i.e., latent) infections, but an estimated 13.7 million have chronic active disease, and 1.8 million people die from TB every year, most of these in the developing world (WHO, 2009). The increasing global burden of the disease, its synergistic action with HIV and the growing problem of drug-resistant TB have spurred interest in studying TB in the laboratory, where long-tailed macaques have become a favored model for studying the disease (Flynn *et al.*, 2003). Research on experimentally infected *M. fascicularis* shows that, given experimental innocula as small as 25colony-forming units, long-tailed macaques show a broad range of response, from latent disease to chronic active disease (Capuano, III *et al.*, 2003). This wide range of disease parallels that seen with human TB infection, and makes this species of macaque well-suited as a model to study human tuberculosis infection (Lin *et al.*, 2009). Additionally, this research on *M. fascicularis* points to an interesting phenomenon which is that it appears that at least some primate taxa are more susceptible to TB than are others. Further, even within a primate genus, response to infection with *M. tuberculosis* has been shown to differ greatly among congeneric species with some (e.g., *M. fasicularis*) able to sustain chronic, slowly progressive disease similar to that which occurs in humans, while others (e.g., *M. mulatta*) succumb rapidly to the disease (Walsh *et al.*, 1996). These differences in response to exposure are important, because animals which quickly succumb to the disease are unlikely to be reservoirs, while those with sufficient resistance to contain the infection at a subclinical level may be able to harbor and transmit the mycobacteria to conspecifics and humans.

Of course, given the widespread use of long-tailed macaques in laboratory settings, the issue of TB infection in this setting is important in and of itself particularly because epizootics in the laboratory can lead to widespread culling of animals and disruption of experimental protocols. In response, since the 1970s, surveillance programs have been instituted to more actively detect and respond to the presence of TB in laboratory primate populations(Roberts and Andrews, 2008). Unfortunately, diagnosis of TB in macaques, as in humans, is bedeviled by the lack of sensitive and specific laboratory tests capable of differentiating exposure from latent and active infection. The intradermal tuberculin skin test (TST) is the test that over the past decades has commonly been employed in laboratory settings. However, recently newer serologic and molecular tests, alone and in combination, show promise as improvements on these traditional approaches (Lerche *et al.*, 2008).

The impact of TB in free-ranging long-tailed populations, especially populations that have frequent and intense contact with human populations where TB is endemic is not well understood. Much of the natural range of *M fascicularis* coincides with hyperendemic areas for TB. Indeed, some of our unpublished data support the logical expectation that synanthropic populations of *M fascicularis* would show evidence of exposure to TB. Yet to be explored is research investigating the incidence and/or prevalence of TB exposure in more remote populations with less contact with humans. One of the many questions that remain is how exposure to TB impacts the health of these populations. This is a particularly difficult question to explore because current methods for detection of exposure to TB require that the animal be sedated, a tuberculin skin test be placed and follow-up at 24, 48and 72 hours. Such an approach is rarely feasible in free-ranging populations. Our research team is working on developing techniques that allow for the noninvasive monitoring of TB exposure in primate populations.

Endemic human respiratory viruses

Here we consider a number of respiratory viruses, endemic in human populations. Endemic human respiratory viruses have been studied extensively in the long-tailed macaques in laboratory settings, but little is known about the impact of these viruses in free-ranging populations. In 2000–2002 we examined serological evidence of infection with a panel of endemic human viruses among six groups of free-ranging long-tailed macaques on the Indonesian island of Bali. The sites were all associated with temples or shrines, but varied in macaque population, amount of human presence and kinds of human contact. A total of 135 animals were captured, sampled, and released, and serum was analyzed by ELISA for evidence of prior infection with Influenza A and B, Parainfluenzas 1, 2 and 3, measles, respiratory syncitial virus (RSV), mumps, and adenovirus. No serological evidence of prior infection with influenza A, influenza B or parainfluenza 1 was found. The most commonly detected antibodies across the populations were to parainfluenza 3 (23.0 percent), adenovirus (33.3 percent), measles (46.7 percent), and mumps (48.1 percent) with seroprevalence of antibodies to varying widely among the sites (Table 7.1).

Of these viruses, measles is considered the most significant threat to macaques. Infections with measles virus can cause pneumonia in macaques and has been associated with spontaneous abortion of fetuses (Renne *et al.*, 1973). Measles has been shown to cause significant mortality and morbidity in recently captured macaques who were likely immunocompromised due to poor nutrition, injuries associated with trapping and handling, overcrowding, and poor husbandry (El Mubarak *et al.*, 2007; Suzuki *et al.*, 1981; Welshman, 1989; Willy *et al.*, 1999). Our research group has measured prevalence of antibodies to measles in primates in a number of countries, including the above cited series in Bali. Interestingly, in contrast to *M. fascicularis* populations in Indonesia, in which serological evidence of infection with measles was found, no evidence of measles infection was detected among the thirty-eight long-tailed macaques sampled from a number of groups in Singapore (Jones-Engel *et al.*, 2006b). We hypothesize that higher levels of measles vaccination in Singapore's human population, in comparison to Indonesia's, is one of the major contributing factors to contrasting prevalence of measles in these respective macaque populations.

The important conclusion to draw from these data, we believe, is that respiratory pathogens endemic in the human population, including several viruses (e.g., RSV, mumps and measles) that are especially prevalent in children, are readily spread to synanthropic primate populations. Further research is needed

Table 7.1. *Antibody prevalence of several endemic human respiratory viruses among six populations of long-tailed macaques in Bali, Indonesia*

Location	N	Flu A[1]	Flu B[2]	Paraflu1[3]	Paraflu2[4]	Paraflu3[5]	Measles[6]	RSV[7]	Mumps[8]	Adenovirus[9]
Alas Kedaton	24	0.00%	0.00%	0.00%	0.00%	58.30%	12.50%	8.30%	79.20%	25.00%
Bedugal	5	0.00%	0.00%	0.00%	80.00%	80.00%	100.00%	100.00%	60.00%	100.00%
Pulaki	22	0.00%	0.00%	0.00%	0.00%	9.10%	45.50%	4.50%	36.40%	36.40%
Sangeh	56	0.00%	0.00%	0.00%	1.80%	19.60%	67.90%	3.60%	47.40%	21.10%
Teluk Terima	9	0.00%	0.00%	0.00%	0.00%	0.00%	66.70%	0.00%	33.30%	33.30%
Uluwatu	19	0.00%	0.00%	0.00%	0.00%	0.00%	5.30%	0.00%	26.30%	57.90%

[1] Influenza A; [2] Influenza B; [3] Parainfluenza 1; [4] Parainfluenza 2; [5] Parainfluenza 3; [6] Measles virus; [7] Respiratory syncitial virus; [8] Adenovirus

to determine the impact of these pathogens on macaque health and the extent to which these viruses can be spread from habituated animals to surrounding groups of unhabituated, "wild" groups.

Vector-borne infections

Malaria

Malaria is the most important vector-borne disease, in terms of global impact, on public health. A tropical/subtropical disease, the geographic extent of endemic malaria overlaps significantly with the natural ranges of primate species. Data suggest that the four protozoan species (*Plasmodium vivax, P. falciparum, P. ovale* and *P. malariae*) that cause most human malaria may have arisen as a result of host switches from non-human primates to human hosts (Escalante *et al.*, 2005; Krief *et al.*, 2010). *Plasmodium spp.* are transmitted by mosquitoes of the genus *Anopheles* (Collins, 1974). Of the hundreds of Anopheles species, about 100 have been shown capable of transmitting "human" malaria. However, much is yet to be learned about the feeding ecology of anopheles, which species feed on humans, which feed on primates, and which feed on both (Garamszegi, 2009).

The most common human malaria in SE Asia is *P. vivax*, shown by phylogenetic studies to have jumped the species barrier approximately 25,000 years ago from an Asian primate (Escalante *et al.*, 2005). Three closely related Plasmodium species (*P.cynomologi, P. inui, P. knowlesi*) associated with Southeast Asian macaque reservoirs, form a phylogenetic cluster with *P. vivax* (Escalante *et al.*, 2005). Until recently, diagnosis of malaria relied on microscopic description of the parasite and was unable to differentiate between *P. vivax, P. malariae* and their close phylogenetic cousins. However, in the past few years advances in malaria diagnosis, using new serological and molecular techniques have enhanced our ability to identify specific *Plasmodium* species (Collins and Barnwell, 2009). These new methods have led to the recognition that some human malaria in Southeast Asia are caused by *P. knowlesi*. The first documented *P knowlesi* cases, in 2004 among Dyaks in Malaysian Borneo, have been followed by other cases on the Malaysian peninsula (Luchavez *et al.*, 2008; Ng *et al.*, 2008; Putaporntip *et al.*, 2009; Vythilingam *et al.*, 2006; Vythilingam *et al.*, 2008). To date, these reports come from areas that contain populations of monkeys as well, especially *M. fascicularis* and *M. nemestrina* (White, 2008). The public health significance of *P. knowlesi* remains unclear. Some reports suggest that *P. knowlesi* malaria is similar in severity, or possibly more severe than *P malariae* (Galinski and Barnwell, 2009). These new findings raise numerous questions. How prevalent is *P. knowlesi* infection in humans and primates? Is *P. knowlesi* infection in humans a recent phenomenon, or has it been taking place, undetected, for a long time? What are the important reservoirs for infection? Which vectors are most important in its transmission? This is an exciting arena for research in the next decade.

Management considerations

The potential impact of cross-species disease transmission on public health and primate conservation should be kept in mind as an important consideration in making decisions regarding management of long-tailed macaque populations. We strongly advocate an approach that emphasizes gathering relevant data in a methodologically rigorous fashion and using this information to guide evidence-based policy making. Such an approach first explores and characterizes the human-primate interface, in terms of geography, human and primate populations, and contexts of interspecies contact. Subsequently, reservoirs of infectious agents, both in human and primates are identified and ongoing serological surveillance is conducted. Based on these data, specific areas and contexts that present high risk for potential cross-species disease transmission can be prioritized. A multidisciplinary approach incorporating the perspectives of different fields of study is necessary to integrate diverse data.

Though, as we point out, few data are available on which to firmly base specific recommendations, we would venture some hypotheses regarding identification of high-risk situations. Certain contexts, by their nature, may constitute particular risks for the transmission of human pathogens to wild primate populations. Pet primates, for example, may be seen as potential conduits for pathogens passing from human reservoirs to wild primates. As we have pointed out, the prolonged, intimate contact pets have with owners, their families, and even their communities provides many opportunities for exposure to infectious agents that infect humans. Children are much more likely than adults to carry and transmit a broad range of infectious agents characteristic of their age group, respiratory viruses and rotaviruses, for example. Children may be less reticent about reaching out and touching a monkey, or, for that matter, about coughing on one. It is also common practice for pet owners to release their pet monkeys when the pets outgrow their appeal–as they generally do as they mature, grow large teeth and become aggressive. These released pets may find their way to groups of free-ranging monkeys that have yet to encounter human pathogens. This situation is a potentially dangerous one for these wild monkeys, particularly with pathogens such as measles virus.

Another likely risk factor for primate conservation is the health status of the human population. For example, a population with low levels of immunization against measles virus is much more likely to sustain an epidemic that can spill over to infect synanthropic, and, by extension, wild primate populations. Human nutrition and infrastructure are two other variables that may have a significant impact on the reservoir of pathogens communicable to macaques, on the one hand, and on the vulnerability of the human population to primate-borne pathogens, on the other.

Certainly, any effective management strategy should carefully address the issue of human-primate contact. That is, how to reduce interspecies contacts, especially those, such as bites, that are most likely to lead to transmission of infectious agents. A thorough understanding of the cultural/economic role of primates in a given community is likely to be indispensable to interventions designed to modify human behaviors. For that matter, knowledge of macaque behavior is equally critical.

References

Abbott, A. 2008. French university under fire for culling macaques. *Nature* **455**: 145.

Altmann, M., Parola, P., Delmont, J., Brouqui, P., and Gautret, P. 2009. Knowledge, attitudes, and practices of French travelers from Marseille regarding rabies risk and prevention. *Journal of Travel Medicine* **16**(2): 107–111.

Boneva, R. S., Switzer, W. M., Spira, T. J., *et al*. 2007. Clinical and virological characterization of persistent human infection with simian foamy viruses. *AIDS Res and Human Retroviruses* **23**: 1330–1337.

Brooks, J. I., Rud, E. W., Pilon, R. G., Smith, J. M., Switzer, W. M., and Sandstrom, P. A. 2002. Cross-species retroviral transmission from macaques to human beings. *Lancet* **360**: 387–388.

Calattini, S., Betsem, E. B. A., Froment, A., *et al*. 2007. Simian foamy virus transmission from apes to humans, rural Cameroon. *Emerging Infectious Diseases* **13**: 1314–1320.

Callahan, M. E., Switzer, W. M., Matthews, A. L., *et al*. 1999. Persistent zoonotic infection of a human with simian foamy virus in the absence of an intact orf-2 accessory gene. *Journal of Virology* **73**: 9619–9624.

Capuano, S. V., Croix, D. A., Pawar, S., *et al*. 2003. Experimental *Mycobacterium tuberculosis* infection of cynomolgus macaques closely resembles the various manifestations of human *M. tuberculosis* infection. *Infection and Immunity* **71**: 5831–5844.

CDR. 2000. Monkeys with herpes B virus culled at a safari park. *Communicable Disease Report* **10**(11): 99–102.

Collins, W. E. 1974. Primate malarias. *Advances in Veterinary Science and Comparative Medicine* **18**: 1–23.

Collins, W. E., and Barnwel, 1 J. W. 2009. *Plasmodium knowlesi:* Finally being recognized. *Journal of Infectious Diseases* **199**: 1107–1108.

Daszak P., Cunningham, A. A., and Hyatt, A. D. 2001. Anthropogenic environmental change and the emergence of infectious diseases in wildlife. *Acta Tropica* **78**:103–116.

Eberle, R. and Hilliard, J. 1995. The simian herpesviruses. *Infectious Agents and Disease* **4**: 55–70.

El Mubarak, H. S., Yuksel, S., van Amerongen, G., *et al.* 2007. Infection of cynomolgus macaques (*Macaca fascicularis*) and rhesus macaques (*Macaca mulatta*) with different wild-type measles viruses. *Journal of General Virology* **88**: 2028–2034.

Elmore, D. and Eberle, R. 2008. Monkey B virus (Cercopithecine herpesvirus 1). *Comparative Medicine* **58**: 11–21.

Engel, G. A., Jones-Engel, L., Schillaci, M. A., *et al.* 2002. Human exposure to herpesvirus B-seropositive macaques, Bali, Indonesia. *Emerging Infectious Diseases* **8**: 789–795.

Engel, G. A., Hungerford, L. L., Jones-Engel, L., *et al.* 2006. Risk assessment: A model for predicting cross-species transmission of simian foamy virus from macaques (*M. fascicularis*) to humans at a monkey temple in Bali, Indonesia. *American Journal of Primatology* **68**: 934–948.

Engel, G. A., Pizarro, M., Shaw, E., *et al.* 2008. Unique pattern of enzootic primate viruses in Gibraltar macaques. *Emerging Infectious Diseases* **14**: 1112–1115.

Escalante, A. A., Cornejo, O. E., Freeland, D. E., *et al.* 2005. A monkey's tale: The origin of *Plasmodium vivax* as a human malaria parasite. *Proceedings of the National Academy of Sciences of the United States of America* **102**: 1980–1985.

Falcone, V., Schweizer, M., and Neumann-Haefelin, D. 2003. Replication of primate foamy viruses in natural and experimental hosts. In *Foamy viruses,* A. Rethwilm (ed.). Berlin: Springer. pp. 161–180.

Favoretto, S. R., de Mattos, C. C., Morais, N. B., *et al.* 2001. Rabies in marmosets (*Callithrix jacchus*), Ceara, Brazil. *Emerging Infectious Diseases* **7**: 1062–1065.

Fiennes, R. 1967. *Zoonoses of Primates.* Ithaca, NY: Cornell University Press.

Flynn, J. L., Capuano, S. V., Croix, D., *et al.* 2003. Non-human primates: a model for tuberculosis research. *Tuberculosis* **83**: 116–118.

Fuentes, A. 2006. Human culture and monkey behavior: Assessing the contexts of potential pathogen transmission between macaques and humans. *American Journal of Primatology* **68**: 880–896.

Fuentes, A. and Gamerl, S. 2005. Disproportionate participation by age/sex classes in aggressive interactions between long-tailed macaques (*Macaca fascicularis*) and human tourists at Padangtegal monkey forest, Bali, Indonesia. *American Journal of Primatology* **66**: 197–204.

Galinski, M. R. and Barnwell, J. W. 2009. Monkey malaria kills four humans. *Trends Parasitol* **25**: 200–204.

Garamszegi, L. Z. 2009. Patterns of co-speciation and host switching in primate malaria parasites. *Malaria Journal* **8**: 110.

Gardner, M. B., Luciw, P., Lerche, N., and Marx, P. 1988. Nonhuman primate retrovirus isolates and AIDS. *Advances in Veterinary Science and Comparative Medicine* **32**: 171–226.

Gautret, P., Schwartz, E., Shaw, M., *et al.* 2007. Animal-associated injuries and related diseases among returned travellers: A review of the GeoSentinel Surveillance Network. *Vaccine* **25**: 2656–2663.

Gonzalez, J. P., Pourrut, X., and Leroy, E. 2007. Ebolavirus and other filoviruses. *Current Topics in Microbiology and Immunology* **315**: 363–387.

Guzman, R. E., Kerlin, R. L., and Zimmerman, T. E. 1999. Histologic lesions in cynomolgus monkeys (*Macaca fascicularis*) naturally infected with simian retrovirus type D: Comparison of seropositive, virus-positive, and uninfected animals. *Toxicological Pathology* **27**: 672–677.

Hahn, B. H., Shaw, G. M., De Cock, K. M., and Sharp, P. M. 2000. AIDS as a zoonosis: Scientific and public health implications. *Science* **287**: 607–614.

Hayami M, Ido E, and Miura T. 1994. Survey of simian immunodeficiency virus among nonhuman primate populations. *Current Topics Microbiology and Immunology* **188**: 1–20.

Heneine, W., Lerche, N. W., Woods, T., *et al.* 1993. The search for human infection with simian type D retroviruses. *Journal of Acquired Immune Deficiency Syndromes* **6**: 1062–1066.

Heneine, W., Switzer, W. M., Sandstrom, P., *et al.* 1998. Identification of a human population infected with simian foamy viruses. *Nature Medicine* **4**: 403–407.

Hilliard, J. K. and Ward, J. A. 1999. B-virus specific-pathogen-free breeding colonies of macaques (*Macaca mulatta*): Retrospective study of seven years of testing. *Laboratory Animal Science* **49**: 144–148.

Hilliard, J. K. and Weigler, B. J. 1999. The existence of differing monkey B virus genotypes with possible implications for degree of virulence in humans. *Laboratory Animal Science* **49** :10–11.

Holmes, G.P., Chapman, L. E., Stewart, J. A., *et al.* 1995. Guidelines for the prevention and treatment of B-virus infections in exposed persons. The B virus Working Group. *Clinical Infectious Diseases* **20**: 421–439.

Huff, J. L. and Barry, P. A. 2003. B-virus (Cercopithecine herpesvirus 1) infection in humans and macaques: Potential for zoonotic disease. *Emerging Infectious Diseases* **9**: 246–250.

Huff, J. L., Eberle, R., Capitanio, J., Zhou, S. S., and Barry, P. A. 2003. Differential detection of B virus and rhesus cytomegalovirus in rhesus macaques. *Journal of General Virology* **84** :83–92.

Jahrling, P. B., Geisbert, T. W., Jaax, N. K., Hanes, M. A., Ksiazek, T. G., and Peters, C. J. 1996. Experimental infection of cynomolgus macaques with Ebola-Reston filoviruses from the 1989–1990 U.S. epizootic. *Archives of Virology Supplement* **11**: 115–134.

Jasmer, R. M., Nahid, P., and Hopewell, P. C. 2002. Clinical practice. Latent tuberculosis infection. *New England Journal of Medicine* **347**: 1860–1866.

Jones, K. E., Patel, N. G., Levy, M. A., *et al.* 2008. Global trends in emerging infectious diseases. *Nature* **451**: 990–993.

Jones-Engel, L., Engel, G. A., Schillaci, M. A., *et al.* 2005. Primate-to-human retroviral transmission in Asia. *Emerging Infectious Diseases* **11**: 1028–1035.

Jones-Engel, L., Engel, G. A., Heidrich, J., *et al.* 2006a. Temple monkeys and health implications of commensalism, Kathmandu, Nepal. *Emerging Infectious Diseases* **12**: 900–906.

Jones-Engel, L., Engel, G .A., Schillaci, M. A., *et al.* 2006b. Considering human-primate transmission of measles virus through the prism of risk analysis. *American Journal of Primatology* **68**: 868–879.

Jones-Engel L, Schillaci M, Engel G, Paputungan P, and Froehlich J. 2006c. Characterizing primate pet ownership in Sulawesi: Implications for disease transmission. In *Commensalism and Conflict: The Human-Primate Interface,* J. D. Paterson, J. Wallis (ed.). Norman, OK: American Society of Primatologists. pp 196–221.

Jones-Engel, L., Steinkraus, K. A., Murray, S. M., *et al.* 2007. Sensitive assays for simian foamy viruses reveal a high prevalence of infection in commensal, free-ranging Asian monkeys. *Journal of Virology* **81**: 7330–7337.

Jones-Engel, L., May, C. C., Engel, G. A., *et al.* 2008. Diverse contexts of zoonotic transmission of simian foamy viruses in Asia. *Emerging Infectious Diseases* **14**: 1200–1208.

Kessler, M. J., London, W. T., Madden, D. L., *et al.* 1989. Serological survey for viral diseases in the Cayo Santiago rhesus macaque population. *Puerto Rico Health Sciences Journal* **8**: 95–97.

Krief, S., Escalante, A. A., Pacheco, M. A., *et al.* 2010. On the diversity of malaria parasites in African apes and the origin of Plasmodium falciparum from Bonobos. *PLoS Pathogens* **6**: e1000765.

Kuller, L., Benveniste, R. E., Watanabe, R., Tsai, C. C., and Morton, W. R. 1992. Transmission of SIVMne from female to male *Macaca nemestrina. Journal of Medical Primatology* **21**: 299–307.

Lerche, N. W., Cotterman, R. F., Dobson, M. D., Yee, J. L., Rosenthal, A. N., and Heneine, W. M. 1997. Screening for simian type-D retrovirus infection in macaques, using nested polymerase chain reaction. *Laboratory Animal Science* **47**: 263–268.

Lerche, N. W., Marx, P. A., Osborn, K. G., *et al.* 1987. Natural history of endemic type D retrovirus infection and acquired immune deficiency syndrome in group-housed rhesus monkeys. *Journal of the National Cancer Institute* **79**: 847–854.

Lerche, N. W., Switzer, W. M., Yee, J. L., *et al.* 2001. Evidence of infection with simian type D retrovirus in persons occupationally exposed to nonhuman primates. *Journal of Virology* **75**: 1783–1789.

Lerche, N. W, Yee, J. L., Capuano, S. V., and Flynn, J. L. 2008. New approaches to tuberculosis surveillance in nonhuman primates. *ILAR Journal* **49**: 170–178.

Li, J. T., Halloran, M., Lord, C. I., *et al.* 1995. Persistent infection of macaques with simian-human immunodeficiency viruses. *Journal of Virology* **69**: 7061–7067.

Lin, P. L., Rodgers, M., Smith, L., *et al.* 2009. Quantitative comparison of active and latent tuberculosis in the cynomolgus macaque model. *Infection and Immunity* **77**: 4631–4642.

Luchavez, J., Espino, F., Curameng, P., *et al.* 2008. Human infections with Plasmodium knowlesi, the Philippines. *Emerging Infectious Diseases* **14**: 811–813.

Marx, P. A., Bryant, M. L., Osborn, K. G., *et al.* 1985. Isolation of a new serotype of simian acquired immune deficiency syndrome type D retrovirus from Celebes black macaques (*Macaca nigra*) with immune deficiency and retroperitoneal fibromatosis. *Journal of Virology* **56**: 571–578.

Meiering, C. D. and Linial, M. L. 2001. Historical perspective of foamy virus epidemiology and infection. *Clinical Microbiology Reviews* **14**: 165–176.

Miranda, M. E., Yoshikawa, Y., Manalo, D. L., *et al.* 2002. Chronological and spatial analysis of the 1996 Ebola Reston virus outbreak in a monkey breeding facility in the Philippines. *Experimental Animals* **51**: 173–179.

Mitsunaga, F., Nakamura, S., Hayashi, T., and Eberle, R. 2007. Changes in the titer of anti-B virus antibody in captive macaques (*Macaca fuscata, M. mulatta, M. fascicularis*). *Comparative Medicine* **57**: 120–124.

Morton, W. R., Kuller, L., Benveniste, R. E., *et al.* 1989. Transmission of the simian immunodeficiency virus SIVmne in macaques and baboons. *Journal of Medical Primatology* **18**: 237–245.

Ng, O. T, Ooi, E. E., Lee, C. C., *et al.* 2008. Naturally acquired human Plasmodium knowlesi infection, Singapore. *Emerging Infectious Diseases* **14**: 814–816.

Normile, D. 2009. Emerging infectious diseases. Scientists puzzle over Ebola-Reston virus in pigs. *Science* **323**: 451.

Ohsawa, K., Black, D. H., Sato, H., and Eberle, R. 2002. Sequence and genetic arrangement of the U(S) region of the monkey B virus (cercopithecine herpesvirus 1) genome and comparison with the U(S) regions of other primate herpesviruses. *Journal of Virology* **76**: 1516–1520.

Putaporntip, C, Hongsrimuang, T., Seethamchai, S., *et al.* 2009. Differential prevalence of Plasmodium infections and cryptic Plasmodium knowlesi malaria in humans in Thailand. *Journal of Infectious Diseases* **199**: 1143–1150.

Putkonen, P., Kaaya, E. E., Bottiger, D., *et al.* 1992. Clinical features and predictive markers of disease progression in cynomolgus monkeys experimentally infected with simian immunodeficiency virus. *AIDS* **6**: 257–263.

Ramos, P. M. and Ramos, P. S. 2002. Human accidents with monkeys in relation to prophylactic treatment for rabies, in the Municipal district of Sao Paulo, Brazil. *Rev Soc Bras Med Trop* **35**: 575–577.

Reimann, K. A., Parker, R. A., Seaman, M. S., *et al.* 2005. Pathogenicity of simian–human immunodeficiency virus SHIV-89.6P and SIVmac is attenuated in cynomolgus macaques and associated with early T-lymphocyte responses. *Journal of Virology* **79**: 8878–8885.

Renne, R. A., McLaughlin, R. and Jenson, A. B. 1973. Measles virus-associated endometritis, cervicitis, and abortion in a rhesus monkey. *Journal of the American Veterinary Medical Association* **163**: 639–641.

Reynes, J. M., Soares, J. L., Keo, C., Ong, S., Heng, N. Y., and Vanhoye, B. 1999. Characterization and observation of animals responsible for rabies post-exposure treatment in Phnom Penh, Cambodia. *Onderstepoort Journal of Veterinary Research* **66**: 129–133.

Richard, A. F., Goldstein, S. J., and Dewar, R. E. 1989. Weed macaques: The evolutionary implications of macaque feeding ecology. *International Journal of Primatology* **10**(6): 569–594.

Ritz, N., Curtis, N, Buttery, J., and Babl, F. E. 2009. Monkey bites in travelers: should we think of herpes B virus? *Pediatric Emergency Care* **25**: 529–531.

Roberts, J. A. and Andrews, K. 2008. Nonhuman primate quarantine: its evolution and practice. *ILAR Journal* **49**: 145–156.

Schillaci, M. A., Jones-Engel, L., Engel, G. A., *et al.* 2005. Prevalence of enzootic simian viruses among urban performance monkeys in Indonesia *Tropical Medicine and International Health* **10**:1305–1314.

Smith, A. L., Black, D. H., and Eberle, R. 1998. Molecular evidence for distinct genotypes of monkey B virus (herpesvirus simiae) which are related to the macaque host species. *Journal of Virology* **72**: 9224–9232.

Suzuki, M., Sasagawa, A., Inayoshi, T., Nakamura, F., and Honjo, S. 1981. Serological survey for SV5, measles and herpes simplex infections in newly-imported cynomolgus monkeys. *Japanese Journal of Medical Science and Biology* **34**: 69–80.

Switzer, W. M., Salemi, M., Shanmugam, V., *et al.* 2005. Ancient co-speciation of simian foamy viruses and primates. *Nature* **434**: 376–380.

Tsai, C. C., Follis, K. E., Snyder, K., *et al.* 1990. Maternal transmission of type D simian retrovirus (SRV-2) in pig-tailed macaques. *Journal of Medical Primatology* **19**: 203–216.

Vythilingam, I., Noorazian, Y. M., Huat, T. C., *et al.* 2008. Plasmodium knowlesi in humans, macaques and mosquitoes in peninsular Malaysia. *Parasites and Vectors* **1**: 26.

Vythilingam, I., Tan, C. H., Asmad, M., Chan, S. T., Lee, K. S., and Singh, B. 2006. Natural transmission of *Plasmodium knowlesi* to humans by *Anopheles latens* in Sarawak, Malaysia. *Transactions of the Royal Society of Tropical Medicine and Hygiene* **100**: 1087–1088.

Walsh, G. P., Tan, E. V., la Cruz, E. C., *et al.* 1996. The Philippine cynomolgus monkey (*Macaca fasicularis*) provides a new nonhuman primate model of tuberculosis that resembles human disease. *Nature Medicine* **2**: 430–436.

Weigler, B. J., Hird, D. W., Hilliard, J. K., Lerche, N. W., Roberts, J. A., and Scott, L. M. 1993. Epidemiology of cercopithecine herpesvirus 1 (B virus) infection and shedding in a large breeding cohort of rhesus macaques. *Journal of Infectious Diseases* **167**: 257–263.

Weir, E. C., Bhatt, P. N., Jacoby, R. O., Hilliard, J. K., and Morgenstern, S. 1993. Infrequent shedding and transmission of herpesvirus simiae from seropositive macaques. *Laboratory Animal Science* **43**: 541–544.

Welshman, M. D. 1989. Measles in the cynomolgus monkey (*Macaca fascicularis*). *Vet Rec* **124**: 184–186.

White, N. J. 2008. Plasmodium knowlesi: the fifth human malaria parasite. *Clinical Infectious Diseases* **46**: 172–173.

WHO. 2009. Global tuberculosis control – epidemiology, strategy, financing. WHO Report, pp 1–28.

Wilkinson, R. C., Murrell, C. K., Guy, R., *et al.* 2003. Persistence and dissemination of simian retrovirus type 2 DNA in relation to viremia, seroresponse, and experimental transmissibility in *Macaca fascicularis*. *Journal of Virology* **77**: 10751–10759.

Willy, M. E., Woodward, R. A., Thornton, V. B., *et al.* 1999. Management of a measles outbreak among Old World nonhuman primates. *Laboratory Animal Science* **49**: 42–48.

Wolfe, N. D., Escalante, A. A., Karesh, W. B., Kilbourn, A., Spielman, A., and Lal, A. A. 1998. Wild primate populations in emerging infectious disease research: the missing link? *Emerging Infectious Diseases* **4**: 149–158.

Wolfe, N. D., Switzer, W. M., Carr, J. K., *et al.* 2004. Naturally acquired simian retrovirus infections in central African hunters. *Lancet* **363**: 932–937.

Part III

Ethnophoresy of long-tailed macaques

8 Macaca fascicularis *in Mauritius: Implications for macaque–human interactions and for future research on long-tailed macaques*

ROBERT W. SUSSMAN, CHRISTOPHER A. SHAFFER
AND LISA GUIDI

Introduction

Macaca fascicularis is an extremely adaptable species that is found throughout the continental and insular Southeast Asia (Gumert, Chapter 1). In the six-teenth century, long-tailed macaques were introduced by Portuguese or Dutch sailors from Sumatra or Java into Mauritius where they have readily adapted to flora entirely different from that of Asia (Sussman and Tatterall, 1986; Cheke, 1987; Kondo *et al.*, 1993; Tosi and Coke, 2007; Blancher *et al.*, 2008).

Throughout their range, long-tailed macaques inhabit dense primary canopy forest, riverine and coastal forest, mangrove and nipa swamp, as well as secondary forest, and disturbed habitats (Medway, 1970; Southwick and Cadigan, 1972; Kurland, 1973; Rijksen, 1978; Rodman, 1978a; Fittinghoff and Lindburgh; 1980; Wheatley, 1980; Crockett and Wilson, 1980; Fooden, 1995; van Schaik *et al.*, 1996; Fuentes *et al.*, 2005; Ong and Richerson, 2008). They are found from sea level to 2000 m (Rowe, 1996; Supriatna *et al.*, 1996), and are "adaptable opportunists" (MacKinnon and MacKinnon, 1980). For example, Crockett and Wilson (1980) found *M. fascicularis* in 22 of 25 habitat types surveyed in Sumatra, in contrast to eight of 25 exploited by *M. nemestrina*.

Long-tailed macaques also are commonly found in habitats disturbed by humans, including urban and agricultural settings. In an early report, Medway (1970) observed that, of all southeast Asian primates, long-tailed macaques consistently and easily utilize secondary forest. Even within their riverine

Monkeys on the Edge: Ecology and Management of Long-Tailed Macaques and their Interface with Humans, eds. Michael D. Gumert, Agustín Fuentes and Lisa Jones-Engel. Published by Cambridge University Press. © Cambridge University Press 2011.

habitat, they seek out naturally disturbed areas. One can expect to see many groups around villages, cultivated fields, and along rivers. In a similar vein, Kurland (1973) reported these macaques to be "clustered around native villages on the rivers, kra macaques often aggressively indulge in and defend the garbage of human habitation" (see also Southwick and Cadigan, 1972). Long-tailed macaque numbers sometimes increase after logging (Fooden, 1995). Richard *et al.* (1989) described both *M. fascicularis* and *M. mulatta* as "weed" species that exploit early succession, riparian and humanly disturbed areas. They suggested that these species exploited naturally occurring ecological "edges" during the Pleistocene, and that their distributions have been importantly influenced by human disturbances which have so greatly expanded the areas of disturbed secondary habitat in eastern Asia (see also van Schaik *et al.*, 1996). They also surmise that these "weed" or, as we prefer to call them (as did Medway, 1970), "edge" species are "camp-followers" of humans.

Although baboons and chimpanzees are the primates most often utilized in developing models of early human evolution and behavior, for a number of reasons, including their adaptability and habitat choices, *Macaca fascicularis* are also a potentially good model for reconstructing how our earliest ancestors may have lived (see Hart and Sussman, 2009). In fact, *M. fascicularis* is among the most utilized species for medical research. This being so, it is surprising and disappointing that there are relatively few detailed, intensive studies of natural populations of long-tailed macaque ecology and behavior. These types of studies have become exceedingly rare in recent years. Furthermore, no long-term research field sites have been established in which identified individuals have been followed over generations for this species.

Thus, there is little information available on the individual life histories of long-tailed macaques, on demography of populations (see, however, Fuentes *et al.*, Chapter 6), or on the variation in ecology and behavior of this species. In a survey of Google Scholar, we found that of 102 recent papers listed on long-tailed macaques, 84 percent were on medical research, 9 percent on reconciliation or agonistic behavior, 6 percent on other social behavior, and less than 1 percent on ecology or conservation. In 425 papers that referred to *Macaca fascicularis* in Primate Lit of the Primate Info Net maintained by the Wisconsin Primate Research Center (WPRC) Library at the University of Wisconsin-Madison over the past six months, only twelve were primary references on behavior or ecology (and five of these were abstracts). Of these, ten were papers on behavior in captive animals and only two related to ecology. Research on the natural history, ecology, behavior, or long-term status of populations in natural habitats related to long-tailed macaque adaptations or adaptability have become essentially non-existent.

We believe this dearth of research is lamentable. Would it not be informative to know which aspects of behavior vary or are changeable in different locations and habitats and which are fixed and unchanging? What aspects of the behavioral repertoire are adaptable and what aspects are species-specific? Obviously, currently, there are not enough data to answer questions such as these. However, in this paper, we propose to use information available on the introduced population of Mauritian long-tailed macaques to provide a possible framework for approaching these types of questions.

It is likely that the population colonizing Mauritius included less than 100 individuals (Lawler *et al.*, 1995), probably being brought to Europe to be sold for display in menageries which were popular at that time. The macaques have done extremely well in Mauritius, even though the environment is radically different from their place of origin. For example, the species composition of the forests in Mauritius is completely different from that in Southeast Asia. As far as we know, the plant species composing the diet of Mauritian macaques differs almost entirely from those of Asian populations.

We will discuss a few aspects of the ecology and behavior of Mauritian macaques in the hope of generating interest in obtaining comparative data. Where they exist, we will point out a few interesting comparisons between this introduced population and those living in Asia. Finally, we will discuss macaque-human interactions in Mauritius, where they are seen as sacred animals, possible tourist attractions, and an export cash crop for medical research, on the one hand, and as agricultural pests, threats to the endemic flora and fauna, and specialty culinary items on the other.

The topography, climate, and vegetation of Mauritius

Mauritius is an island situated a few degrees north of the Tropic of Capricorn. It is only 62 km long and 45 km wide but has a remarkable climatic and habitat diversity because of dramatic local variation in relief. There is an elevated and undulating central plateau that yields gradually in the east, and more sharply on the other sides of the island, to a coastal plain that is relatively narrow in the west and south, and wider elsewhere. The northern portion of the island is flat and low-lying, and is the only major region that lacks the spectacular erosional remnant massifs that dramatically rise in places to over 800 m. Rainfall is highest on the plateau, peaking at about 5,000 mm/year, and is lowest in the north and west, where it often is below 1,000 mm/year. Seventy percent of the rain falls in the cyclone season, between December and April. Temperatures are higher on the coast than on the plateau (by about 2^0 C), with an average daily maximum on the coast of around 30^0 C and an average daily minimum on the plateau of about 16^0 C.

This varied topography and climate has given rise to an enormous variety of vegetation and habitat diversity, though much of the native vegetation has already been destroyed (Lorence and Sussman, 1986; Safford, 1997). This wide spectrum of native vegetation was characterized in a seminal paper by Vaughan and Wiehe (1937), in which note was made of the extremely high proportion of endemic and Mascarene species present in all indigenous Mauritian plant communities. The upland regions, including the central and southeastern plateaus and mountains above 370 m support evergreen plant habitats of five major types. These include marsh, heath, and thicket formations, and stratified climax and cloud forest. The lowland is divided into two zones. A narrow arid zone (< 1,000 mm/year of rainfall) on the western coastal plain naturally supported a palm savanna, and a moister zone covering the remainder of the lowlands (1,000–2,500 mm/year) that supports several native plant communities, including dry semideciduous forests and thickets, and shrubby beach vegetation. Forests at lower altitudes are less clearly stratified than those of the uplands, are less dense in trees and, though still with a closed canopy, support more growth at ground level.

Much of the indigenous, native vegetation has been cleared for plantations or has been invaded by largely exotic species, which are rapidly supplanting them (Vaughan and Wiehe, 1937; Lorence and Sussman, 1986). Depending upon location and climate, upland forest has been replaced by guava (*Psidium cattleianum*), privet (*Ligutrum robustum*) and similar exotics, or by forests of ravenala (*Ravenala madagascariensis*). Secondary regrowth in lowland areas has produced tree savanna or thorn scrub characterized almost completely by exotic species. Native associations have mostly disappeared, though they survive in a few isolated patches around the island. In Figure 8.1, we give a broad indication of the distribution of the various vegetal types in Mauritius.

In the mid 1980s, somewhat over half of the island, some 105,000 ha was devoted to intensive cultivation, primarily sugarcane. Savanna covered about 5,000 ha; scrub around 39,500 ha; and indigenous scrub with some tropical evergreen formations interspersed about 8,000 ha. Indigenous natural forests accounted for about 2,000 ha and forest plantations for a further 9,000 ha. Other environments capable of supporting monkeys occupied a few thousand hectares at most. Hunting plantations occupied about 28,000 ha of both forest and savanna/scrub lands. At that time, indigenous forest covered only about 5.6 percent of the island's surface (Sussman and Tattersall, 1986). How much this has changed in the past 20 years is not known but these figures are similar to those reported in le Roux (2005). He states that in 1999 just over 50 percent of the land was under sugarcane cultivation and 31 percent was classified as forest and scrub. Lutz and Holm (1993) reported that between 1965 and 1985, the

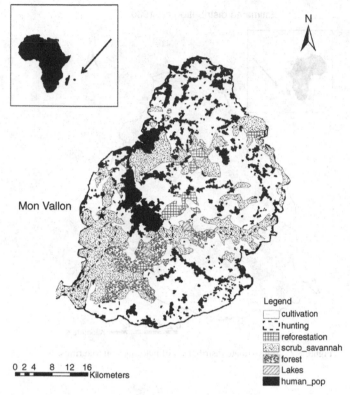

Figure 8.1. Distribution of vegetation types in Mauritius.

area used for growing sugarcane decreased by 15 percent, to cover 45 percent of the total land area. During the same period, urban land use increased by 215 percent, thereby significantly increasing its share of the total island land use. Another significant change was the rapid growth, especially in the late 1980s, of the tourist industry. Some 50 beach hotels now occupy almost 30 km of the coastline, or about 2.9 km². This is a small amount of total island land space, but a very scarce land resource that included rare endemic mangrove vegetation. Lutz and Holm go on to state that, compared to these, other land use changes during this period were minor.

Macaques in Mauritius

The macaques are located in three principal areas in Mauritius, broadly corresponding to three major massif complexes, and areas contiguous with them.

Estimated distribution ca. 1980

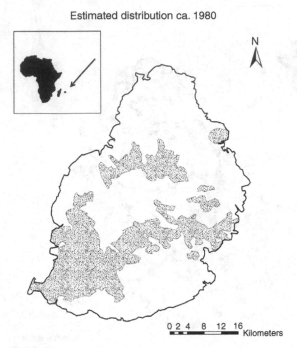

Figure 8.2. Approximate distribution of macaques on Mauritius.

The massif complexes include the "Port Louis Group" in the northwest, close to the capital city, the "Black River group" in the southwest, and the "Bambous group" in the west (Sussman and Tattersall, 1986; Bertram and Ginsberg, 1994). Bertram and Ginsberg (1994) estimated that approximately four-fifths of the areas occupied by the monkeys are connected to one another and provide corridors for the monkeys (Figure 8.2). The monkeys are most numerous and wide-spread in the southwestern portion of the island, an area centering on the Black River group of massifs. Much of this area consists of degraded thicket and savanna formations but on the higher elevations and on the steeper slopes vestiges of indigenous primary forest remain. As elsewhere, the long-tailed macaques on Mauritius are extremely adaptable and inhabit most of the habitat types in which Asian populations are found. The distribution of the macaques on Mauritius is not uniform and different habitats support different densities of monkeys.

Macaca fascicularis in Mauritius are relatively small with long tails, and are sexually dimorphic. Mean weight of adult females was 4.45 kg (3.63–5.67 kg range, N=16) and that of adult males was 7.68 kg (6.80–9.30 kg range, N=8).

Adolescent males weighed an average of 6.09 kg (4.99–7.03 kg range, N=12). Average body length (i.e., base of the tail to nose) for females was 49.67 cm; tail length 51.33 cm and for males was 53.25 cm with a tail length of 57.75 cm. The largest male trapped was 57 cm long with a tail of 65 cm. From a distance the Mauritian macaques appear gray in color. On closer examination the dorsal surfaces are gray with brown and black highlights on the top of the head, back and dorsal surfaces of the legs, arms, and tail. Eyelids are "flashed" as part of facial gestures. All individuals possess a vertical tuft or "top notch" of hair on the head and older females have a white beard (Figure 8.3 a and b). Face and body markings vary such that individuals can be discerned.

From the mid 1980s to the mid 1990s, the population of long-tailed macaques on the whole island of Mauritius was estimated to be around 35,000–40,000 individuals (Sussman and Tattersall, 1986; Bertram and Ginsberg, 1994). However, beginning sometime in the late 1980s and early 1990s, a large number of feral long-tailed macaques began being trapped and shipped to the UK and the US for medical research. Export of these animals has continued to the present, although many of the animals now being exported are the product of controlled breeding populations (see below). Due to trapping for export and to fill captive breeding colonies, the current population of monkeys in Mauritius is unknown, though recent estimates indicate it may as low as 8,000 animals (Guidi and Sussman, 2009). Furthermore, we do not know the precise locations or schedule of trapping activities and, thus, the present distribution and density of the monkeys in various locations on the island needs to be reassessed. We are currently working with three of the macaque research and breeding laboratories (Bioculture Mtius Ltd., Biodia Co. Ltd., and Noveprim Ltd.) to examine the current status of the feral population. Until very recently, in the absence of mammal competitors or of predators, and with their success in human disturbed habitats, the monkeys had thrived in Mauritius.

Our comparative study of the Mauritian macaques

Sussman, his students, and colleagues have studied the macaques on Mauritius over several different time periods. Sussman and Tattersall conducted three separate short-term studies in 1977, 1984, and 2005. Sussman did a relatively long-term study of the ecology and behavior of one group of macaques in western Mautitius for 20 months in 1979–1980. Jamieson (1998) conducted a twelve-month study on diet, foraging and social behavior at the same site in 1980–1981. Guidi did a preliminary reconnaissance study of the distribution and status of the Mauritian monkey population in August 2009 in preparation for her Ph.D. thesis research on the island macaques.

Figure 8.3. (a) Male Mauritian long-tailed macaque; (b) Female Mauritian long-tailed macaque.

The results reported here are mainly from the studies of Sussman and Jamieson, many of which derive from yet unpublished data. Our main study site was Mon Vallon, located at the western base of the Montagne du Rempart, in western Mauritius. The site forms a part of a private reserve owned by Médine Estates Ltd. The reserve is bordered by fields of sugarcane. At the time of the early studies, the reserve was a mosaic of lowland vegetation types interlaced with savanna. To the south, a steep ridge rises and falls away again, the ground is very rocky, and a *Leucaena-Furcraea* thick formation occurred. This was generally an open scrub with a thick undergrowth of aloes and with scattered trees about 4–5 meters in height. At the foot of the southern slope of the hill, and along streams and tributaries running through the site, there was

Figure 8.3. *(cont.)*

evergreen forest, with trees not exceeding 10 meters in height. The canopy was closed and the forest floor was relatively open and shaded. The same species of plants found in the forest were also found in scattered distribution in the savanna and thicket. At the edges of these vegetation types, and scattered within them, were substantial patches of thorn scrub. The plant species were a mix of indigenous and introduced plants, with a number of commercial and ornamental tree species scattered throughout the study area. Other sympatric mammals at the site included deer, wild pigs, mongoose, rabbits, a variety of rodents, and the only endemic mammal, fruit bats (*Pteropus spp*). There were no natural predators of the macaques, but feral dogs were observed on occasion. See Sussman and Tattersall (1981) and Jamieson (1998) for more detailed descriptions of the site.

In 2005, we found that much of the research site had been converted into a captive breeding colony operated by Biodia Co. Ltd., a subsidiary of Médine Estates, Ltd. Even though when we conducted our study the site was in many ways "man-made", it was relatively undisturbed and used seasonally for organized deer hunts. By 2005, Mon Vallon had been converted and there were no free-ranging monkeys remaining at the site.

In order to follow and observe the animals in this diverse and often impenetrable vegetation, we trapped members of two groups, tranquillized the animals, examined and weighed and measured them, and then placed radio collars on 25 individuals (ten from one group and fifteen from the other). Trapping techniques are described in detail in Jamieson, 1998. Other individuals were marked with Chinese hair dye. Radio collaring individuals of different ages and sexes in both resident groups enabled us to follow the groups throughout the day. Data were collected on the ecology and behavior of the animals using five-minute scan samples of all individuals in site. Changes in the location of the animals were noted and, even when the animals were in thick vegetation and out of site, location of the group under observation was obtained from the radio collared individuals. *Ad lib* notes were taken of all occurrences of social interactions or of any unusual behaviors (e.g., interactions with deer, human interference, intergroup interactions). Data were also collected on nearest neighbor and on grooming partners when sex or individual identifications could be determined. The data were collected in all-day sessions or partial-day sessions but were tabulated weekly with the goal of distributing the sampling periods evenly over the day and month. Since groups were very large and moved fairly quickly, census data were collected by two individuals or with a tape recorder as the groups moved through open areas or crossed roads or paths.

Some results and comparisons

Group size and structure

The two groups studied in Mauritius contained at least 85 (south group) and 72 (north group) individuals (Table 8.1). The groups were multi-male/multi-female with a sex ratio of around one male to three females. The sex ratio in some parts of the island, however, is now likely different due to selective capture of females over males. The mean adult to immature ratio is one adult to one to two young. Members of the large groups would leave their sleeping site together but soon split into subgroups throughout the day. They would then

Table 8.1. *Group composition, and mean sex and age ratios of Macaca fascicularis in Mauritius and Asia.*

| | | | | | | Ratios | | |
| | | | | | | Adult | Adult: | |
AM	AF	Adol	J	I	Total	Sex	Immat.	Reference
3.4	8.4	-	7.8	4.2	23.8	1:2.7	1:1.0	Southwick and Cadigan (1972)
2.8	5.0	-	8.0	2.8	18.5	1:1.7	1:1.4	Kurland (1973)
2.5	7.0	2.0	4.5	5.0	21.0	1:2.8	1:1.2	Aldrich-Blake (1980)
7.0	18.0	3.0	20.0	18.0	66.0	1:2.6	1:1.6	Koyama et al., (1981)
3.0	10.0	2.0	13.0	2.0	30.0	1:3.3	1:1.3	Wheatley (1982)
1.7	7.5	-	3.5	3.3	16.0	1:4.6	1:0.7	Khan and Wahab (1983)
4.5	7.6	2.1	7.9	2.7	24.8	1:1.7	1:1.0	van Schaik et al., (1983)
3.3	10.3	1.7	11.0	8.3	34.7	1:3.1	1:1.9	Koyama (1984)
8.0	**23.0**	**5.0**	**39.0**	**10.0**	**85.0**	**1:2.9**	**1:1.6**	**This study. Group S**
6.0	**20.0**	**4.0**	**36.0**	**5.0**	**72.0**	**1:3.3**	**1:1.6**	**This study. Group N**

AM = Adult males.
AF = Adult females.
SAM = Subadult (i.e., adolescent) males.
J = Juveniles
I = Infants

reconvene before returning to their evening sleeping sites. These subgroups were made up of matrilineal groups with some attached males, or young, adolescent and/or subordinate male subgroups (Sussman and Tattersall, 1981; Jamieson, 1998). Over a year-long period, Jamieson found that subgroup size and number correlated with resource availability and with reproductive cycle.

The Mauritian macaque group sizes are among the larger group sizes observed for *Macaca fascicularis*, which range from < 10 to > 85 (Table 8.2). As can be seen from this table, this is one of the characteristics of long-tailed macaques that is highly variable. Generally, the largest groups are found in savanna and thorn scrub vegetation, as in Mauritius, or in urban forests, urban parks, and temples throughout Malaysia and Indonesia. In many urban habitats, the monkeys are fed by people. The smallest groups are found in swamp and mangrove forests. Groups living in secondary forest appear to live in larger groups than those in found in primary forests (Crockett and Wilson, 1980).

Although group size is highly variable from site to site and in different habitat types and ecological conditions, group structure is phylogenetically

Table 8.2. *Variations in home range size (km²) and daily ranges (in meters) of Macaca fascicularis.*

Home Range	Daily Range	Group Size	Reference
2.0	—	± 30	Furuya (1965)
0.8	—	40–50	Poirier and Smith (1974)
0.8	400–1000	16	Kurland (1973)
0.7–1.0	—	16	Rodman (1978a)
0.462	1400	17.0 m	MacKinnon and MacKinnon (1980)
0.35*	150–1500 (760 m)	23	Aldrich-Blake (1980)
1.25	1869	30.0 m	Wheatley (1982)
0.42	—	27.4 m	van Schaik and van Noordwijk (1985)
0.33	—	30–40	Lucas and Corlett (1991)
0.50	—	8–60	de Ruiter and Geffen (2009)
0.57	—	**85**	**This Study**
1.17	—	**72**	**This Study**

* = approximate
m = mean

conservative. As in the Mauritian macaques, groups are multi-male/multi-female with a sex ratio averaging around one adult male to around three adult females (Table 8.1). Other aspects of social structure that are relatively fixed among these macaques are: female philopatry, male emigration from natal groups, and some solitary males and all-male groups. Breaking into subgroups throughout the day also is a common feature among long-tailed macaques throughout their range (Kurland, 1973; Poirier and Smith, 1974; Aldrich-Blake, 1980; Fittinghoff and Lindburgh, 1980; Mackinnon and MacKinnon, 1980; van Schaik *et al.*, 1983; van Schaik and van Noordwijk, 1986).

The conservative features of relatively large, multi-male groups may be related to the predator prone edge, open, and secondary habitats where the majority of naturally occurring long-tailed macaques are found. Males provide a sentinel role. The ability to break up into subgroups is highly adaptable, given the mosaic character of many of these habitats, with a diverse array of resources spread over a variety of types of "resource packages." Matrilines, with male friends, and all-male groups provide natural mechanisms for subgroup formation. The large variation in group size is likely to be related to local resources, environmental conditions related to relatively recent historical events, and the macaque populations' demographic history correlated with these events. Now, for example, group sizes are likely to be very small in parts of Mauritius, due to intense capture rates. The ability of these animals to maintain populations with groups of very different sizes speaks to their ability to inhabit such a great variety of habitats.

Table 8.3. *Mauritian macaques nearest neighbors n=3048*

	F/I	F	I	J	M	SAM
F/I	31.78*	17.76	1.4	40.19*	7.48	1.40
F	3.01	20.43	2.76	41.60*	30.20*	1.75
I	0.00	12.50	25.00*	37.50*	12.50	12.50
J	1.68	16.21*	0.76	68.96*	10.70	1.68
M	5.39	40.60*	0.69	27.27*	22.27	3.16
SAM	0.00	22.22*	0.00	44.44*	13.89	19.44

* Higher than predicted by chance.

Social organization – interactions within groups

Patterns of social interaction among different sexes and age classes of individual Mauritian long-tailed macaques can be seen in nearest neighbor and grooming data. Besides juveniles who stay close to all classes of individuals, mothers
with infants remain near one another, and adult females without infants are
most frequently seen close to adult males (Table 8.3). However, 70 percent of
mutual grooming occurs among females or females and juveniles (Table 8.4).
These patterns reflect the matrilineal core of groups in long-tailed macaques,
which have been referred to as "mother clubs" (Fady, 1969).

This type of social organization is characteristic of *Macaca fascicularis*.
As found elsewhere, Mauritian groups are organized around matrilines, with
mothers, daughters, and sisters maintaining close social ties (van Schaik and
van Noordwijk, 1988; Jamieson, 1998; de Ruiter and Geffen, 1998). There
is a dominance order among females. Males migrate, both from their natal
group and after they reach adulthood (Wheatley, 1982; Jamieson, 1998; van
Noordwijk and van Schaik, 2001, Sussman, personal observation). Males
are dominant to females at the individual level but groups of closely related
females can have some level of dominance over males (matrilines create interesting dynamics in a group). In the groups that we studied there was one dominant or two codominant males at any particular time.

The role of dominant male changed hands a number of times in both study
groups and these changes correlated with both migration and aggressive encounters. However, the individual taking over the dominant position could not be
predicted by his previous status or behavior. For example, in south group, two
males, Patch and Gorilla, were codominant at the beginning of the study and for
many months. These two males had very different personalities. Patch was very
social and interacted with adult females and younger animals, often grooming and playing with juveniles. Gorilla was much less social and had a more

Table 8.4. *Mutual grooming among Mauritian macaques.*

	#	%
F-J	72	47%
F-F	35	23%
J-J	26	17%
M-J	9	6%
F-M	7	5%
SAM-J	2	1%
SAM-M	1	0.05%
SAM-SAM	1	0.05%
Total	n = 153	100%

aggressive demeanor. Halfway through the study, Patch was injured (cause unknown) and did not continue to participate in dominance interactions. After this, Gorilla was harassed more by lower ranking males and by one immigrating male in particular. After a few weeks and while still dominant, he left the group of his own volition. A younger adult male, #1, took over dominance. This male had not been one of the males harassing Gorilla and, in fact, he had a very calm, nonaggressive, permissive personality. Rank below the dominant, central males was neither consistent nor stable. van Noordwijk and van Schaik (2001) hypothesize that "transfer by males is a self-chosen option with a high probability of improving a male's reproductive opportunities" and that "males show sophisticated assessment and decision-making." We agree with this assessment but would add that many factors seem to affect migration patterns at any given time. Long-term studies of individual life histories would be needed to assess how migration patterns relate to lifetime reproductive patterns.

It appears that the general social organization of these macaques is a relatively fixed trait. However, particular patterns of social interaction, introducing variation around this theme, appear to be related to specific personalities of individuals, to demographic conditions, to environmental conditions and perturbations, and to interactions with humans. Unfortunately, unlike with Japanese macaques and the introduced rhesus monkeys in Cayo Santiago, Puerto Rico, there have been few long-term studies involving known genealogies for long-tailed macaques.

Activity cycles

At Mon Vallon, feeding was the most common macaque activity (30 percent of observation time, with foraging an additional 2 percent) followed by moving (23 percent), resting (22 percent), grooming (13 percent), other (primarily social behavior 5 percent), and travel (5 percent) Overall, the Mauritian macaque activity patterns were very consistent throughout the day. There was a slightly bimodal feeding pattern with increases in feeding from 5 a.m. to 7 a.m. and 2 p.m. to 5 p.m.. However, feeding never represented less than 20 percent of the hourly macaque activity budget. Macaques spent significantly more time feeding during the wet season (December–April) than during the dry season (Wilcoxon W=33, p=0.42). Otherwise, they were consistent in their activity patterns throughout the year.

The Mauritian study groups returned to the same sleeping sites every night for the entire study. The sleeping site of the south group was at the westernmost point of the monkey's range, on the side of a hill overlooking a seasonal marsh (see Jamieson, 1998). It was covered with large rocks and medium sized trees in which the macaques groomed, rested and slept. They slept on the hill where the ground cover was most dense. Access to this area was difficult for people because of the loose volcanic rock, which undermined footing, and the thick tangle of thorny vines and small, often spiny tree saplings. The animals would spend time grooming and in social activities prior to leaving the sleeping site in the morning and, especially, after arriving at the sleeping site each evening.

Generally, *Macaca fascicularis* activity cycles are quite variable from site to site and are affected by food resources, habitat, human activity, and possibly group size (Ashmore, 1992). MacKinnon and MacKinnon (1980) observed an activity cycle similar in Malayan long-tailed macaques, with their monkeys exhibiting high levels of feeding and moving throughout the day. At both the Mauritian and Malayan study sites, food resources were patchily distributed over a wide area. At other sites in Malaysia (Aldrich-Blake, 1980) and Angaur (a.k.a. Ngeaur) Island (Poirier and Smith, 1974), long-tailed macaques were observed to feed more during early and late afternoon and to have more intense resting periods during midday when heat is most intense.

The amount of time spent in various activities differs in some of the study sites (Table 8.5). *M. fascicularis* spends more time feeding than in other activities in most sites. However, at his site in Indonesia, Wheatley (1980) observed his animals to spend most of their time traveling and moving. The long-tailed macaques living along Naaf River in Bangladesh spend little time moving (Khan and Wahab, 1983). At this location, the monkeys spent much of their time sitting in mangrove trees waiting for low tide as they hunt for food along the shoreline. Bernstein (1968) studied groups that lived near a Catholic school and its refuse bins. Here, the monkeys' activity patterns corresponded to the school schedule, and their schedule changed in relation to human activities. At Mon Vallon, the animals would also often modify their normal schedule in response to human behaviour. For example, they would rest and hide in dense

Table 8.5. *Activity budgets of Macaca fascicularis*

Feed	Move	Travel	Rest	Groom	Play	Other	Total	Reference
28	—	30	42	—	—	—	100.0	Rodman (1973)
35.0	20.0	—	34.0	12.0	—	—	101.0	Adrich-Blake (1980)
32.0	23.0	5.0	22.0	13.0	—	5.0	100.0	Sussman and Tattersall (1981)
13.0	—	45.0	42.0	—	—	—	100.0	Wheatley (1982)
55.0	15.4	—	13.6*	6.0	10.0	—	100.0	Khan and Wahab (1983)

* Combined values for rest (3.3%) and sit (10.3%) categories.

vegetation if deer hunting parties were close by. They also would raid the sugarcane fields only in early morning or late evening, when the fields were not being tended and when no hired human monkey-guards were present

As mentioned above, throughout its range, *M. fascicularis* has been described as a riverine "refuging" species (Hamilton and Watt, 1970), departing and returning to a central place, usually near a river, each day (Fittinghoff and Lindburgh, 1980; Wheatley, 1978, 1980, 1982; Sussman and Tattersall, 1986; van Schaik *et al.*, 1996). Among these macaques, although activity cycles are quite variable from location to location, refuging is a common characteristic. This behavior is quite rare among primates and further research into the factors affecting activity cycle and refuging in this species would be a worthwhile endeavor. It is another feature that makes these macaques a potentially interesting model for early human behavioral evolution.

Locomotion and choice of forest strata

In Mon Vallon, the monkeys were very versatile in their locomotion and use of different forest strata. They were mostly terrestrial, with over 70 percent of time spent on or just above the ground during all activities except feeding. All traveling (group displacement) was done on the ground. This was necessary because of the absence of contiguous forest canopy. In addition, over 50 percent of all movement and grooming occurred on the ground. On the other hand, the monkeys fed in all levels of the forest and over 70 percent of feeding was in small and larger canopy trees. They also were frequently seen in water searching for food (e.g., snails and water plants), swimming, and playing.

In most of SE Asia, especially where sympatric with pig-tailed macaques, long-tailed macaques are almost exclusively arboreal. Elsewhere they are primarily arboreal but readily are able to exploit terrestrial habitats. In Malaysia and Indonesia, for example, *M. fascicularis* characteristically travels and flees in the trees (Furuya, 1965; Kurland, 1973; Rodman, 1978a, 1978b, 1991; Aldrich-Blake, 1980; Wheatley, 1980, 1982; Crockett and Wilson, 1980). Wheatley (1980) observed long-tailed macaques to spend 97 percent of the time in the trees while foraging, and Crockett and Wilson (1980) found these macaques in the trees in 71 percent of encounters with them. Even in these sites, however, long-tailed macaques forage on the ground along rivers and on the mudflats of swamps and the sea (Crockett and Wilson, 1980; Wheatley, 1982; Khan and Wahab, 1983; van Schaik *et al.*, 1996). At other sites, as in our Mauritian site, where continuous canopy is not available, long-tailed macaques spend a great deal of time on the ground (e.g., Poirier and Smith, 1974; Spencer, 1975). Also, as in our site, *M. fascicularis* is often found living in close proximity to water and is an accomplished swimmer (Spencer, 1975; Poirier and Smith, 1974; Fittinghoff and Lindburgh, 1980; Son 2003). Long-tailed macaques have even been observed catching and eating fish (Stewart *et al.*, 2008).

Diet and feeding behavior

The Mauritian macaques exploited more than 55 plant species during the study and ate a variety of different food items. They were predominantly frugivorous, with fruit (17 percent), pods (10 percent), and seeds (11 percent) accounting for 38 percent of feeding time. Other important food sources were leaves (including petioles and buds) (22 percent), flowers (9 percent), sugarcane (6 percent), and grass (5 percent). The macaques also ate insects, bark, snails, mushrooms, exudates, grass, water plants and molasses. Six species (*Leucaena leucocephala* [14 percent], *Acacia concinna* [13 percent], *Tamarindus indica* [7 percent], sugarcane [7 percent], *Mangifera indica* [6 percent], and *Psidium sp* [5 percent]) made up 50 percent of the diet (Figure 8.4).

The monkeys were also highly selective in their choice of food items. Of the 27 species for which selection ratios could be calculated, the macaques selected fourteen more often than would be expected based on their relative density. The macaques exploited a few species intensively throughout the study (e.g., *Acacia concinna* and *Leucaena leucocephala*) but most species only made up a high portion of the diet during one or two months. The average number of species consumed each month was 22 and the maximum was 31. An average of 4.23 species made up over 60 percent of the diet each month. Dietary diversity (Sussman, 1987) ranged from 1.9–2.8 with an average of 2.4. The macaques increased dietary breadth during the dry season with the number

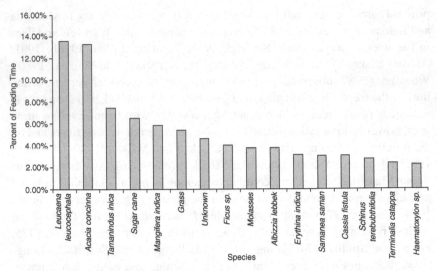

Figure 8.4. Main species consumed by Mauritian macaques.

of species making up more than 50 percent (Wilcoxon W=20, p=0.02) and 60 percent (W=22.5, p=0.05) of the diet significantly higher in the dry season than wet season.

While fruit made up a high percentage of the macaque diet throughout the study, the consumption of leaves, flowers, grass, and cultivated food items (i.e., sugarcane and molasses) appeared to be related to availability. Data on food availability were only available for seven months, making significant correlations difficult to detect. During these months, percent folivory was significantly correlated with the availability of young leaves at the p<0.10 level (Spearman's rho=0.68, p=0.09). The monkeys spent a significantly higher percentage of time feeding on flowers (W=28, p<0.01), and stems (W=30, p=0.01) during the wet season and grass during the dry season (W= 20, p=0.04).

In sum, the Mauritian macaques were eclectic feeders, exploiting a large number of different plant species and a variety of food items. However, they were also highly selective, concentrating on only a few species at any one time and showing a marked preference for certain species. The monkeys were highly frugivorous and were able to eat large amounts of fruit regardless of availability by exploiting pods, seeds, mature, and immature fruit. They supplemented their diet with a diverse array of food items and fed intensively on leaves, flowers, grass, and cultivated crops as these became available. Overall, this dietary pattern is very similar to that reported in the few comparable studies of wild *Macaca fascicularis* populations in Southeast Asia. The species is mostly frugivorous, with fruit making up from 40 percent to over 80 percent of its diet, except in highly provisioned populations or highly disturbed environments (Table 8.6). Like the Mauritian macaques, other populations are eclectic feeders that exploit a variety of different food items. In addition, they show a high level of short-term selectivity, eating from a large number of species but concentrating on a few at any one time.

A twelve-month study of the diet of *M. fascicularis* in a Borneo swamp forest by Yeager (1996) offers an excellent comparison to our study of Mauritian macaques. The Borneo macaques had an average dietary diversity of 1.91 and showed a preference for thirteen of nineteen species for which she could calculate selection ratios. The monkeys ate from more than thirty-three plant species but relied on five species (*Ganua motleyan, Diospyros ainguyi, Licania splenens, Eugenia sp.* and *Garcinina sp.*), accounting for more than 60 percent of their diet. Although the plant species eaten by the Borneo population were completely different from those exploited by the Mauritian macaques, the types of items eaten, the degree of dietary diversity, and the degree of selectivity are very similar. We believe that these features of *M. fascicularis* feeding ecology represent a species-specific dietary pattern (Sussman, 1987) and can be related to its wide biogeographical dispersion.

Home and day-ranging behavior

The average day range for the study groups in Mauritius was 3428 meters (1132–7696 m) and the average home range was 0.76 km² The straight-line distance groups spent from their sleeping site averaged 374 m (0–1044 m), with a maximum average distance of 656 m. The amount of time per day spent away from the sleeping site averaged 11.7 hours (8.8–14.0 hours) (Jamieson, 1998). As stated above, the groups returned to the same sleeping sites every day throughout the eighteen-month study period.

Movement of the groups to sites away from the sleeping sites appeared to be related to a number of factors. Fruit availability correlated with site choice only during periods when fruit was least abundant. Other factors, such as human activity, ripeness of sugarcane, amount of shade, proximity to water, resting or sleeping locations, or to other groups, influenced choice of sites visited throughout the day. Seasonal changes in subgroups, interactions and dispersion, male migration and group fission were related more to changes in social behavior associated with the macaque reproductive cycle than to environmental factors *per se* (Jamieson, 1998).

One of the most interesting things about long-tailed macaque behavior is their propensity for returning to the same sleeping site each evening. This gives

Table 8.6. *Percentage of plant parts in the diet of various long-tailed macaque populations. Data below "this study" represent populations in highly disturbed environments (after a recent fire) or in locations where the diet includes a high proportion of provisioned foods.*

Study	Fruit	Seed	Pod	Total Fruit	Leaf	Grass	Flower	Insect	Prov.	Other/?
Supriatna et al., 1996				50.0	16.7					
Rijksen 1979				60.0	35.0					
Lucas and Corlett 1991*				77.0	12.0		11.0			
Yeager 1996				66.7	17.2		8.9	4.1		
Wheatley et al., 1996 (Kalimantan)				87.0	2.0	2.0	3.0	4.0		
This study	**18.9**	**13.0**	**10.3**	**42.2**	**25.7**	**5.1**	**7.0**	**2.4**		**18.7**
Berenstein 1986 pre-fire				65.0	9.0		26.0	<1.0		
Berenstein 1986 post-fire				14.0	46.0		<1.0			
Lucas and Corlett 1991				44.0	8.0		7.0		14.0	27
Wheatley et al., 1996 (Bali)				18.0	5.0	11.0	2.0	12.0	42.0	
Son 2004				28.0	12.0	4.0		33.0	25.0	
Fuentes et al., 2005				12.0				7.4	69.0	10.3

groups a reliable, presumably secure and safe place to return from their daily activities. It enables groups to split into subgroups during the day and reunite each evening. A great deal of the social activity takes place each morning and evening at these sites. However, this characteristic of ranging also restricts the size of the home ranges and the pattern of day range use, since the animals must schedule their day in a way that enables them to return to this site each evening. Average day ranges of between 150–1,900 meters and home ranges ranging from 0.42–2.0 km² have been reported from the few sites for which they are recorded, outside of Mauritius (see Table 8.2).

The ranging behavior of long-tailed macaque groups is one of the most intriguing aspects of their behavior. Such strict refuging behavior is unusual among primates. How widespread is this behavior among long-tailed macaques and their closest relatives? What are the factors that determine when refuging occurs or not? What other animals typically exhibit this behavior and under what circumstances? We believe that this is a subject that deserves a great deal of further study.

Interactions between the macaques and humans on Mauritius

Macaques and humans arrived on Mauritius relatively late and both have inhabited the island for approximately the same amount of time. The macaque population has grown with the human population. In fact, Carié (1916) attributed the abandonment of Mauritius by the Dutch in 1712 to the damage caused to their plantations by the rats and monkeys.

The current relationship between these two mammals is complex. On the one hand, a portion of the population considers the monkeys sacred and, in certain areas, food offerings are left for the monkey god. This is especially true around the Hindu temple at Grand Bassin where the local feral monkeys are protected and fed. The monkeys are of considerable interest to tourists and tourism is one of the major economic industries of Mauritius. Mauritians are aware of the extreme decrease in the monkeys in recent years. One man in particular – a Port Louis resident – was somewhat distraught over the fact since, until recently, he made extra income taking the tourists to see them. The monkeys also are kept as pets. This is a problem for monkeys and humans alike. Just as with other primates, the macaque adults are unruly and dangerous. It is at this point that they are invariably released and, being unable to join other populations, they remain in the towns, stealing food from humans. Some of these individuals are captured by locals and killed for their meat. Since these monkeys are tame, they are easy targets. However, for the most part the monkeys live in remote habitats and do not interact with the local people.

In the more natural habitats, the monkeys are seen as destructive pests. They raid crops, help to destroy indigenous vegetation, help spread invasive introduced plant species (such as guava and privet), and are thought to be a major contributor to the possible extinction of indigenous species of birds (Gleadow, 1904; Owadally, 1980; Cheke, 1987; Jones and Owadally, 1988; Carter and Bright, 2002; Global Invasive Species Data Base, 2007; Bunbury *et al.*, 2008a). In fact, the macaques have long been implicated as one of the major factors leading to the destruction and near extinction of the highly endangered pink pigeon (*Columba mayeri*) (Durrell, 1973; Swinnerton, 2001). They are thought to opportunistically feed on the eggs of the pigeon as well as any adults they are able to capture. Most references to the pink pigeon suggest that predation by macaques and other introduced mammals is one of the greatest threats to their survival. However, very few, if any systematic studies have been conducted to determine the rate at which predation occurs or the degree to which it affects the pigeon population.

Although it is highly probable that predation of the pink pigeon by the macaques does occur on occasion, it is unlikely to be one of the major threats facing these rare birds. In a three-year study on the island, Bunbury *et al.*, (2008b) found that over half of all pink pigeon deaths were due to parasite infection and only 9 percent were due to predation by macaques, rats, mongooses and feral cats. Given the number of studies conducted on parasite infection (i.e., Bunbury *et al.*, 2007, 2008a; Swinnerton *et al.*, 2005), it is clearly a threat to the remaining pigeon population and is perhaps exacerbated by a lack of genetic diversity due to inbreeding depression (Swinnerton *et al.*, 2004). Complete destruction of the macaques on the island of Mauritius would therefore not ensure the continued survival of the pink pigeon and is unjustified as the main argument for their capture and export as biomedical subjects.

Throughout the island, the monkeys are hunted as pests and for food and, in recent years, they have been poached and sold to some of the less reputable biomedical labs. Monkey curry (Carri de Jocot or Curry No. 2) is a popular dish, often served as a specialty meal during weekend fairs at Catholic Churches (referred to as Fancy Fairs, Figure 8.5), though we have never observed monkey on a restaurant menu. Most people in Mauritius rarely see or interact with the monkeys and there have been no detailed, systematic studies of the interface between the two species or on the attitudes of the general population towards the monkeys. We intend to conduct such a study.

Recently, the macaques have taken on a new and different role in Mauritius. Their value for medical research has been discovered. Since the early 1980s, the monkeys have been exported for this purpose and are a valuable commodity. It appears that the Mauritian macaques are relatively free from natural

Figure 8.5. An advertisement from a Mauritian newspaper giving notice of a weekend fair at a Catholic church featuring the local specialty of curry No. 2 or monkey curry.

infections by various pathogenic agents (Matsubayashi, 1992). In 2008, of the 28,091 primates imported into the US for medical research 26,512 (94.3 percent) were long-tailed macaques. The second leading exporter of this species is Mauritius (behind China). The UK also imports a relatively large number of long-tailed macaques, mostly originating from Mauritius. In 2008, 5,396 long-tailed macaques were imported into these two countries. A similar number, entered the US and UK in 2007 (Thirlray, 2008, 2009). The country expects a total of around 10,000 monkeys to be exported each year for an annual income to Mauritius of about 20 million USD (Pappiah, 2001). In Mauritius, there are two major companies that only export animals born in breeding facilities. However, other exporters are interested in the more immediate rewards and export wild-caught animals. The effect this has had on the natural population over the last fifteen years is unknown but expected to be drastic. This, of course, is not good for the long-term sustenance of this new national economic product.

Thus, the status of the Mauritian macaques is currently questionable. Given their negative influences on the island, there is little enthusiasm to protect them. However, their religious importance, their value for tourism, their research interest as a model for ecology, biology, and anthropology, their humane treatment, and the importance of maintaining a viable population for breeding purposes for sustainable export necessitate some compromise and a long-term, feasible plan to maintain some natural populations of this intriguing and important primate. Furthermore, complete eradication of the monkeys would be next to impossible as well as extremely costly. A much more sound strategy is to control the population.

This is a valuable, interesting, extremely social and intelligent animal from which we can learn a great deal. This knowledge should not be restricted to our interest in our own health, but also to our understanding of non-human and human behavior, ecology and evolution. We realize that a solution to saving some "natural" populations on this island will not be simple or easy, but we hope that a solution to this problem will be attempted and achieved.

Acknowledgements

Financial support was provided by NSF grants BNS-7916561 and BNS-7917931, Explorers Club of New York, the Boise Fund, Sigma Xi, the Scientific Research Society, the Richard Lounsbery Foundation, the National Geographic Society, the Frederick Voss Fund of the American Museum of Natural History, and Washington University. We are grateful to the Médine Sugar Estate Co. Ltd., the Cie Sucrière de Bel Ombre Ltd., Bioculture Ltd., Noveprim Ltd., Biodia Ltd., Patricia Koenig, Sandra Lefin, David Lagane, Eric Baboo and Nada Padayatchy for their assistance and helpful advice. We thank two anonymous reviewers and the editors for their excellent suggestions. We also thank Cassie Snyder for her assistance in the preparation of this manuscript.

References

Aldrich-Blake, F. P. G. 1980. Long tailed macaques *(Macaca fascicularis)*. In *Malayan Forest Primates: Ten Years's Study in Tropical Rain Forest.* D. J. Chivers (ed.). New York: Plenum Press. pp. 147–166.

Ashmore-DeClue, P. C. 1992. Macaques: An adaptive array (a summary and synthesis of the literature on the genus *macaca* from an ecological perspective. Ph.D. thesis, Washington University, St. Louis, MO.

Berenstein, L. 1986. Responses of long-tailed macaques to drought and fire in Eastern Borneo: A preliminary report. *Biotropica* **18**: 257–262.

Bernstein, I. S. 1968. Social status of two hybrids in a wild troop of *Macaca irus*. *Folia Primatol.* **8**: 121–131.

Bertram, B. and Ginsburg, J. 1994. *Monkeys in Mauritius: potential for humane control*. Report to the Zoological Society of London commissioned by the RSPCA.

Blancher, A., Bonhomme, M., Crouau-Roy, B., Terao, K., Kitano, T., and Saitou, N. 2008. Mitochondrial DNA sequence phylogeny of 4 populations of the widely distributed cynomolgus macaque (*Macaca fascicularis fascicularis*). *Journal of Heredity* **99**: 254–264.

Bunbury, N., Barton, E., Jones, C. G.., Greenwood, A. G., Tyler, K. M., and Bell, D. J. 2007. Avian blood parasites in an endangered columbid: Leucocytozoon marchouxi in the Mauritian Pink Pigeon Columba mayeri. *Parasitology* **134**: 797–804.

Bunbury N., Jones C. G., Greenwood A. G., and Bell D. J. 2008a. Epidemiology and conservation implications of *Trichomonas gallinae* infections in the endangered Mauritian pink pigeon. *Biological Conservation* **141**: 153–161.

Bunbury, N., Stidworthy, M., Greenwood, A., *et al.* 2008b. Causes of mortality in free-living Mauritian pink pigeons Columba mayeri, 2002–2006. *Endangered Species Research* **9**: 213–220.

Carié, P. 1916. L'acclimatation a l'ile Maurice. Premiere partie. Mammiferes et Oiseaux de Maurice. *Bulletin de la Societe Nationale D'Acclimatation de France*. Maretheaux, Paris.

Carter, S. P. and Bright, P. W. 2002. Habitat refuges as alternatives to predator control for the conservation of endangered Mauritian birds. In *Turning the Tide: The Eradication of Invasive Species*, C.R. Veitch and M.N. Clout (eds.) Switzerland: IUCN SSC Invasive Species Specialist Group. pp. 71–18.

Cheke, A. S. 1987. An ecological history of the Mascarene Islands with particular reference to extinctions and introductions of land vertebrates. In *Studies of Mascarene Island Birds*. A.W. Diamond (ed.). Cambridge University Press. pp. 5–89.

Crockett, C. M. and Wilson, W. L. 1980. The ecological separation of *M. nemestrina* and *M. fascicularis* in Sumatra. In *The Macaques: Studies in Ecology, Behavior and Evolution*, D. G. Lindburgh (ed.). New York: Van Nostrand Reinhold. pp. 148–181.

de Ruiter, J. R. and Geffen, E. 1998. Relatedness of matrilines, dispersing males and social groups in long-tailed macaques (*Macaca fascicularis*). *Proceedings. Biologial Sciences/The Royal Society* **265**: 79–87.

Durrell, G. 1977. *Golden Bats and Pink Pigeons*. New York: Simon and Schuster.

Fady, J. C. 1969. Les jeux sociaux: Le compagnon de juex chez les jeunes. Observations chez *Macaca irus*. *Folia Primatologica* **11**: 134–143.

Fittinghoff, N. A. and Lindburgh, D. G. 1980. Riverine refuging in East Bornean *M. fascicularis*. In *The Macaques: Studies in Ecology, Behavior and Evolution*, D. G. Lindburgh (ed.). New York: Van Nostrand Reinhold. pp 182–214.

Fooden, J. 1995. Systematic review of Southeast Asian longtail macaques. *Fieldiana (Zoologica)* **81**: 1–206.

Fuentes, A., Southern, M., and Suaryana, K. 2005. Monkey forests and human landscapes: Is extensive sympatry sustainable for *Homo sapiens* and *Macaca fascicularis* on Bali. In: *Commensalism and Conflict: The Primate–Human Interface.* J. Patterson and J. Wallis (eds.). Norman, OK: The American Society of Primatoligists Publications. pp. 168–195.

Furuya, Y. 1965. Social organization of crab-eating macaques. *Primates* 6: 285–336.

Global Invasive Species Data Base. 2007. issg database: Ecology of *Macaca fascicularis*. Available at www.issg.org/database/species/ecology.asp?si.

Gleadow, F. 1904. Report on the forests of Mauritius with a preliminary working plan. Port Louis, Mauritius: Government Printer.

Guidi, L. M. and Sussman, R. W. 2009. Survey and recensus of the long-tailed macaques (*Macaca fascicularis*) of Maurtius. Paper delivered at the Midwest Primate Interest Group Annual Meeting, October 2009.

Hamilton, W. J. III, and Watt, K. E. F. 1970. Refuging. *Annual Review of Ecology, Evolution and Systematics* 1: 263–286.

Hart, D. and Sussman, R. W. 2009. *Man the Hunted: Primates, Predators, and Human Evolution* (expanded edn.). Boulder, CO: Westview.

Jamieson, R. W. 1998. The effects of seasonal variation in fruit availability on social and foraging behavior of *macaca fascicularis* in Mauritius. Ph.D. thesis, Washington University, St. Louis, MO.

Jones, C. G. and Owadally A. W. 1988. The life histories and conservation of the Mauritius Kestrel *Falco purictatus*, Pink Pigeon *Colomba mayeri* and Echo Parakeet *Psittacula eques*. *Proceedings of the Royal Society of Arts and Sciences of Mauritius* 5: 80–131.

Khan, M. A. R. and Wahab, M. A. 1983. Study of eco-ethology Of The crab-eating macaque, *Macaca fascicularis*, in Bangladesh. *Journal of the Asiatic Society Bangladesh* 9: 101–109.

Kondo, M., Kawamoto, Y., Nozawa, K., *et al.* 1993. Population genetics of crab-eating macaques (*Macaca fascicularis*) on the island of Mauritius. *American Journal of Primatology* 29: 167–182.

Koyama, N., Asnan, A., and Natsir, N. 1981. Socio-ecological study of the crab-eating monkeys in Indonesia. *Kyoto University Overseas Research Report of Studies on Indonesian Macaque. I.* Kyoto University Primate Institute, pp. 1–10.

Koyama, N. 1984. Socio-ecological study of the crab-eating monkeys at Gunung Meru, Indonesia. *Kyoto University Overseas Research Report of Studies on Asian Nonhuman Primates.* 3:17–36.

Kurland, J. A. 1973. A natural history of Kra macaques (*Macaca fascicularis*, Raffles, 1821) at the Kutai Reserve, Kalimantan, Timur, Indonesia. *Primates* 14: 245–262.

Lawler, S. H., Sussman, R. W., and Taylor, L. L. 1995. Mitochondrial DNA of Mauritian Macaques (*Macaca fascicularis*): An example of the founder effect. *American Journal of Physical Anthropology* 96: 113–141.

le Roux, J. J. 2005. Soil erosion prediction under changing land use on Mauritius. Master of Science (Geography). University of Pretoria, Pretoria.

Lorence D. H., and Sussman, R. W. 1986. Exotic species invasion into Mauritius wet forest remnants. *Journal of Tropical Ecology* 2: 147–162.

Lucas, P. W. and Corlett, R. T. 1991. Relationship between the diet of *Macaca fascicularis* and forest phenology. *Folia Primatologica* **57**: 201–215.

Lutz, W. and Holm, E. 1993. Mauritius: population and land use. In *Population and Land Use in Developing Countries: Report of a Workshop*. National Research Council, Washington, DC, National Academy Press. pp. 98–105.

MacKinnon, J. R. and MacKinnon, K. S. 1980. Niche differentiation in a primate community. In *Malayan Forest Primates*, D. J. Chivers (ed.). New York: Plenum Press. pp. 167–189.

Matsubayashi, K., Gotoh, S., Kawamoto, Y., *et al.* 1992. Clinical examinations on crab-eating macaques in Mauritius. *Primates* **33**: 281–288.

Medway, Lord. 1970. The monkeys of Sundaland. In *Old World Moneys: Evolution, Systematics and Behavior*. J. R. Napier and P. H. Napier (eds.), New York: Academic Press. pp. 513–553.

Ong, P. and Richardson, M. 2008. *Macaca fascicularis ssp. fascicularis*. IUCN 2009. IUCN Red List of Threatened Species. Version 2009.1. www.iucnredlist.org.

Owadally, A. W. 1980. Some forests pests and disease in Mauritius. *Revue d'agriculture sucerie de l'ile Maurice* **59**: 76–10

Pappiah, H. 2001. Forestry outlook studies in Africa. Mauritius: Ministry of Natural Resources and Tourism.

Poirier, F. E. and Smith, E. O. 1974. The crab-eating macaques of Anguar island, Paulau, Micronesia. *Folia Primatologica* **22**: 258–306.

Richard, A. F., Goldstein, S. J., and DeWar, R. E. 1989. Weed macaques: the evolutionary implications of macaque feeding ecology. *Intlernational Journal of Primatology* **10**: 569–594.

Rijksen, H. D. 1978. *A Field Study on Sumatran Orang utans (Pongo pygmaeus abelli Lesson 1827)*. Wageningen, The Netherlands: Beenman and Zonen.

Rodman, P.S. 1973. Synecology of Bornean Primates, I. A Test forInterspecific Interactions in Spatial Distribution of Five Species. *American Journal of Physical Anthropology* **38**: 655–659.

1978a. Diets, densities, and distributions of Bornean primates. In *The Ecology of Arboreal Folivores*. G. G. Montgormery (ed.). Washington DC: Smithsonian Institute Press. pp. 465–478.

1978b. Food distribution and terrestrial locomotion of crab-eating and pig-tailed macaques in the wild. (Abstract) *American Journal of Physical Anthropology* **47**: 157.

1991. Structural differentiation of microhabitats of sympatric *Macaca fascicularis* and *Macaca nemestrina* in East Kalimantan, Indonesia. *International Journal of Primatology* **12**: 357–375.

Rowe, N. 1996. *The Pictorial Guide to the Living Primates*. Hampton, NY: East Pogonias Press.

Safford, R. J. 1997. A survey of the occurance of native vegetation remnants on Mauritius in 1993. *Biological Conservation* **80**: 181–188.

Southwick, C. H., and Cadigan, F. C. 1972. Populations studies of Malaysian primates. *Primates* **13**: 1–18.

Spencer, C. 1975. Interband relations, leadership behavior and the initiation of human-oriented behavior in bands of semi-wild free-ranging *Macaca fascicularis*. *Malaysian Nature Journal* **29**: 83–89.

Son V. D. 2003. Diet of *Macaca fasicularis* in a mangrove forest. *Vietnam Laboratory Primate News* **42**: 1–5.

2004. Time budgets of *Macaca fascicularis* in a mangrove forest. *Vietnam Laboratory Primate News* **43**: 1–4.

Stewart, A.-M. E., Gordon, C. H., Wich, S. A., Schroor, P., and Meijaard, E. 2008. Fishing in Macaca fascicularis: A rarely observed innovative behavior. *International Journal of Primatology* **29**: 543–548.

Supriatna, J., Yanuar, A., Martarinza, *et al.* 1996. A preliminary survey of long-tailed and pig-tailed macaques (*Macaca fascicularis* and *Macaca nemestrina*) in Lampung, Bengkulu, and Jambi provinces, southern Sumatera, Indonesia. *Tropical Biodiversity* **3**: 131–140.

Sussman, R. W. 1987. Species-specific dietary patterns in primates and human dietary adaptations. In *The Evolution of Human Behavior: Primate Models*. W. G. Kinzey (ed.), New York, SUNY Press. pp. 151–179.

Sussman, R. W. and Tattersall, I. 1981. Behavior and ecology of *Macaca fascicularis* in Mauritius: A preliminary study. *Primates* **22**: 192–205.

1986. Distribution, abundance, and putative ecological strategy of *Macaca fascicularis* on the island of Mauritius, southwestern Indian Ocean. *Folia Primatologica* **64**: 28–43.

Swinnerton, K. J. 2001. Ecology and conservation of the pink pigeon Columba mayeri on Mauritius. *Dodo* **37**: 99.

Swinnerton, K. J., Groombridge, J. J, Jones, C. G., Burn, R. W., and Mungroo, Y. 2004. Inbreeding depression and founder diversity among captive and free-living populations of the endangered pink pigeon Columba mayeri. *Animal Conservation* **7**: 353–364.

Swinnerton, K. J., Greenwood, A. G., Chapman, R. E., and Jones, C. G. 2005. The incidence of the parasitic disease trichomoniasis and its treatment in reintroduced and wild Pink Pigeons Columba mayeri. *Ibis* **147**: 772–782.

Thirlray, H. 2008. US and UK primate imports. *IPPL News* **35**: 17.

2009. US and UK primate imports. *IPPL News* **36**: 14–16.

Tosi, A. J. and Coke, C. S. 2007. Comparative phylogenetics offer new insights into the biogeographic history of *Macaca fascicularis* and the origin of the Mauritian macaques. *Molecular Phylogenetics and Evolution* **42**: 498–504.

van Noordwijk, M. A., and van Schaik, C. P. 2001. Career moves: Transfer and rank challenge decisions by male long-tailed macaques. *Behaviour* **138**: 359–95.

van Schaik, C. P., Noordwijk, M. A., van, Boer R. J. de, and Tonkelaar, I. de. 1983. The effect of group size on time budgets and social behaviour in wild long-tailed macaques (*Macaca fascicularis*). *Behavioral Ecology and Sociobiology* **13**: 173–181.

van Schaik, C. P., and Noordwijk, M. A. van. 1985. Evolutionary effect of the absence of felids on the social organization of the macaques on the island of Simeulue (*Macaca fascicularis fusca*) (Miller, 1903). *Folia Primatologica* **44**: 138–147.

1986. The hidden costs of sociality: intra-group variation in feeding strategies in sumatran long-tailed macaques (*Macaca fascicularis*). *Behaviour* **99**: 296–315.

1988. Scramble and contest in feeding competition among female long-tailed macaques (*Macaca fascicularis*). *Behaviour* **105**: 77–98.

van Schaik, C.P., van Amerongen, A., and van Noordwijk, M.A. 1996. Riverine refuging by wild Sumatran long-tailed macaques (*Macaca fascicularis*). In *Evolution and Ecology of Macaque Societies*. J. E. Fa and D. G. Lindburgh (eds.) Cambridge University Press. pp. 160–181.

Vaughan, R. E. and Wiehe, P. O. 1937. Studies on the vegetation of Mauritius: A preliminary survey of the plant communities. *Journal of Ecology* **25**: 289–343.

Wheatley, B. P. 1978. Foraging patterns in a group of long-tailed macaques in Kalimantan Timur, Indonesia. In *Recent Advances in Primatology, vol 1: Behavior*, D. J. Chivers and J. Herbert (eds.). New York: Academic Press. pp. 347–349.

1980. Feeding and ranging of East Bornean *Macaca fascicularis*. In *The Macaques: Studies in Ecology, Behavior and Evolution*. D. G. Lindburgh (ed.), New York: Van Norstrand. pp. 215–246.

1982. Energetics of foraging in *Macaca fascicularis* and *Ponao pvamaeus* and a selective advantage of large body size in the orang-utan. *Primates* **23**: 348–363.

Wheatley, B. P., Putra, D. K. H., and Gonder, M. K. 1996. A comparison of wild and food-enhanced long-tailed macaques (*Macaca fascicularis*). In *Evolution and Ecology of Macaque Societies*. J. E. Fa and D. G. Lindburgh (eds.). Cambridge University Press. pp. 182–206.

Yeager, C. P. 1996. Feeding ecology of the long-tailed macaque (*Macaca fascicularis*) in Kalimantan Tengah, Indonesia. *International Journal of Primatology* **17**: 51–62.

9   The support of conservation programs through the biomedical usage of long-tailed macaques in Mauritius

NADA PADAYATCHY

Introduction

Mauritius is a 1,865 m² oceanic island found within the Mascarene archipelago in the Indian Ocean some 900 km east of Madagascar around latitude 20°15' S and longitude 57°30'E with the highest point of the island being at 828 m above sea level (Saddul, 1995). (Figure 9.1).

Although found on Arab mariner charts as early as 1300 AD and visited since then by sailors of various origins, such as the Portuguese, it was only colonized by the Dutch in 1638 AD (Cheke and Hume, 2008). After the Dutch, the island was colonized by the French followed by the British before attaining independence in 1968. Mauritius has had a Republic status since 1992 and a population of 1.26M (CSO, 2007), making it the eighteenth most densely populated country in the world. The official language is English, but the local dialect, the Mauritian Creole, is closer to French, making the latter a more used language than English. The Mauritian population is ethnically very diverse with people of Indian origin (68 percent), descended mostly from the indentured laborers from India composed of people from different religious backgrounds – Hindus, Muslims, Tamils, Telegus, Marathis, etc. The next largest ethnic group is composed of Creoles (27 percent) – people descended mostly from slaves brought over from Madagascar and Africa. People of Chinese origin, mostly descended from Chinese immigrants from Guangzhou compose 3 percent of the population and the remaining 2 percent are made of people of European descent (mostly from France) and some other minorities. Traditionally, the economy of Mauritius was based mostly on sugarcane, but this has now diversified to other sectors such as tourism, export, services, etc.

Monkeys on the Edge: Ecology and Management of Long-Tailed Macaques and their Interface with Humans, eds. Michael D. Gumert, Agustín Fuentes and Lisa Jones-Engel. Published by Cambridge University Press. © Cambridge University Press 2011.

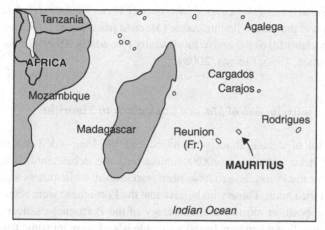

Figure 9.1. Map of the south west Indian Ocean showing location of Mauritius.

Mauritius has a very rich biodiversity and together with Madagascar and the surrounding islands it forms part of one of the world's biodiversity hotspots. Mauritius is famous for being the only home of the Dodo (Cheke and Hume, 2008). Following the colonization of the island, the Dodo rapidly disappeared due to the consequences of human actions. The Dodo was not the only Mauritian victim of man's actions. Of the fourteen endemic species of birds, eight have gone extinct with the remaining six surviving ones all threatened according to IUCN (2008). Out of three species of fruit bats, only one remains. Of the seventeen species of reptiles, six are extinct, including two species of tortoises and among the remaining eleven, six are confined to small islets around Mauritius. The same scenario, where humans have radically and detrimentally changed the environment, applies for the island's terrestrial invertebrates and flora.

Human action such as forest destruction are among the culprits for the demise of the Mauritian native fauna and flora. Forest destruction through hardwood harvesting began in 1638 as soon as the island was colonized by the Dutch and continued through the French occupation as they cleared forests to make way for agriculture. The most recent survey of indigenous vegetation cover remaining in Mauritius completed in the mid 1990s points to only about 2 percent of the original forest remaining and even those 2 percent consist of highly degraded forest (Page and D'Argent, 1997). The reason for this degradation is due mostly to invasive plant species introduced by man with at least 21 species now considered to be seriously threatening Mauritian biodiversity (Florens, 2008). Invasive alien animals also represent a major threat to native fauna and flora. Mauritius has thirteen species of invasive alien species all of which are believed to impact upon native wildlife. The major ones being feral cats (*Felis catus*), rats (*Rattus* sp.), Java deer (*Cervus timorensis*), feral pigs (*Sus scrofa*) and the long-tailed macaque (*Macaca fascicularis*). All have been recognized as harmful to the native biodiversity (Owadally, 1979; Cheke *et al.*, 1984; Quammen, 1996; Florens, 2008).

The introduction of *Macaca fascicularis* to Mauritius

There is a lot of debate about who introduced the long-tailed macaque to Mauritius. Cheke and Hume (2008) summarized the debate, indicating that some consider the Portuguese to have been responsible, while others argue that it was in fact the Dutch. Those who believe that the Portuguese were responsible base their supposition mostly on the strategy of the Portuguese sailors at that time, which was that when they found a suitable island, to plant some fruit trees and leave livestock to multiply as a source of food for future visits. It has also been debated whether the term livestock would include monkeys which in turn gave rise to the rumor that Portuguese were fond of monkey meat. If indeed the Portuguese did introduce the long-tailed macaque to the island, this introduction would have taken place in the second quarter of the sixteenth century.

Those who support the idea that the Dutch were responsible for the initial introduction of monkeys argue that studies of contemporary manuscript accounts of the first Dutch visit in 1598 makes no mention of monkeys being present on the island. The only mammals reported to be present in 1598 were bats. Another account dated from 1602 reports the only four-legged animal present to be cats. However, some authors (Cheke and Hume, 2008) believe this to be a misprint or transcription error. The Dutch term for cat – *Katten* is very similar to the term for rat – *Ratten*. It might be possible that the 1602 report made mention of rats rather than cats, especially given that there was no mention of cats in subsequent reports for more than 100 years following this report. By 1606, it was reported that rats and monkeys were the only four-legged animals to be present, indicating that (i) there might have been a mistake in 1602 in translating rats and cats and (ii) that monkeys were introduced somewhere between 1602 and 1606 by the Dutch.

While the literature is full of contradictions regarding whether the Portuguese or the Dutch introduced the long-tailed macaque to Mauritius, of lesser debate is the origin of these macaques. Although the long-tailed macaque is widely distributed (i.e., Maleyo-Indonesian and Indochinese regions), it is generally agreed that the founder population in Mauritius comes from Java, Indonesia (Sussman and Tattersal, 1981; Cheke and Hume, 2008). This is supported by historical considerations as well as a close examination of the physical characteristics of the different populations, namely pelage coloration and hair tufts which are consistent with the Javan population (Sussman and Tattersal, 1981). Furthermore, DNA analysis in the mid 1980's has confirmed this origin as well as establishing that the founder population was very small, consistent with an unintended introduction such as escaped or abandoned pets (Lawler *et al.*, 1995). It is worthwhile to note that a more recent study explored the alternative hypothesis of the Mauritian macaque being of Sumatran origin by a phylogenetic analysis but due to a very limited sample size from monkeys of Javan origin, they did not reject the hypothesis that Mauritian monkeys originate from Java (Tosi and Coke, 2006).

However, regardless of who introduced them or the geographic origin of Mauritius' long-tailed macaque, this generalist has since thrived in the country. Mauritius offered all the habitat types that these macaques are known to occupy, primary, secondary, coastal, mangrove, riverine and savannah forest types. It was even reported that one of the reasons the Dutch left the island in 1710 was due to damages inflicted on crops and food stocks by rats and allegedly monkeys (Carie, 1916). Even after the French settlement in 1722, reports of damage by monkeys were still being related by the French settlers with one report by Baron Grant in 1741 going as far as to say that native birds were becoming scarce in the woods "as the monkies, which are in great numbers, devour their eggs" (Cheke and Hume, 2008).

After 1810, when the British took over, the island went through a major change. There was a boom both in terms of population as well as in agriculture, with sugarcane establishing itself as the driving force behind the island's economy. Massive deforestation occurred throughout the island over the years resulting in wildlife (both native and exotic) to be confined in smaller and smaller areas. In 1859, George Clark commented on the rarity of birds and blamed the monkeys for their destruction (Cheke and Hume, 2008.). In 1880, the first report to be done on Mauritian forests, assessing their condition and recommendations for future management, Richard Thompson of the Indian Forest Service, wrote that monkeys ate eggs and young of birds and in addition, stated that monkeys "devour and throw down the unripe seeds of all the principal and important forest trees, so that it is scarcely possible to secure ripe seed that will germinate" (Cheke and Hume, 2008).

The only attempt to study the distribution and abundance of *Macaca fascicularis* was carried out between 1977 and 1984 and it was concluded that the estimated 25,000–35,000 population of long-tailed macaque lived over the 45,000 ha of available and suitable habitat with a variation in the densities in different habitat types (Sussman and Tattersal, 1986). These figures were revised upwards by Bertram and Ginsberg (1994), who used different statistical assumptions estimating the population at 40,000. But, most recently have been reassessed by Sussman *et al.* (Chapter 8) to possibly be below 10,000 individuals.

Perceptions *of* Macaca fascicularis *in Mauritius*

Throughout their existence on the island, the long-tailed macaque has been perceived as a pest, from the early settlers (the Dutch, the French and the British) to the existing population, because they raid crops and even damage properties. However, the raiding of crops or damaging of properties occurs mostly in areas bordering the forests. Most of these lands are considered as marginal lands and it is very often people with relatively low incomes that cultivate these lands. Hence, they are affected by the damage inflicted by monkey troops more than other farmers on less marginal lands. It has been estimated that long-tailed macaques cause US$1.5 million in agricultural damage per year (Bertram and Ginsberg, 1994), which represents about 0.01 percent of the island's GDP.

Macaques affect several types of farms in Mauritius. In one region of the island, known as the ex-tea belt, where previous tea plantations have been replaced by sugarcane plantations and are adjacent to riverine and secondary forest habitats, more than 40 percent of small planters have abandoned their plot of land due to monkey damage (M. Boodnah, Manager of Farmers Service Corporation, Curepipe Branch, personal communication). It is not only sugarcane that suffers from monkey raids. Due to the flexible diet of long-tailed macaques and depending on the predominant food source available, they may feed on other crops such as bananas, maize, vegetables, and fruit trees (Figure 9.2). In some deer rearing estates, where the deer are given supplementary food, the monkeys have developed a commensal relationship with the deer. They live in close proximity with the herds and also feed on the deer food, which is a mixture of grains and molasses (pers. obs.)

Apart from being perceived as a pest to agriculture, the long-tailed macaque is devastating to the native fauna and flora. For centuries, they have been recognized as predators of birds' nests, eating eggs and young and sometimes even attacking incubating adults (McKelvey, 1976; Jones *et al.*, 1992). Their role in preventing regeneration of native trees through the predation of immature fruits (Figure 9.3), as well as contributing to the seed dissemination of invasive exotic plants have also been documented (Florens, 2008).

It is likely that early settlers hunted monkeys for food. This practice continued through the years, with escaped/ex-slaves using monkeys as a food source. In Mauritius, a large proportion of forest areas (excluding Nature reserves) are leased privately for deer rearing and game shooting. Prior to the 1980's monkeys were considered as game and were regularly shot for

Figure 9.2. Feral monkeys feeding on crops.

food. This practice has since been greatly reduced as the lease owners prefer using the monkeys as a resource (trapped for monkey breeding farms) rather than shooting them. Today, there is still a fraction of the population that will eat monkey meat. In the 1980s it was quite common for fancy-fairs (i.e., a family event organized by the local community, usually the church, as a fund-raising event) to sell dishes prepared from game, including deer, wild pig and monkey. Some old cookbooks of Mauritian cuisine even featured recipes for cooking monkey meat. However, today, the practice of eating monkey meat has almost disappeared from Mauritian culture (see Sussman *et al.*, Chapter 8).

Although not very common, there are people keeping long-tailed macaques as pets. In fact, subsection 17 of the Wildlife and National Parks Act of 1993, specifies that "a person may keep as a pet an individual specimen of any species of wildlife as may be prescribed for the purposes of this subsection." Long-tailed macaques are usually taken very young from the wild and kept on a leash or in a cage in the yard of the person. Very often, when a pet macaque reaches puberty, their behavior changes and they can become aggressive. Unfortunately, most of the pet owners find that the easiest solution is to release the long-tail macaques back into nature where they are unable to adapt and hence live a life of commensalism in close contact with humans.

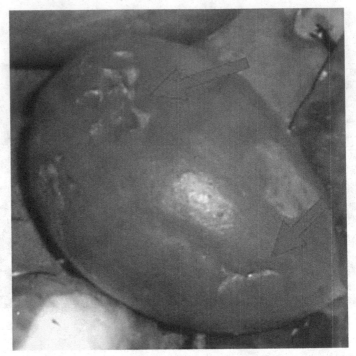

Figure 9.3. Evidence of chewing by long-tailed macaques on immature endemic fruits.

The way the local population looked at long-tailed macaques underwent a major change in 1985 with the establishment of Bioculture (Mauritius) Ltd (BCM), a local private company involved in the capture, breeding and export of long-tailed macaque for biomedical research in Europe and USA. The idea of using the Mauritian macaque population for research had already been evoked in 1983 (Kavanagh, 1983), two years before Bioculture (Mauritius) Ltd established itself. Conditions established by the government for the exploitation of this natural resource were very clear – export would be allowed only to companies involved in biomedical research, not military or space research. In addition the company was expected to abide by existing international norms regarding general animal welfare. Bioculture (Mauritius) Ltd signed agreements with private land owners, lease-land holders and large/small planters to trap monkeys on their land with a payment paid to the owner for every animal caught. Many people decided that instead of being just a pest, the monkey was suddenly turning into a resource, with the result that most people stopped shooting monkeys as game, but preferred rather, to allow BCM to come and trap them, thus earning themselves additional revenue. Small planters who for years had suffered from damage to their crops inflicted by monkeys found themselves in a win-win situation from the humans' perspective. First by trapping monkeys, BCM helped to control the population in the area, hence less damage to the crops. Second, BCM paid for every monkey caught, which in some ways, compensated for the damaged crops. In 1989, a second permit was delivered to another company, nowadays known as Noveprim. For more than ten years, these two companies were the only ones in the field, and then the industry suddenly mushroomed with one joint-venture of Bioculture (Mauritius) Ltd, Biodia Ltd operating on its own and the government giving permits to three other companies.

The attractiveness of the Mauritian long-tailed macaque as a potential candidate for biomedical research is due to its unique viral profile and naturally occurring specific pathogen free (SPF) status (i.e., free from Herpes-B, SIV (Simian immunodeficiency viruses), SRV(Simian retrovirus) and STLV1 (Simian T-Lymphotropic virus Type 1). This status is very important under research conditions as it confers a safer working environment for the researchers. The naturally occurring SPF status would suggest that animals brought to Mauritius were very young, not yet sexually mature, and would hence support the theory of the animals being transported as pets which were either accidentally or deliberately released on the island, rather than a deliberate introduction for meat source which would have included a mature stock of animals.

Interestingly, from 1985, when trapping of monkeys started, until 2005, an estimated maximum of 4,000 animals were removed on an annual basis. Without clear studies of the population, there was no noticeable impact on the remaining population, as the proportion of male vs. female, adult vs. juvenile appeared fairly constant. This would suggest that the population might have been even higher than the estimated 40,000 in 1994. However, in the years 2005–2008, the island witnessed an unprecedented wave of trapping of monkeys – mostly due to expanding new companies. Whether this has had an impact on the monkey population dynamics remains to be studied (see Sussman, Chapter 8, for a preliminary assessment), What is certain is that it has been noticed that in general the macaques have become neo-phobic and trap-shy. Since 2009 however, the trapping of long-tailed macaques has undergone a major slowdown with some companies not trapping at all. This was mostly due to economical reasons with the global financial crisis.

Macaca fascicularis, the conservationist

It is ironic that a species that has for decades has been maligned as one of the major culprits in the downfall of Mauritius' native fauna and flora, is today the behind-the-curtain savior of the same native fauna and flora. Today, the majority of Mauritius' native forest remnants are found within the Black River Gorges National Park, a 6,574 ha area proclaimed in 1994. The creation of the National Park came together with the creation of the National Parks and Conservation Service, a government unit that manages the park, offshore islets of conservation value, and various other conservation projects.

To help them in their endeavor, the government created the National Parks Conservation Fund. This fund is used to run conservation projects in Mauritius such as forest restoration programs which weed out exotic plants, as well as maintaining fenced Conservation Management Areas within the park. This fund is fueled by money from the macaque industry through the payment of annual trapping fees as well as a levy of US $70 on each animal exported. This levy alone accounts for more than a US $250,000 annually (and is expected to increase with the recent increase of the levy to $100/exported animal as well as with the establishment of other breeding companies). There is no other source of funding of this magnitude available for conservation initiatives on Mauritius, making the macaque-export industry fundamental to conservation. Putting a stop to this industry would cripple long-term conservation in Mauritius.

Apart from the conservation aspect, the impact of the macaque-export industry on the Mauritian society cannot be understated. It contributes to the economy through export, it provides jobs, both direct and indirect, and it supports local production by purchasing locally grown vegetables and fruits and though their Corporate Social Responsibility (CSR) Programs, helping the community in various ways. For instance, Bioculture (Mauritius) Ltd has a CSR program that encompasses a variety of projects at different levels (i.e., community, local, and international) ranging from sponsoring community socio-cultural events, long-term support of sports clubs, schools and local NGOs to direct involvement in conservation projects in the region.

Conservation projects funded by a pest species – a breeder's perspective

Bioculture (Mauritius) Ltd, established in 1985 as the first macaque breeding industry in Mauritius, is recognized as one of the world's leaders in the supply of naturally occurring SPF *Macaca fascicularis* to the biomedical industry. It is regarded as having extremely high standards of animal care and welfare. The company is ISO 9001:2000 certified, UK Home Office approved as well as being accredited by AAALAC International (Figure 9.4).

Founded and run by dedicated conservationists, the company recognizes that the backbone of its business is derived from a natural resource and, as such, felt

Figure 9.4. Bioculture (Mauritius) Ltd logo and certifications.

morally obligated to put back some of the profits derived from this resource into conservation. BCM funds and manages five conservation projects in the region, namely two forest restoration programs in Mauritius, one re-wilding project on the nearby island of Rodrigues and two forest conservation programs in Madagascar through its own Non-Governmental Organization (NGO). To date, BCM employs more than 80 persons directly in these conservation projects.

Mauritius project

BCM owns two blocks of native forests each of approximately 100 acres. One, a moist tropical hardwood forest in the south east of the island and the other, a dry tropical hardwood forest dominated by Mauritian Ebony located in the south west. These blocks of forest, found outside the range of the National Park, are not represented in the Government Conservation Management Areas and were in a state of severe degradation. Invasive alien species were smothering the forest, preventing natural regeneration. BCM worked together with a local NGO, the Mauritian Wildlife Foundation (more specifically with its Flora Division), as well as the Ecology and Conservation Department of the University of Mauritius, towards a forest restoration program. These external bodies provided the expertise in plant identification, trained BCM staff and are consultants on the project. BCM also opened the door to University of Mauritius undergraduate students to carry out various studies on weeding techniques at our conservation sites.

The BCM forest restoration program consists of weeding – manually removing the invasive alien species by cutting, uprooting and ring-barking then applying herbicide to cut stumps to prevent regrowth (Figure 9.5). Seeds and seedlings of native plants are collected and propagated in in-situ nurseries before being planted back in the forest once they have reached a certain size.

The restoration work carried out by BCM has been noted by both governmental agencies and conservation bodies, with international experts coming to witness the work done. The BCM forest restoration program has been used as a case study for how the private sector can carry out conservation work and, based on BCM data and perfected techniques, larger scale weeding will be carried out in other blocks of native forests.

In recognition to the commitment of Bioculture (Mauritius) Ltd towards conservation, in 2008, it has been invited to be a member of the National Threatened Native Plants committee, to advise on national strategies for conservation of native plant diversity. The forest restoration program is still being carried out and it is hoped that in the near future, these blocks of forests will become showcases that can be used for educational purposes for school children and the population in general.

Rodrigues project

Rodrigues, a 104 km² island is a dependency of Mauritius found 574 km east of the latter. Just like Mauritius, it had a high level of biodiversity. Of particular relevance was the high densities of giant tortoises – historically the highest density ever recorded (Francois Leguat mentioned large aggregations of 2,000–3,000 tortoises), hence earning the island the nickname of "Tortoise Island." The tortoise population has been extensively exploited, mostly by the French settlers in the Mascarene in the eighteenth century. This, together with habitat destruction and the effect of introduced rats, decimated the tortoise population in less than a century. Ecological disaster continued with the open forests of Rodrigues being destroyed over the years by deforestation,

Figure 9.5. (a) Mauritian forest invaded with exotics,
(b) immediately after weeding and (c) a restored forest several years after weeding.

Figure 9.5. (cont.)

fires, and overgrazing resulting in severe soil erosion. The consequence today is that the native fauna and flora of Rodrigues is hanging on to existence by the tip of the fingers. However, since the early 2000s, the island has received much international exposure, in terms of the efforts being made towards restoration work, making it a flagship site for ecological restoration of devastated islands.

BCM has a lease on 20 hectares of degraded forest found in the south west of the island where it set up the Francois Leguat Giant Tortoise and Cave Reserve. The team there is recreating a parcel of the almost completely lost lowland forest of Rodrigues by planting native trees with the help of the Rodrigues Branch of the Mauritius Wildlife Foundation. Aldabra Giant tortoises, *Dipsochelys elephantina* and Radiata tortoises, *Astrochelys radiata* have been introduced on the reserve.

BCM also built a museum on the site, with galleries focusing on the history of the island, the natural history of Rodrigues and the limestone geology of the region. An educational centre also makes available to researchers and students, rare and historical documents on the history of Rodrigues.

Madagascar project

Because they are using free-ranging primates as a resource for their business, BCM wanted to return some of their profits derived from these monkeys into primate conservation. However, rather than just contributing financially to the cause, BCM wanted to be a key player in the primate conservation field, doing hands-on conservation. Since 2002, BCM set up and funds a NGO, Biodiversity Conservation Madagascar, employing more than thirty locals to conserve the natural habitats of the primates of Madagascar. The NGO manages two forested areas: one 2,500 hectares forest in the east of the island, since 2002 and the other, a 14,000 hectares forest in the west, since October 2009.

The BCM strategy for conservation involves the concept of "Conservation Payments." Simply put, Conservation Payment is giving the local population incentives NOT to go in the forest to carry out destructive activities. Conservation Payments takes the form of employment opportunities, technical advice on agriculture, power generators, computers for local administration, mobile phones, and even a television set for better access to information. BCM also carried out health campaigns for the benefit of the population whereby free medical consultation and free medication were made available. It is taking the local population on board as a stakeholder in the conservation process, encouraging them to value the forest, value the lemurs and be proud of its conservation status.

Furthermore, BCM has included an educational aspect to its conservation activities. BCM funds scholarships through the Malagasy Primate Society – GERP. These scholarships are made available to Malagasy University students carrying out lemur related studies at Sahafina. To date, more than 25 students have benefited from these scholarships.

Conclusion

Since being introduced to Mauritius more than 300 years ago, *Macaca fascicularis* established themselves really well. We believe that the macaque breeding industry has converted this perceived pest into a resource whose exploitation has contributed to the local economy. Furthermore, conditions (i.e., levies and trapping fees) attached to the exploitation of these feral primates allows the harvesting of these animals to subsidize conservation projects through the National Parks and Conservation Fund, sustaining the very existence of the National Park in Mauritius as well as actively funding conservation of the highly threatened biodiversity in the region, with a major focus on primate conservation in Madagascar.

Primate breeding companies for biomedical research are generally regarded by conservation primatologists as "the big bad wolf." Bioculture (Mauritius) Ltd has shown that this need not be the case. An ethical and responsible company can operate in a sustainable way by utilizing a primate resource in a balanced manner and also direct part of its profit to the conservation of other threatened species of primates. The primate breeding industry and primate conservation need not be mutually exclusive.

References

Bertram, B. and Ginsberg, J. 1994. Monkeys in Mauritius: Potential for humane control. Unpublished report commissioned by the Royal Society for the Prevention of Cruelty to Animals.

Carié, P. 1916. L'acclimatation a l'ile Maurice. Premiere partie. Mammiferes et Oiseaux de Maurice. *Bulletin de la Societe Nationale D'Acclimatation de France.* Maretheaux, Paris.

Cheke, A. S., Gardner, T., Jones, C., Owadaly, A., and Staub, F. 1984. Did the Dodo do it? *Animal Kingdom* **87**: 4–6.

Cheke, A. S. and Hume, J. 2008. *Lost land of the Dodo.* T and AD Poyser, London.

CSO. 2007. Mauritius in figures 2006. Central Statistics Office, Ministry of Finance and Economic Development.

Florens, V. F. B. 2008. Ecologie des forets tropicales de l'Ile Maurice et impact des especes introduites envahissantes. PhD thesis, Universite de la Reunion.

IUCN. 2008. 2007 IUCN red list of threatened species. www.iucnredlist.org, accessed 27 July 2010.

Jones, C., Swinnerton, K. J., Taylor, C. J., and Mungroo, Y. 1992. The release of captive-bred Pink Pigeons *Columba mayeri* in native forest on Mauritius. A progress report July 1987–June 1992. *Dodo* **28**: 92–125.

Kavanagh, M. 1983. *A Complete Guide to Monkeys, Apes and Other Primates.* New York: Viking Press. p. 213.

Lawler, S. H., Sussman, R. W., and Taylor, L. L. 1995. Mitochondrial DNA of the Mauritian Macaques (*Macaca fascicularis*): An example of the founder effect. *American Journal of Physical Anthropology* **96**: 133–141.

McKelvey, D. S. 1976. A preliminary study of the Pink Pigeon. *Mauritius Inst. Bull.* **8**: 145–175.

Owadally, A. W. 1979. The Dodo and the Tambalacoque tree. *Science* **203**: 1363–1364.

Page, W. and D'Argent, G. A. 1997. A vegetation survey of Mauritius (Indian Ocean) to identify priority rainforest areas for conservation management. IUCN/MWF report, Mauritius.

Quammen, D. 1996. *The Song of the Dodo.* Pimlico, London.

Saddul, P. 1995. *Mauritius: A Geomorphological Analysis.* Moka Mauritius: Mahatma Gandhi Institute.

Sussman, R. W. and Tattersal, I. 1981. Behaviour and ecology of *Macaca fascicularis* in Mauritius: A preliminary study. *Primates* 22(2): 192–205.

 1986. Distribution, abundance and putative ecological strategy of *Macaca fascicularis* on the Island of Mauritius, southwestern Indian Ocean. *Folia Primatologica* 46: 28–43.

Tosi, A. J. and Coke, C. S. 2006. Comparative phylogenetics offer new insights into the biogeographic history of *Macaca fascicularis* and the origin of the Mauritian macaques. *Molecular Phylogenics And Evolution* 42: 498–504.

10 Ethnophoresy: The exotic macaques of Ngeaur Island, Republic of Palau

BRUCE P. WHEATLEY

The only non-human primate population in Oceania is on the Island of Ngeaur in the Republic of Palau. This population of macaques, *M. fascicularis*, was the result of ethnophoresy, the dispersal of animals by humans. German colonialists introduced about a half-dozen macaques from Indonesia around 1909. Islanders view these macaques not only as agricultural pests to be hunted, but also, in a limited sense as an increasing part of their natural heritage. Ngeaur is a useful case study for ethnoprimatology to better understand what causes communities to choose conservation, control, or eradication of commensal macaques, which will be important as human and non-human primate interactions continue to increase around the globe. Ethnoprimatologists can also offer counsel to Palauan communities living near macaques by providing guidance on management strategies. Population surveys of the macaques on Ngeaur have estimated the population to be slightly less than 1,000 individuals, which is twice the size of the human population. This chapter discusses how the local communities can control their macaque population by limiting macaque dispersal into the southern part of Ngeaur where taro gardens are located and onto the other islands of Palau. Totally eradicating the macaques may not be necessary nor economically in Palau's best interest. Dispersal management will be a more cost effective approach than eradication efforts, which can be expensive to implement. Macaques could also be used to draw economic gains from tourism, providing a resource to local communities. Proper management and utilization of macaques could allow for finances being available for use on other species that threaten human health, conservation of endangered species, and limiting the loss of biodiversity in Palau.

Heinsohn (2003) has proposed the term ethnophoresy to refer to animal dispersal via human vessels. Specifically, he uses the term "ethnotramp" to refer to any species such as *Macaca fascicularis* that is regularly translocated and dispersed by humans. Such translocation might be suspect to any land

Monkeys on the Edge: Ecology and Management of Long-Tailed Macaques and their Interface with Humans, eds. Michael D. Gumert, Agustín Fuentes and Lisa Jones-Engel. Published by Cambridge University Press. © Cambridge University Press 2011.

beyond the exposed continental shelf in Southeast Asia or Sundaland which once connected Sumatra and Borneo (Tjia, 1970). While the term "ethnophoresy" is new, the concept of human translocation of species is old. In Alfred Russel Wallace's (1869) 1859 journeys, for example, he discussed how local Malayans often kept tame pets such as monkeys and carried them around in their boats. He found it difficult to imagine how *Macaca nigra* could have reached the Island of Batchian (Bacan) by any natural means of dispersal without passing over the narrow straits to Gilolo, Halmahera. We need to be careful however, because, as Fooden (1991) points out, complete land bridges during glacial events are not always necessary for this macaque's dispersal as apparently occurred in its colonization of the Philippine archipelago. Macaques are good swimmers and rafting between islands is another possibility.

Macaca fascicularis is not only extensively distributed throughout the Southeast Asian archipelago, but it is also one of the most widespread non-human primates (see Gumert, Chapter 1). It is, therefore, an interesting and sometimes difficult question to determine the extent of the connection between the distribution of this species and its human dispersal. It has been a resident on some islands such as Java for over a million years (Hooijer, 1952), but on other islands such as Mauritius (Sussman and Tattersall, 1981), and Ngeaur (Poirier and Smith, 1974), it is a relative newcomer.

Although the most eastern point on the globe inhabited by a macaque during Wallace's time was *M. nigra* on Bacan, it has now been eclipsed by *M. fascicularis* and by the same process – man. Furthermore, *M. fascicularis* is now the only established macaque population in all of Oceania in Micronesia on the island of Ngeaur (a.k.a., Anguar), one of sixteen states in the Republic of Palau (Figure 10.1). Palau is about 500 miles east of Mindinao and about 700 miles southeast of Guam. Ngeaur is a raised atoll or limestone reef of 3.2 sq. mi. (Figure 10.2). The nearest population of macaques is over 500 miles away in the Republic of the Philippines. The somewhat recent human dispersal of this population of long-tailed macaques is an excellent example of how humans have facilitated the adaptive success of this primate. This species dispersed many thousands of years ago throughout the Indonesian archipelago and the means of their dispersal can, in many cases, only be conjectured. The long-tailed macaque has benefited from their close association with humans in some areas while suffering from this association in others. The species ranges from commensal to completely non-commensal in forest habitats. This chapter examines the macaque population of Palau since its introduction and the various issues relating to its management or attempted eradication.

The study of the more encompassing interactions between humans and non-human primates is called ethnoprimatology (Sponsel, 1997). Sponsel (1997) appears to have been one of the first to draw attention to this informative

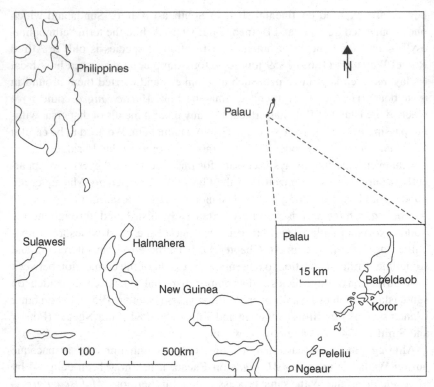

Figure 10.1. Location maps for the Republic of Palau and for the island of Ngeaur.

area of study which has witnessed tremendous growth since its inception. Ethnoprimatology explores our many cultural connections to non-human primates (Wheatley, 1999a), especially such broadly commensal ones such as *M. fascicularis*. Animals do not live in physical environments alone but also in social and cultural environments. Thus it is important to understand all of these environments in order to explain the human response to animals such as the long-tailed macaques. The species can be seen, for example, as a pest and hunted as on Ngeaur or as a food source among the Penan of Borneo (Labang and Medway, 1979). The macaques of Bali, while seen as crop raiders, also have a long history of association with and integration into the Hindu culture (Wheatley, 1999a). In contrast to Bali, the species has less integration into the culture of Ngeaur, Palau. Many people on Ngeaur appear to view the macaques as general pests, especially to agricultural crops, but there are some Islanders who view the animals as part of their heritage (Farslow, 1987; Marsh-Kautz and Wheatley, 2004).

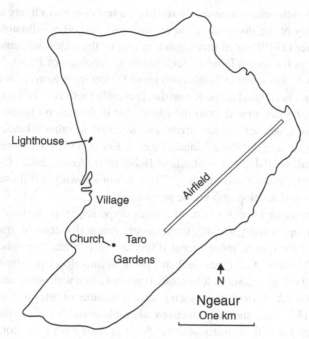

Figure 10.2. Map of Ngeaur Island.

Palauans are justifiably proud of their environmental heritage and are intimately concerned with preserving their unique flora and fauna. Article VI of the Palaun Constitution, for example, calls for the "conservation of a beautiful, healthful, and resourceful natural environment" and one of the government's eight ministries is the Ministry of Natural Resources, Environment, and Tourism. The famous Rock Islands, enclosed by a barrier reef that makes up most of the archipelago's hundreds of islands is protected by many agencies and regulations. The Bureau of Marine Resources is one such agency that regulates and restricts harvesting and collecting of fish, sponges, and corals. The non-profit group CEDAM (Conservation, Education, Diving, Awareness and Marine Research) declared the reefs of Palau as one of the seven underwater wonders of the world in 1989. The Rock Islands are in the process of being nominated to the United Nation's World Heritage listing (Marsh, 2009). The community in Ngeaur and its' local government have shown some interest in understanding the monkeys in many ways, including funding and training students in primatological research (Marsh-Kautz and Wheatley, 2004). The national and state governments, however, have also repeatedly attempted to control the monkeys, including eradication.

The Palauan archipelago of over 350 islands is a test case and a living laboratory in the study of the dispersal of these macaques and ethnoprimatology in general. Kramer (1919) stated that Ngeaur is one of the oldest settlements in Palau. Ngeaur is the fourth largest island in the archipelago of Palau. Ngeaur Island is about 11 km south of Peleliu and about 60 km from Koror, the capital of Palau. Ngeaur is isolated to the extent that presently there is only a ferry that runs every other week to and from the island. Air flights are no longer available. Some macaques have been carried and released to other islands in the archipelago, including the Rock Islands, Peleliu, Koror, and at least one of the ten states, Airai, on Palau's largest island Babeldaob (Anon, 2008; Crombie and Pregill, 1999; Poirier and Smith, 1974). It is not known yet if these introduced animals will develop into viable populations.

Unlike many other islands where macaques dispersed in prehistoric times, Palau offers an opportunity to study the relatively recent dispersal of macaques to Ngeaur and their even more recent dispersal by humans throughout the archipelago. Primates of all types are being brought into closer proximity with humans around the globe. Such close contacts are not often welcomed and Palau is probably no different from many such places. A number of relevant questions can be asked. How have such introductions taken place and how have the communities responded to these introductions? What options have these communities considered and are there other options that can be explored? How successful have these responses been? Can primate conservation succeed in Ngeaur? Can the people of Ngeaur live comfortably with these macaques? Ngeaur thus offers us a complex case study by which we can better understand what causes communities to choose conservation, control, or eradication of commensal macaque populations. It is only when all the facts about the humans and non-human interactions are known that difficult choices can be made.

Introduction of macaques to Palau

The Germans bought the Caroline Islands from Spain in 1899 after the Spanish-American war. Copra was the most profitable trade item until phosphates were mined in 1908. Ngeaur's large phosphate deposits are the result of millions of years of bird guano. These deposits were estimated to be two and a half million tons of high grade ore (Palau Community Action Agency, 1977). A consortium of German banks and other firms, called the *Deutsche Sudsee Phosphat Aktien Gesellschaft*, organized in Bremerhaven in 1908 and was granted mining rights. The population of Ngeaur was estimated at 150 people in 1909 (Kramer, 1919) and a train ran on the island carrying the ore to waiting steamships.

Many Palauans consider the unique presence of *M. fascicularis* in their country to be a dubious honor. While the island dispersal of many non-volant animals and especially monkeys is generally of unknown origin, the introduction of *M. fascicularis* to Palau is reasonably well known. Although Chinese, Malayan, and Spanish traders were known to carry monkeys on their ships (Kramer, 1919), the animals were probably introduced to Ngeaur around 1909, during the German administration of Palau (Poirier and Smith, 1974). These authors state that German phosphate mining engineers brought two monkeys to the Island. Ngeaur elders, however, interviewed in English and Japanese stated that five animals were brought via steamship from Indonesia (Wheatley *et al.*, 1999b). Another informant told us that her mother remembers two monkeys that the Germans had as pets in the hospital on Ngeaur. Support for more than two introduced monkeys also comes from Matsubayashi *et al.* (1989) who note that the genetic diversity of the present population is too great for only two animals. The genetic studies of these researchers also suggest an Indonesian origin such as Sumatra or Java for the macaques. Matsubayashi *et al.* (1989) also found significantly smaller body sizes in the animals on Ngeaur compared to Indonesian populations and an interesting grayish coat-color variant. There is no mention of monkeys in Kramer's ethnography of Ngeaur (Kramer, 1919).

The macaques were freed, perhaps when the Germans lost their colony to Japan during World War I and departed from the region, and have prospered ever since. The reason why they were introduced is unknown. Several reasons were suggested during interviews. Perhaps they were simply pets or even used to test medicines, but they were not needed as a surrogate canary for mining phosphates as many have claimed. Luskin (1989) for example, states that: "This subterranean operation brought the macaques to the island as proverbial guinea pigs for testing mine air quality." The mining operations on Ngeaur were a simple process of clearing the jungle, removing the dead vegetation, breaking up the ore and shoveling it into mining cars (Otsuki, 1916). It wasn't until 1936, decades after the monkey introduction is thought to have occurred, that mining operations went sub aerial.

The macaque is a serious pest in Palau

A government report published in 2002 stated that "The macaque is a serious pest in Palau" (Anon, 2002). A more recently published and widely circulating pamphlet in Palau is entitled "Invasive Mammal: Monkey (*Macaca fascicularis*, crab-eating macaque)". The pamphlet notes that the Republic of Palau's Division of Fish and Wildlife Protection considers macaques as a serious threat to the farms, health, and ecosystem of Palau. It further asserts that these "monkeys have caused major socio-economic and environmental damage to Angaur" by "destroying farm crops and gardens in Angaur" and that they prey on crabs and birds in Palau. The pamphlet mentions that another potential threat from macaques is the potential for disease and that the B-virus carried by the macaques "is often lethal to humans" (Anon, 2008). Kemp and Burnett (2003) have also stated that the B-virus is fatal to humans in about 80 percent of cases, although they neglect to mention that cases of this rare disease in humans has only occurred in the laboratory (see Engel and Jones-Engel, Chapter 7 for a discussion). Moreover, the B-virus strain is different in *M. fascicularis* from that in *M. mulatta* of which all known cases of transmission to humans has occurred (Smith *et al.*, 1998).

The contention that macaques cause damage is overemphasized and it has a long history. The Chief Conservator of Palau, for example, asserted in 1977 that macaques were damaging, not only the crops on Ngeaur but also the bird populations (Farslow, 1979). Kemp and Burnett (2003) state that "it is likely (although not yet reported) that the Angaur macaques prey on native birds, small vertebrates and eat/destroy native fruits." Carter (2007) makes a similar assertion. These claims could possibly be validated because long-tailed macaques have been seen to do so elsewhere in Southeast Asia. Macaques in Ngeaur are said to dig up and to eat taro especially if other food sources are not available. Local interviews also established that the animals also may eat and damage many other crops such as betel nut, oranges, papaya, and pandanus (Wheatley *et al.*, 2002). The monkeys are also said to eat the endangered *bekai* eggs (*Megapodius laperouse*), as do the people. They may also destroy the nests of pigeons, Belau fruit doves, (*Ptilinopus pelewensis*) and the Micronesian starling (*Aplonis opaca*) which nests in betel nut trees, but at this point these claims are not documented. Another threat stated in the IUCN/SSC Invasive Species Specialist Group, ISSG Database (Carter, 2007) is that they probably help to disperse alien plant species and even damage electrical wiring. Lastly, some local people have worried about the safety of their children with macaques nearby.

1994 Ethnoprimatological survey

In 1994, the author, two faculty members from the University of Guam, Dr Stephenson and Dr Kurashina, and nine students conducted research on Ngeaur (Wheatley, *et al.*, 1999b). Our primary research topic was the hunting of macaques by the local people. We conducted a line transect survey to estimate the size of the macaque population. Using detailed maps of the island, we divided up into pairs and slowly walked along 62,360 meters of roads and trails. We collected 61 samples of over a dozen variables that were recorded the first minute of contact and half-dozen variables every minute over the next ten minutes. The mean detection distance was 41.8 meters giving a transect area of 521 hectares. The number of macaques detected was 286. The population

Table 10.1. *Summary of Ngeaur macaque population*

Reference and Year	Estimated Population Size, Individuals	Number of Troops	Average Troop Size	Number of Macaques Shot or Trapped
1973[1]	480–600	>7	40–50	
1981[2]	825–900	15–20	52	94
1986[3]				66
1987[4]		10–20		70
1994[5]	350–400	27	11	3
1999	600–700			83

[1] see Poirier and Smith, 1974; [2] Farslow, 1987; [3] Kawamoto, *et. al.*, 1988; [4] Matsubayashi, *et.al.*, 1989; [5] Wheatley, *et. al.*, 1999b; [6] Marsh-Kautz and Singeo, 1999.

estimates using this method was therefore 401 individuals. Another method that we used was to take the estimated number of troops that were detected during our survey, 27, and the total number of individuals per troop, about thirteen individuals at the 99 percent confidence interval about the mean that we counted before the macaques detected us to give us an estimated 344 individuals. We concluded that our survey estimated about 400 macaques on Ngeaur which is considerably more than the 150 estimated people that reside on the island (Marsh-Kautz and Wheatley, 2004) (Table 10.1).

Another conclusion from our 1994 study (Wheatley *et al.*, 1999b) was that human hunting had a different effect on the macaques than non-human predators had on the macaques reported elsewhere. We found no significant correlation between the total number of animals in the troop and the detection distance. Larger troop sizes did not seem to avoid predators better. Not only were there fewer macaques per troop, but also, there were twice as many troops as reported since Farslow's (1987) and Matsubayashi's *et al.* (1989) studies (Table 10.1). Our interviews of hunters found that they preferred to hunt larger sized troops because they were more successful. Table 10.1 also lists the number of macaques shot or trapped on Ngeaur during the studies listed; three macaques were shot during our study (Wheatley *et al.*, 1999b). Natural selection appears to favor smaller-sized troops and quieter animals since Poirier and Smith's 1974 study.

A number of local people on Ngeaur that we interviewed in 1994 were contemptuous of monkeys. Some people hated them and wanted them killed. "We hunt monkeys because they are bad," they would say, or "everything monkeys do is bad." In general, however, most people probably leave the macaques alone as Poirier and Farslow (1984) mentioned. Poirier and Smith (1974) even state that their interviews revealed that the local people "had no desire to rid the island of monkeys" but "only to prevent crop damage." Several men told us that children especially seemed to enjoy the macaques. Several women also mentioned that monkeys were cute and smart and another woman said that she liked monkeys.

A survey conducted in Ngeaur by Marsh-Kautz and Singeo (1999) lists the monkey as the top-ranked pest causing the most damage and as the most difficult pest to control. Other pests listed are the shrew, rat, mouse, fruit fly, and taro viruses. Probably the most serious problem caused by these macaques is the damage to the taro gardens that are primarily in the swampy southern part of the island. Taro and other root crops are the staple agricultural foods of Palau. The taro gardens are small plots but give Palauan women great prestige in this matrilineal culture. "It is their life," they say, "One cannot breathe without taro." Women use taro to meet household and ceremonial food requirements as well as exchange obligations. The people typically roast, bake or boil taro corms. One woman said that the macaques ate the tender green shoots which spoiled the whole taro plant. They also stated that the macaques watched them plant which may attract them later when the plants were unguarded. A chief mentioned that the collective farms set up to help feed tourists were destroyed by macaques.

While macaques will damage gardens and crops, the extent of the damage is unknown. Poirier and Farslow (1984), Poirier and Smith (1974), and Farslow (1987) all state that reports of extensive crop damage were unsubstantiated. Although some troops of macaques range in the southern swampy part of the island where most of the taro gardens are located, most of the monkey population ranges in the northern part of the Island (Farslow, 1987). In both areas, the macaques primarily feed in the forests (Farslow, 1987).

Eradication programs

Ngeaur has had numerous eradication programs to hunt, kill, and trap macaques. They are not eaten or sold as bush meat: they are buried or said to be fed to pigs or crocodiles (Wheatley et al., 1999b). Some of the captured young are kept as pets or sold (Figure 10.3). The macaques were killed for sport, food, and medicinal purposes during the Japanese occupation (Poirier and Smith, 1974). A law passed in 1975 called for the complete eradication of the macaques as pests (Whaley, 1992). A year-long field study on the macaques of Ngeaur, by Farslow (1987) estimated that over 90 animals or 10% of the population were killed from 1980 to 1981. Farslow also describes two occasions when he was pinned down by ricocheting bullets from hunters of monkeys in the northern

Figure 10.3. A captive macaque on Ngeaur.

part of the island. The constitution of 1979 banned firearms and the eradication campaign ended, however, people continue to shoot and trap the animals. About 83 animals were shot between 1997 and 1999 and another attempted eradication program started in June of 2001(Wheatley *et al.*, 2002). The animals are killed to remove them as threats to garden areas and as sport. The lighthouse troop in the northern part of the island was most actively hunted despite the lack of gardens there (Farslow, 1987). About 70 animals were trapped on Ngeaur in 1986 for biological evaluations to assess the feasibility of establishing an experimental animal line (Matsubayashi *et al.*, 1989). Trapping appeared widespread in 1994. We were informed in 1994 that male baby monkeys were sold for $100. The government requires export permits for female macaques. Farslow (1987) reports that pet macaques were sold in Koror during his field study. He further mentions that people also hunt fruit bats and sell them off island.

Some non-lethal methods of control have been attempted. Monkey repellents of hot peppers and bilimbi (*oterbekai*) fruits were applied to trees and other plants for a few months but were ineffective after being washed away by the rain. According to informants, the monkeys eat *oterbekai* fruits rendering the control ineffective (Marsh-Kautz and Singeo, 1999). During their 1999 research, Marsh-Kautz and Singeo heard of several other local attempts to keep the animals out of gardens and away from houses. These include ringing large metal bells, placing scarecrows in gardens, and tying up dogs in the gardens.

The community apparently felt that those measures had little effect on the animals compared to trapping, firing weapons and shooting them.

Palau's National Invasive Species Committee, NISS, began "Operation Counter-Invasion" in 2006 (Anon, 2007). This project was developed by local communities and multiple organizations after initiation by the President of the Republic of Palau. It received $100,000 from the government of Taiwan. The two highest priority species identified by this committee were a vine (*Merremia peltata*) and (*Macaca fascicularis*). An expert in invasive mammal eradication, Mr. Karl Campbell, according to the summary report, submitted his recommendations "leading to eventual eradication of macaques from the Republic" (Anon, 2007). The Ministry of Justice also improved the enforcement of existing laws prohibiting transport of monkeys to other islands. Several other projects were also mentioned in this report that focused on preserving Palau's marine ecosystems. The NISS also began a Macaque Sterilization Project in 2007 that completed a pet macaque sterilization project in November of 2009 (Miles, 2009). The third phase will be the eradication campaign on Ngeaur.

Benefits of macaques

It is difficult to find any references to the benefits that the macaques offer on Ngeaur, but there were a few individuals who said that they were a good tourist attraction. Several other informants mentioned that macaques are intelligent, funny, and can provide companionship (Marsh-Kautz and Singeo, 1999). One of the chiefs told us that monkeys should not be viewed as a special natural resource, but he admitted that even he might keep a few monkeys around for tourists. This chief wanted to encourage more tourism on Ngeaur. He especially pointed to the additional employment opportunities that tourists provide for the local people such as being tour guides, preparing food, and renting the guest houses. Tourists also buy food and other supplies at the store. A sakura-kai or community organization exists to handle tourism and provide food for tourists. Tourists from many different countries signed the guestbook at the beachside guest house and mentioned that they enjoyed seeing the monkeys. Researchers drawn to the island to study monkeys have added to these visitor numbers. Tourists of all types have engaged in mutual beneficial interactions with the local people thus connecting all in a more global world (Marsh-Kautz and Wheatley, 2004). A survey conducted on Ngeaur reported by Marsh-Kautz and Singeo (1999) found that while most gardeners dislike macaques, some of the younger adults are becoming more accepting of macaques. Half of the respondents however, wanted the macaques eradicated. We were also told of the many requests by other islanders for baby monkeys which suggest the possibility that baby macaques are, or have been at times, a cash crop.

The macaques are a dispersal agent of the Palaun apple or *rebotel*, the tropical almond or *meiich* (*Terminalia catappa*), and *Macaranga carolinensis* (Farslow, 1987 and Wheatley *et al.*, 2002). We were shown how to crack open *meiich* during our fieldwork in 1994. The local people also made some candy from the nuts and shared it with us.

Introduced species

Many species besides macaques have been introduced either intentionally or inadvertently to Ngeaur and to Palau in general. Palau has been acquiring new species ever since it emerged from the ocean some 30 million years ago (Crombie and Pregill, 1999). Other introduced species to Ngeaur besides macaques include rodents *Rattus rattus*, *R. exulans*, and *R. norvegicus* and *Mus musculus* and the Asiatic musk shrew, *Suncus murinus*, (Crombie and Pregill, 1999; Pregill and Steadman, 2000). Several invasive species of fruit flies threaten the fruit crops such as mountain apple, carambola, guava and bananas (Anon., 2002). Epidemic outbreaks of dengue fever possibly from the introduced mosquitos *Aedes aegypti* and *A. albopictus* occur periodically such as in 1988, 1995 (Ashford *et al.*, 2003) and 2007 (Mesekiu's News, 2007). These diseases also pose a threat to many other species of wildlife. The determination of native species as opposed to invasive species can be somewhat tricky as van Leeuwen *et al.*, (2008) found on the Galapagos Islands using pollen samples. These researchers reclassified six presumed introduced species as native after finding pollen samples that predated European contact.

M. fascicularis is tagged as "one of the 100 worst invasives in the world" (Lowe *et al.*, 2000). Recently, however, because of losses due to extreme trade pressures, *M. fascicularis* will probably be categorized among the "primates still widespread but rapidly declining" (Eudey, 2008). Ninety-nine percent of all imported primate species into the US are *M. fascicularis* (Thirlway, 2009). Most of these animals are used for biomedical research.

Discussion

The simplest and most cost-effective management priority is to prevent invasive species from gaining a foothold in a country (McNeely *et al.*, 2001). The macaques of Ngeaur have spread throughout the island and have lived there for over one hundred years. The macaques are very prolific. Farslow (1987) estimates 150 new infants per year based on his population estimate of 900 animals. His estimated population density of 108 animals per square kilometer is high compared to that of some other populations. If we assume that there are 1,000 or so animals on the island it would seem extremely time consuming to exterminate every macaque. An example might be instructive. An experimental release of a single Norway rat onto a 9.3 hectare island took almost five months to trap (Russell *et al.*, 2005). This was despite intensive trapping efforts and the fact that the rat's position was known at all times via a radio transmitter. Farslow (1987), our surveys and the statements by local people show that the population of monkeys is much higher on the northern part of the Island, north of the ruins of the phosphorus refinery. The macaque population has tenaciously survived numerous calamities. German and later, Japanese phosphate mining as well as warfare and extensive military occupation have damaged the island and destroyed many plants and animals (Lorenza, 1999). In the 1930s over 3,000 people lived on the Island. The macaques survived saturation bombing during World War II and several major typhoons, including typhoon Louise in 1964. One man interviewed had admiration for the ability of monkeys to survive the war. He also said that it would be a shame if there were no more monkeys on Ngeaur after all that they went through.

Macaques can survive severe droughts and forest fires such as those in the early 1980s in eastern Borneo (Berenstain, 1986). They can hide and live in the numerous coral nooks and crannies on Ngeaur Island. Furthermore, the animals have apparently adapted to hunting and trapping pressures by ranging in smaller-sized troops (Wheatley *et al.*, 1999b) and by remaining silent and out of sight (Farslow, 1987; Wheatley *et al.*, 1999b). Trapping is also not as effective as it was when it was introduced (Marsh-Kautz and Wheatley, 2004). If trapping continues it is to be hoped that humane concerns will prompt close attention to trapped animals. We heard a number of stories about the death of trapped animals because the traps were not checked as frequently as they should have been to ensure that food and water were being provided. It is worth remembering that the Indian ban on the export of rhesus macaques was imposed for humanitarian reasons (Eudey, 1995). The local people of Ngeaur are sensitive about the possibility of chemicals, whether they are poisons or birth control hormones getting into the environment or their food supply. Thus, all eradication measures up to now have failed. The eradication of monkeys from Ngeaur may thus be very costly and very time consuming. Complete eradication may even be impossible.

The various anonymous reports and pamphlets including the public awareness campaign, Operation Counter-Invasion against monkeys and a vine (Anon, 2007), almost sounds like a military offensive. The message in its entirety is exaggerated perhaps as an effort to frighten people into controlling this macaque better. In an earlier paper, Wheatley *et al.*, (2002) noted the association between colonial powers, especially the Germans and macaques that some local residents mentioned. Why don't the Germans come back and take the monkeys away? They brought them here said one resident. It seems possible that the macaques are a symbol of past colonial resentment and the extensive environmental damage done during colonial times from which the people of Ngeaur still must contend with on a daily basis. The negative economic and health effects of European contact were noted by Kramer (1919). One conservation officer in Koror is reported to have said that "the monkeys were brought in by foreigners and are now being protected by foreigners such as Farslow" (Whaley, 1992). "He got his Ph.D. from the monkeys, but what about the people of Angaur? They get nothing."

Eradication is often feasible only when colonization is at an early phase (Clout and Lowe, 1996). This early colonization appears to be the case for the macaques on some of the smaller islands of Palau. Another priority might be to better protect Babeldaob since it is the largest island in the archipelago accounting for about 80 percent of Palau's total land mass. Therefore, the emphasis against the illegal transport of macaques from Ngeaur to other islands in Palau in the Invasive Mammal pamphlet (Anon, 2008) appears to be a good idea. The steep penalty of $500 for illegally exporting a monkey appeared to be an effective deterrent according to informants in 1999. Such fines and the close vigilance of residents on other islands in Palau, so far, have dampened the spread of these macaques to other Islands. Control and containment on Ngeaur may thus be the only realistic possibility. Some interviewees did accept the idea of a co-existence between them and macaques if the latter population could be controlled (Marsh-Kautz and Wheatley, 2004). Younger adults especially considered macaques to be part of the island's heritage in the survey by Marsh-Kautz and Singeo (1999).

There is one study on the extent of damage to subsistence gardening by the macaques despite Kemp and Burnett's (2003) statement that "No ecological studies on the impact of *M. fascicularis* in Ngeaur have apparently been conducted. That study was conducted by Farslow (1987) in 1980 and 1981. He and other primate researchers such as Poirier and Farslow (1984), Poirier and Smith (1974) state that reports of extensive crop damage are unsubstantiated. Farslow (1987) says that the macaques are not your "typical crop raiders" as there are no commercial farms, but this does not make the damage that they do to the gardens as any less serious from the people's point of view, especially as most still live a largely subsistence lifestyle.

Most of the taro gardens were undamaged during Farslow's study period and damage was mostly on the periphery of the gardens. Only 2 percent of the macaque damage to taro involves damage to the root system; the rest of the damage consists of breaking off leaves and stalks (Farslow, 1987). As previously mentioned, however, taro is of the highest priority to the local women. They say taro is the mother of our life. Farslow (1987) reports damage to the taro crops and betel nut to be the most serious problem and that most of the damage was on the southern third of the island. Evidence of only one small taro leaf was found eaten by a macaque during our study in 1994 (Wheatley *et al.*, 2002) and this plant was one of about a thousand plants located under the animal's sleeping tree in the swamp. Raw taro is generally considered inedible as it is acrid and causes an intense burning and itching sensation. It is doubtful that macaques eat much raw, mature taro. The young and newly planted taro is at most risk because it is said to be sweet and tender without the itchy irritation to the mouth.

Prevention of damage to taro is one of the highest concerns to the people of Ngeaur and effective controls could focus on those crops mostly found in the swamps. One informant mentioned that chickens actually do more damage to taro than macaques because of the scratching around the plants. Dogs are reportedly used on Ngeaur to chase macaques out of the gardens. Another important crop that macaques are alleged to damage is betel nut. Some local people stated that in some years, all of the betel, especially when young was eaten by macaques. According to Farslow (1987) the macaques receive the most hostility from the local people because the animals eat betel nut. Ngeaur is famous for its high quality betel nut as far away as Yap and Guam. It is considered a culturally valuable crop and increasingly, a profitable cash crop. Garden crops grown near houses such as corn, yams, tomato, beans, watermelon, sugarcane and squash were not eaten by feral monkeys during Farslow's (1987) study. In addition, he says that most of the crops such as papaya and banana were inaccessible to the animals. At present, there are no studies, besides that of Farslow (1987), that document the percentage of agricultural foods or natural foods consumed by Ngeaur macaques. Two of the favored foods of monkeys (Poirier and Smith, 1974; Farslow, 1987) were in fruit during the time of our 1994 study: the Palauan apple, *Eugenia javanica* and the tropical almond, *Terminalia catappa*. Both of these species fruit several times a year (Farslow, 1987): thus there is a continual supply of food for the animals in the forests.

Some of the other introduced species on Ngeaur also have a significant impact on the socio-economy and the environment. Crombie and Pregill (1999) state that they believe that *Rattus rattus* is a substantial predator of herpetofauna and birds. Rats and starlings do more damage to crops than macaques according to Farslow (1987). Rats were seen by Farslow (1987) chewing off large pieces of cassava roots as well as eating all cultigens and fruits. Carter (2007) identifies rats as the main predator on the endangered and endemic bird, the Mauritius fody with macaques as the second major predator. The large, green tree skink eats ripe papaya on Ngeaur (Crombie and Pregill, 1999). Starlings also eat papaya. Farslow (1987) states that "no papaya fruit left to ripen on the tree is free from attack by starlings." He also mentions that fruit bats eat bananas. Monitor lizards were apparently introduced to Ngeaur Island to control rodents and "they are notorious among the residents for eating chicken eggs and chicks" (Farslow, 1987). These large, six-foot long lizards (*Varanus*) are increasing in numbers and one informant said that they are important predators of birds on Ngeaur and elsewhere. Farslow (1987) recommends a study of the effects of lizard predation on birds. We were told that the shrew eats taro, cassava, and sweet potato. There is no documentation as far as we know of macaques eating crabs and birds in Palau. Neither Farslow (1987) nor Poirier and Smith (1974) ever saw monkeys prey on crabs, birds or their nests. Wheatley *et al.* (2002) previously reported that according to local people, even starving macaques would not eat the land crabs. The statement that Ngeaur macaques probably disperse alien plant species as mentioned in the Invasive Species Specialist Group database (Carter, 2008) and by Kemp and Burnett, (2003) is also unsupported. A study of all the endangered and exotic plants and animals that need protection, not just on Ngeaur, but also on the other islands of Palau is sorely needed. For example, if Ngeaur were found to have a high percentage of endemics or a high level of biodiversity then sometimes eradication might be more understandable (Clout and Lowe, 1996).

The most significant effect on natural biodiversity is habitat loss (Mittermeier, 1987; Lee and Macdonald, 1996). A National Report to the United Nations Convention to Combat Desertification (Anon, 2002) states that after climate change, the most serious threat to the Republic of Palau is land degradation caused by population growth and development. This report specifically refers to the 53-mile road on Babeldaob which has caused significant vegetation loss including the dredging of reefs. Mature forests are being cleared there and siltation is also occurring (Pregill and Steadman, 2000). The construction of new roads and hotels should be ecologically free from erosion and siltation. The various entities responsible for threatening biodiversity could be held accountable for any damage. The State of Hawaii, for example, began imposing fines on tour companies for damaging coral reefs several years ago. The money can be used to help restore reefs, for example by re-attaching broken coral colonies. According to Dr Soule (2008), the founder of the Society for Conservation Biology, roads and development are by far the biggest of all environmental problems, even bigger than climate change.

The Palauan government showcases its marine resources and has formed a Marine Invasives Survey Team (Anon, 2007). An important endemic turtle nesting beach on Ngerchur Island was recently protected by the eradication of feral pigs. Exotic turtles could also be a serious threat to endemic freshwater fish and Crombie and Pregill (1999) state that further import of these turtles should be prohibited and their presence eradicated.

Recommendations for management

This paper has attempted to discuss the scale and priorities of conservation concerns on Ngeaur. Resources such as money and time are often limited in dealing with introduced species. Can years of effort and expense towards eradication of monkeys be justified when other, more dangerous species to human health and to the economy exist? For species not causing substantial harm, the most cost-effective method might be to exert some control over their numbers and practice the LTL approach- Learn to live with 'em (Davis, 2009). It is difficult for primatologists to understand how such interesting animals cannot be viewed as a unique resource. They are the only macaque population in all of Oceania. They could be promoted as a tourist attraction to a much greater extent than they are. Tourists are likely to be upset, however, and perhaps at some risk at seeing the hunting of macaques and dead macaques as we were during our brief stay in 1994. The macaques are, nevertheless, not generally viewed as a unique resource.

Management of the macaque population on Ngeaur Island is also a conclusion reached by Farslow (1987) and he suggests, as only one of several options, the trapping and selling of some macaques. Respondents to the survey by Marsh-Kautz and Singeo (1999), favored continued hunting and the selling of macaques to foreigners. Who should be the managers? A good place to start might be with the community-based sakura-kai organization. This organization could use tourist and perhaps government money to organize conservation efforts and macaque management and it could compensate some of the local people who suffer crop damage.

The best antidote for monkey-free gardens may be the ubiquitous presence of humans and dogs, guarding the fields, and preventing the macaque's access to the fields; a sort of human fence that would be on-guard for monkeys. While labor-intensive, it is still the most effective means based on my experience in Bali and Kalimantan. Other primatologists such as King and Lee (1987) also found this to be effective in Malawi against crop-raiding vervet monkeys. Perhaps the best that one could do on Ngeaur is, as humanely as possible, get the macaques out of the swampy taro producing areas in the entire southern third of the island. Keeping the monkeys north of the village using guarding tactics, dogs, and residences might be effective in preventing the animals' return to the south. The southern half of the island could then be used for gardening, betel nut trees, and taro. Perhaps too, a bitter variety of taro can be planted around the periphery of the gardens so that monkeys might learn to avoid eating taro. The farmers of Bali used this approach, although the monkeys learned to use their teeth to peel away the most toxic part of the cassava before they ate it (Wheatley, 1999a). Monkeys north of the village, near the lighthouse area would be much more difficult to eradicate because of the extensive coral outcroppings and this is the area where tourists can view them. Tourists should not be allowed to feed or interact with monkeys because this will increase the potential for human and macaque conflict. Keeping these macaques as pets should also be discouraged because of the potential for disease transmission as well as other concerns.

Conclusion

The macaques of Ngeaur in the Republic of Palau are a good example of ethnophoresy. They were brought to Ngeaur Island during the German occupation sometime around 1909. The macaques are viewed as a serious pest in Ngeaur despite some claims to the contrary. The macaque population exceeds the human population on Ngeaur. The damage that macaques do can be exaggerated while the damage that other animals do can be ignored or minimized. The allegations, unsupported assertions, and anonymous reports are repeated as if they were facts. One needs to ask, why assertions such as these are regularly made and who might profit from them.

Ethnoprimatology has a conservation application because it examines the interactions between humans and other primates. This growing field is at the forefront of many conservation efforts and we may need to study other species to better understand human-animal problems and to offer more rational management decisions. Through dialogue we should be able to find common ground among the conservationists of Palau in protecting its endangered species and in promoting environmental concerns to preserve the beauty of Palau. The success of any conservation project and the future of the macaques of Ngeaur will ultimately be up to the people of Ngeaur themselves.

Acknowledgements

I am grateful to the community of Ngeaur for their generous hospitality. Their receptiveness to our studies on their Island and their sharing of information with us were of immense help. I also thank Kelly G. Marsh for helping me access updated information about the macaques in Palau and for her review of an earlier version of this chapter. My thanks are also due to the anonymous reviewers.

References

Anon. 2002. National Report to the United Nations Convention to Combat Desertification, April 2002. Office of Environmental Response and Coordination, Office of the President of the Republic of Palau, 15 pp.

Anon. 2007. "Operation Counter-Invasion." Interim Implementation Plan for the Palau National Invasive Species Strategy. Summary Report August, 2007. Prepared by the Palau National Invasive Species Committee. A committee of the National Environment Protection Council.

2008. Invasive mammal: Monkey (*Macaca fascicularis*), crab-eating macaque. www.palau.biodiv-chm.org/.

Ashford, D. A., Savage, H., Hajjeh, R., *et al.* 2003. Outbreak of dengue fever in Palau, Western Pacific: Risk factors for infection. *American Journal of Tropical Medicine and Hygiene* **69**(2): 135–140.

Berenstain, L. 1986. Responses of long-tailed macaques to drought and fire in eastern Borneo: A preliminary report. *Biotropica* **18**(3): 257–262.

Carter, S. 2007. *Macaca fascicularis* (mammal) 2007. ISSG database: Ecology of *Macaca fascicularis* and impact information for *Macaca fascicularis*, pp. 1–3.

Clout, M. and Lowe, S. 1996. *Biodiversity loss due to biological invasion: Prevention and Cure*. Conserving Vitality and Diversity. IUCN, Montreal. pp. 29–40.

Crombie, R. and Pregill, G. 1999. A checklist of the herpetofauna of the Palau Islands (Republic of Belau), Oceania. *Herpetological Monographs* **13**: 29–80.

Davis, M.A. 2009. *Invasion Biology*. New York: Oxford University Press.

Eudey, A. A. 1995. Southeast Asian primate trade routes. *Primate Report* **41**: 33–41.

2008. The crab-eating macaque (*Macaca fascicularis*): widespread and rapidly declining. *Primate Conservation* **23**: 129–132.

Farslow, D. 1987. The behavior and ecology of the long-tailed macaque (*Macaca fascicularis*) on Angaur Island, Palau, Micronesia. Ph.D. dissertation, Ohio State University.

Fooden, J. 1991. Systematic review of Philippine macaques (Primates, Cercopithecidae: *Macaca fascicularis* subspp.). *Fieldiana. Zoology n. s*, **64**: 1–44.

Heinsohn, T. 2003. Animal translocation: Long-term human influences on the vertebrate zoogeography of Australasia (natural dispersal versus ethnophoresy). *Australian Zoologist*. **32**(3): 351–376.

Hooijer, D. A. 1952. Fossil mammals and the Plio-Pleistocene boundary in Java. *Koninklijke Nederlandse Akademie van Wetenschappen, Proceedings, Series B* **55**: 436–443.

Kawamoto, Y., Nozawa, K., Matsubayashi, K., and Gotoh, S. 1988. A population-genetic study of crab-eating macaques (*Macaca fascicularis*) on the Island of Angaur, Palau, Micronesia. *Folia Primatologica* **51**(4): 169–181.

Kemp, N. J. and Burnett, J. B. 2003. A biodiversity risk assessment and recommendations for risk management of long-tailed macaques (*Macaca fascicularis*) in New Guinea. Final report. Indo-Pacific Conservation Alliance: Washington.

King, F.A. and Lee, P.C. 1987. A brief survey of human attitudes to a pest species of primate-*Cercopithecus aethiops*. *Primate Conservation* **8**: 82–84.

Kramer, A. 1919. Ethnography (II): Micronesia (B). In *Results of the South-Pacific expedition 1908–1910*, G. Thilenius (ed.). Hamburg: L. Friedrichsen & Co. Translation in 2002 by C. H. Petrosian-Husa, Republic of Palau, Bureau of Arts and Culture.

Labang, D. and Medway, L. 1979. Preliminary assessments of the diversity and density of wild mammals, man and birds in alluvial forest in the Gunong Mulu National Park, Sarawak. In *The Abundance of Animals in Malesian Rain Forests*, A. G. Marshall (ed.). Hull, UK. pp. 53–66.

Lee, G. and Macdonald, I. 1996. *Conserving Vitality and Diversity*, Foreword., IUCN, The World Conservation Union. pp. v–vi.

Lorenza, P. 1999. The effects of foreign culture and school on Angaur, Palau, 1899–1966. MA thesis. Micronesian Area Research Station, University of Guam, Mangilao.

Lowe S.J., Browne, M., Boudjelas, S., and DePoorter, M. 2000. IUCN/SSC Invasive Species Specialist Group (ISSG), Auckland, New Zealand.

Luskin, D. 1989. Islander. *The Pacific Sunday News Magazine* **1** October: 5–9.

Marsh, K. G. 2009. Palau National Registrar of Historic Places Training Workshop, Final Report. *Obis ra Ibetel a Cherechar* (Bureau of Arts and Culture), Ministry of Community and Cultural Affairs, Republic of Palau.

Marsh-Kautz, K. G. and Singeo, Y. 1999. Ngeaur community perceptions of their island's *mongkii (Macaca fascicularis)*: An island under siege. Roughdraft. Report prepared for the community of Ngeaur. Privately circulated.

Marsh-Kautz, K. G. and Wheatley, B. 2004. Connecting local and international: Exploring the pressing mongkii (monkey) issues of Ngeaur, Beluu er a Belau (Angaur, Republic of Palau). In *The Challenges of Globalization*, Lan-Hung Nora Chiang, J. Lidstone, and R. Stephenson (eds.). Maryland: University Press of America, Inc. pp. 141–157.

Matsubayashi, K., Gotoh, S., Kawamoto, Y., . Nozawa, K., and Suzuki, J. 1989. Biological characteristics of crab-eating monkeys on Angaur. *Primate Res* **5**: 46–57.

Mesekiu's News. 2007. Update on Dengue Fever. Palau Community College, *Weekly Newsletter* **9** (28) July 13.

McNeely, J. A., Mooney, H. A. Neville, L. E., Schei, P., and Waage, J. K. (eds.) 2001. A global strategy on invasive species. Gland, Switzerland and Cambridge, UK: IUCN.

Miles, J.E. 2007. Macaque Sterilization Project Overview, November 2007.

Miles, J. E. 2009. Project Progress. Pacific Invasive Initiative, The News, November 2009. Available online at www.issg.org

Mittermeier, R. A. 1987. Effects of hunting on rain forest primates In: *Primate Conservation in the Tropical Rain Forest*, C. W. Marsh and R. A. Mittermeier (eds.) New York: Alan R. Liss Inc. pp. 109–146.

272   *Bruce P. Wheatley*

Palau Community Action Agency. 1977. *A History of Palau*, Vol. Two: Traders and Whalers, Spanish Administration and German Administration. Koror, Palau.

Poirier, F. and Farslow, D. 1984. Status of the crab-eating macaque on Angaur Island, Palau, Micronesia. *Primate Specialist Group Newsletter* 4: 42–3.

Poirier, F. E. and Smith, E. O. 1974. The crab-eating macaques (*Macaca fascicularis*) of Angaur Island, Palau, Micronesia. *Folia Primatol.* 22: 258–306.

Pregill, G.K. and Steadman, D.W. 2000. Fossil vertebrates from Palau, Micronesia: A resource assessment. *Micronesica* 33 (1/2):137–152.

Russell, J. C., Towns, D. R., Anderson, S. H., and Clout, M. N. 2005. Intercepting the first rat ashore. *Nature* 437: 1107.

Smith, A., Black, D., and Eberle, R. 1998. Molecular evidence for distinct genotypes of monkey B virus (*Herpesvirus simian*) which are related to the macaque host species. *Journal of Virology* 72: 9224–9232.

Soule, M. 2008.Thinking anew about a migratory barrier: roads. *The New York Times*, 14 October. p. D3.

Sponsel, L. E. 1997. The human niche in Amazonia: explorations in ethnoprimatology. In: *New World Primates: Ecology, Evolution, and Behavior*. W. Kinzey (ed.). Hawthorne, New York: Aldine de Gruyter. pp. 143–165.

Sussman, R. and Tattersall, I. 1981. Behavior and ecology of *Macaca fascicularis* in Mauritius: a preliminary study. *Primates* 22: 192–205.

Thirlway, H. 2009. US and UK primate imports. *International Primate Protection League News* May 36(1): 14–16.

Tjia, H.D. 1970. Quaternary shore lines of the Sunda Land Southeast Asia. *Geologie en Mijnbouw* 49(2): 136–144.

van Leeuwen, J., Froyd, C., van der Knapp, W., *et al.* 2008. Fossil pollen as a guide to conservation in the Galapagos. *Science Magazine* 322(5905): 1206.

Wallace, Alfred Russel. 1869. (new ed. 1962). *The Malay Archipelago*. New York: Dover Publications, Inc.

Whaley, F. 1992. Palau wants monkeys off its back. *Pacific Daily News*, 6 December, 1, 4.

Wheatley, B. P. 1999a. *The Sacred Monkeys of Bali*. Prospect Heights, Illinois: Waveland Press, Inc.

Wheatley, B.P., Stephenson, R., and Kurashina, H. 1999b. The effects of hunting on the Long-tailed Macaques of Ngeaur Island, Palau. In: *The Nonhuman Primates*, P. Dolhinow and A. Fuentes (eds.). Mountain View, California: Mayfield Publishing. pp.159–163.

Wheatley, B. P., Stephenson, R., Kurashina, H., and Marsh-Kautz, K. G. 2002. A Cultural Primatological Study of *Macaca fascicularis* on Ngeaur Island, Republic of Palau, In *Primates Face to Face*, A. Fuentes and L. D. Wolfe (eds.). Cambridge University Press. pp. 240–253.

Part IV

Comparisons with rhesus macaques

11 India's rhesus populations: Protectionism versus conservation management

CHARLES H. SOUTHWICK
AND M. FAROOQ SIDDIQI

Introduction

The rhesus monkey *(Macaca mulatta)* population of India has long been one of the classic non-human primate populations of the world. Scientific field studies of abundance were not undertaken until the 1950s, but leading naturalists in India estimated their numbers in the millions "in the United Provinces alone [Uttar Pradesh] the monkey population– in my opinion–is not less than 10 million..." (Corbett, 1953). Other naturalists referred to "vast hordes of rhesus roam large parts of India..." (Sanderson, 1957). Rhesus were abundant in forests, agricultural areas, villages, roadsides, parks, temples, and limited areas in towns and cities.

Although the truly sacred monkey of classical Hinduism is the Common or Hanuman langur, rhesus are also considered sacred animals, representatives of the Monkey God, Hanuman. Hanuman holds a high status in Hindu religion since the Monkey God played a key role in the reunion of Rama, the incarnation of Lord Vishnu, and his beloved wife, Sita. Sita had been kidnapped by the Demon King, Ravana, and taken to the island of Lanka (Ceylon). The defeat of Ravana by Hanuman and his troop of monkeys and reunion of Rama and Sita symbolized the victory of Good over Evil. This is a central concept in Hindu philosophy, and still a very common belief today. It is celebrated every year throughout India in the story of the Ramayana, and the song, dance, and classical drama of the Ramlila. Hence, Hanuman is one of the most popular Hindu deities, and monkeys are traditionally honored and worshipped in classic Hinduism (Lutgendorf, 2007).

In the past 50 years, several forces, demographic, economic, cultural, and ecologic, have all impacted the traditional monkey–human relationship in

Monkeys on the Edge: Ecology and Management of Long-Tailed Macaques and their Interface with Humans, eds. Michael D. Gumert, Agustín Fuentes and Lisa Jones-Engel. Published by Cambridge University Press. © Cambridge University Press 2011.

India. As a result, both human and rhesus populations have shown various stages of change. After a brief description of the methods of this study, these stages are summarized in the Results.

Methods

This field study began in 1959 in association with Aligarh Muslim University in Aligarh, U.P. (i.e., Uttar Pradesh, formerly the United Provinces), north central India. Aligarh is a city in the rich agricultural area of the Gangetic basin, 130 km southeast of New Delhi. Our first objectives were to obtain field data on the abundance, group sizes, sex and age ratios, and habitat distribution of rhesus monkeys in India, primarily in Uttar Pradesh, but extending to the neighboring states of Delhi, Rajasthan, Madhya Pradesh, Haryana, Punjab, Himachal Pradesh, Bihar, and West Bengal. We undertook systematic surveys of forests, agricultural areas, roadsides, canal banks, villages, towns, cities, temples, and public areas such as parks, archaeological sites, and railway stations. Much of our field coverage was on foot, but we also traveled by bicycle, motor scooter, automobile, and occasionally by train or boat to cover greater distances.

In Aligarh District, we selected a representative sample of 20 groups in various typical habitats to census three times annually: In February, before the birth season; in June and July, after the birth season; and in November, after the monsoon and before winter. We did not include a local Hindu Temple site in these annual censuses. Three annual censuses were maintained for 25 years – for the remainder of this 50-year study, we reduced annual censuses to just one a year, in June and July.

Throughout the 50 years of this project, our goals have focused on population abundance, group sizes, sex and age ratios of groups, geographic and habitat distribution, population trends, and ecologic relations with other animals and humans. Several short-term studies of social behavior and ethology were also done, but the emphasis in this paper is on population ecology. Similar field research on rhesus was also undertaken in neighboring countries including Nepal, Burma [Myanmar] Bangladesh, and China. Details on ecological and behavioral field methods are given in other papers listed in the references (Southwick, Beg and Siddiqi, 1961a, 1961b, 1965; Southwick *et al.*, 1980; Southwick and Siddiqi, 2001).

Results

The introduction lists four types of influences affecting rhesus monkey populations in India: demographic, economic (and political), cultural, and ecologic.

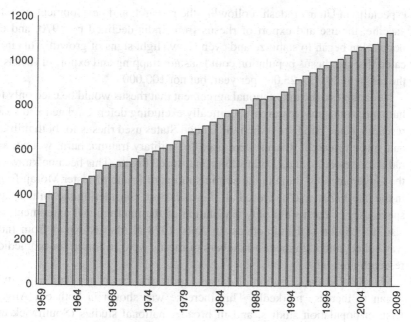

Figure 11.1. Human population growth in India between 1959 and 2009 in millions of people.

In demographic terms, the human population of India has increased from 360 million in 1959 to 1.14 billion in 2009, more than a three-fold increase (Figure 11.1). Even in 1953, Jim Corbett recognized a monkey problem by noting "ten million monkeys living on crops and garden fruit present a very major problem." Nonetheless, because of the sacred status of monkeys, combined with the modest aspirations of Indian villagers, most villagers seemed to accept this problem without too much complaint. For many, it was a normal part of their lives.

In response to economic factors, rhesus populations began to decline in the late 1950s with increasing use of rhesus in biomedical research throughout the world. India had conflicting views about exporting monkeys, but a politically acceptable agreement was signed that assured India that monkeys would be used for humane biomedical research with notable benefits for human health, especially vaccine development for polio. For many years in the late 1950s and early 1960s India exported more than 100,000 rhesus monkeys annually, and as many as 200,000 in some years (Bennett, Abee and Henrickson, 1995; Blum, 1994). This clearly had a negative effect on rhesus populations, and our surveys showed declining numbers of rhesus throughout northern India, especially in Uttar Pradesh. Following the research and development of polio vaccine, the use and export of rhesus from India declined by 1970, and the population began to stabilize and even show slight signs of growth. This indicated that the rhesus population could sustain trapping and export of rhesus in the range of 20,000–25,000 per year, but not 100,000.

Nevertheless, the international agreement that rhesus would be used only for humane biomedical research, specifically excluding defense-related and space research, came into focus when the United States used rhesus for both military and space research. Rhesus were used for military trauma, burn, wound, and radiation research, and also in sub-orbital space flights. This became known to the Government of India in the administration of Prime Minister Moraji Desai in the late 1970s, and he closed all rhesus export, considering the military and space uses of rhesus not only a violation of the international agreement, but also a violation of Hindu ethics. In April 1978 all rhesus export from India was halted, although modest numbers of rhesus were still used for biomedical research in India.

Shortly after the cessation of rhesus export, the rhesus populations of India began to increase markedly. This increase was shown in both our Aligarh District population studies, and in broader national studies (Southwick and Siddiqi, 1988).

Aligarh District population

In the 1960s, our Aligarh District population declined from 403 rhesus in 22 groups to 175 in only ten groups. These were mainly rural agricultural and village groups in an area of approximately 10,000 sq. km., and did not include Aligarh temple groups. In the 1970s, with reduced trapping for export, this population remained at ten groups, but increased to 275 monkeys. Since 1979, however, this population has increased almost three-fold to 744 (Figure 11.2), even though it has declined in the number of groups to just eight. The average group size increased from 18.3 in 1962 to 93.0 in 2008. The reason for the loss of groups, from 22 to ten, and subsequently from ten to eight, accompanied by a great increase in group size was that some groups were no longer tolerated by local people and were trapped or driven away, whereas the other groups were tolerated, even protected and fed, and given free-rein to increase.

Since 1979, the annual increase of our Aligarh population has averaged 4.2 percent, varying from a yearly decrease of 10.2 percent to an increase of 21.8 percent. In the 1960s when trapping and export of rhesus were common, the annual rate of population change was negative, -8.4 percent, varying from +12.3 percent in just one year to –19.7 percent.

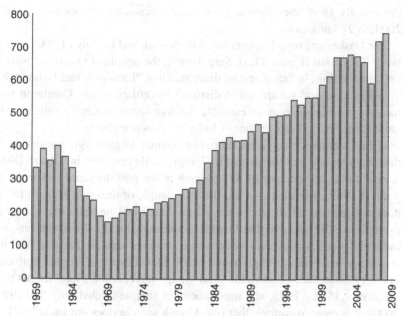

Figure 11.2. The rhesus macaque population in Aligargh between 1959 and 2008

A separate group in our study area, the Qasimpur group, is not included in the above numbers because it originated as a transplant of 20 monkeys into our area. In 1982, an unhappy villager whose mango grove outside our main study area was being heavily damaged by a group of more than 100 rhesus monkeys pleaded with us to remove some of these monkeys to reduce his crop losses. He claimed rhesus were taking one-fifth of his crop. We agreed to remove some, not realizing how difficult it would be to find a place to put them. The trapping was no problem, but finding a suitable area where villagers or neighboring people were willing to accept even a small number of rhesus proved to be frustrating and difficult. After a year's field effort, we found a forest patch of 4 ha along a canal, 2 km. from a village known as Qasimpur where the canal authorities, the owners of the woodlot, and the villagers of Qasimpur were willing to have 20 rhesus relocated in the forest patch. We did this in September of 1983. For several months, one of us (MFS) distributed food for these monkeys in an effort to keep them at the forest release site. The monkeys ranged along the canal bank, however, and finally after several months took up residence at a bridge crossing in the village of Qasimpur. This group of 20monkeys began reproducing within a few months, and for the next six years, the birth rates of adult females in the group were 90 to 100 percent. By 1994, the original group of 20 rhesus had grown to 89, and by 2004, to 227 monkeys.

The Qasimpur group has continued to expand, and by July of 2008, it numbered 258 rhesus (Figure 11.3). Surprisingly, the people of Qasimpur were not yet complaining. In fact, some of them said that "Hanuman had brought these monkeys to us and we are not to disturb." Nevertheless, this Qasimpur translocation provided a dramatic example that translocation is not a valid management procedure to limit or control India's rhesus population.

If the Qasimpur group is added to our original Aligarh eight census groups, the total population of rhesus that we have recently censused in Aligarh District is now 1,002, an increase of 473 percent in the past 40 years (since the low point of 1969). We must also add that our sample of rhesus in Aligarh District does not include all rhesus in the Aligarh area and neighboring towns. For example, we did not cover the Aligarh temple area, local railroad stations, local industrial plants, or several towns, where rhesus groups are present. An intensive rhesus study in the Aligarh area by Dr. Ekwal Imam of Aligarh University in the 1990s included eight additional groups and counted a total rhesus population of 1,337 by 1995, at a time when our sample totaled only 540 (Imam, 2000). It is clear, therefore, that our Aligarh sample does not include all rhesus in the area, although we feel it is representative of rhesus populations and trends where rhesus occur in India, because it contains typical habitats and human population patterns.

Discussion

It is apparent that several forces have combined in India to alter the ecological relationships of human and non-human primates. With the combination of rapid human population growth and rapid rhesus population growth, interacting with forest destruction and conversion, plus the commensal tendency of rhesus to adapt to human villages, towns and cities, rhesus groups are now often considered pest problems. Both agricultural losses and health and safety issues are increasingly evident, including occasional serious attacks on humans by rhesus.

Agricultural problems have been highlighted in rice, wheat, and soybean areas, where agribusiness has taken a strong stand against rhesus, but individual villagers are even more vulnerable to crop depredations. Fruit orchards are especially susceptible to rhesus damage: mango, papaya, bananas, apples, pears, and peaches. In a farm associated with Aligarh University on the edge of the city, the farm manager felt that 20–25 percent of his mango, papaya, and banana crop was lost to rhesus monkeys, although we have no specific data

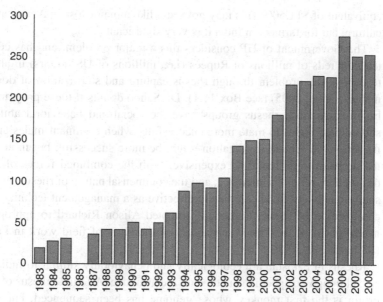

Figure 11.3. The rhesus macaque population growth at Qasimpur Canal Site after 20 macaques were introduced in 1983.

to confirm this. His estimates of crop loss were similar to those of the owner of the mango grove north of Aligarh from which we removed 20 monkeys to establish the Qasimpur group. We do not have estimates of field crop losses in Aligarh District, but time-sample behavioral observations of roadside groups at Chatari, north of Aligarh, showed that 10 percent of their total feeding time was spent on field crops, primarily wheat, grams or pulses (i.e., a type of bean), and sugarcane, whereas 83 percent of their feeding time was spent on food handouts from passersby on the road, which came from human food sources (Siddiqi and Southwick, 1988). These food handouts, from pedestrians, cart drivers, and occasional motorists, relieved pressure on adjacent field crops.

In the Himalayan foothills of Himachal Pradesh (HP), northwest India, where orchard crops are prominent, rhesus are considered a serious problem. Research on agricultural crop losses by S. K. Sahoo in 46 farmland sites of six Districts in HP showed District crop losses varied from 11.4 percent to 30.9 percent, averaging 19.4 percent. Major horticultural crops damaged were apple, pear, cherry, plum, almond, walnut and apricot (Sahoo, 2005). Major field crops fed upon by monkeys, primarily rhesus, but also including common langurs, were maize, wheat, potato, vegetables and pulses. Monetary losses for these 46 farms were estimated at 5.23 lakhs (520,000 Rupees), or a US dollar equivalent of $11,500. This may not seem like a major loss in American agriculture, but for farmers in India it is very significant.

The Government of HP considers this a major problem, and has committed hundreds of millions of Rupees (i.e., millions of US dollars) in attempts to control the problem through rhesus capture and sterilization of dominant males (Sahoo, 2005) (see Box 11.1). Dr. Sahoo doubts if these programs will be effective since rhesus groups have the social and behavioral ability for subordinate males to mate more successfully when dominant males are sterile. Efforts at female sterilization might be more successful, but in any case, the problem is difficult and expensive. With the combined forces of habitat destruction, religious tolerance, and the commensal nature of rhesus, no single solution is likely to be completely effective as a management technique. This combination of rhesus traits has prompted Alison Richard to appropriately name rhesus as the "weed macaque" on the basis of field work in Pakistan (Richard *et al.*, 1989).

Rhesus in Pakistan are ecologically and behaviorally similar to Indian rhesus, although no specific genetic data are available. A female rhesus of Indian origin is the first monkey whose genome has been sequenced, but little is known of the biogeography of rhesus genetics. Certainly, genetic differences occur throughout the wide geographic range of rhesus, from Afghanistan and Pakistan to China andSoutheast Asia. In local and regional areas, genetic variation is maintained by the tendency of some males to leave natal groups and enter others, a process we have witnessed, but we regret we do not have quantitative or genetic data on either the frequency of this or the genetic consequences. In our early field work, we trapped and tattooed 50 monkeys in groups outside our study area, only to find that villagers captured these individuals and kept them as pets because they had "received a mark from Hanuman." Unfortunately, we had to give up this field technique.

A secondary problem of what is known as the "weed macaque" or "monkey menace" is the fact that rhesus give primates a bad name. At least half of the non-human primate species in India are rare and endangered – those that are obligate forest dwellers and do not possess the adaptive ability or commensal tendency to live in village, town or agricultural environments. This includes species such as Phayre's langur, capped langur, golden langur, pig-tailed, stump-tailed, and lion-tailed macaque, and hoolock gibbon. These, and other species, require protected forest habitats, in much the same way as the tiger, leopard, and rhino. With rhesus as a prominent symbol of non-human primates, a concern is that forest conservation efforts for other primates may be weakened. It is very important that the public be made aware of the separate issues of problem primate species and endangered primate species in terms of management or conservation.

Box 11.1   Managing human–macaque conflict in Himachal, India

Sandeep K. Rattan

Himachal Pradesh is a small state in the northern region of India that has more than six million people spread over a geographical area of 55,673 km^2. The state boasts a rich biodiversity due to extreme variation in elevation ranging from 350 m to 6975 m above mean sea level. The climatic conditions vary from semitropical to semiarctic resulting in great variety of floral and faunal species. Although there is a lot of geographical separation of species, the rhesus macaque (*Macaca mulatta*) has made its presence felt in almost all parts of the state. In a survey conducted during the year 2003–4, the rhesus macaque population was estimated to be over 317,000[1]. For Himachal Pradesh managing this large population of macaques has never been easy. This conflict is more than a century old, but in the past, the macaque population was concentrated in forests and a few townships. At that time, the species enjoyed a status of respectable wildlife. Unfortunately, over the years this has taken a downward spiral and today they are close to being declared vermin. The massive destruction of natural habitat due to deforestation, construction activities, and depletion of natural food sources has forced more macaques to move near human dwellings. Continual feeding by the millions of tourists and religious devotees visiting the state annually has further aggravated the situation. As a result, macaques have fast adapted to an easier way of life and have left their natural habitats to forage in garbage dumps and along highways.

Human-macaque conflict is highly seasonal and location specific. During and immediately after the monsoons, the natural habitat has more than enough feeding opportunities for macaques, especially in rural areas and thus instances of conflict are minimal. In contrast, during lean months of the year conflicts rise sharply since natural food is scarce. In urban areas, there is direct conflict between humans and macaques, with many instances of aggression against humans, pilfering, damage to property, etc. In rural areas, conflict is more indirect and projected more at massive crop raiding and farm destruction. Economic losses from this conflict are extreme for an area where the dependence of people on agriculture and horticulture reaches as high as 66.71 percent[2]. This conflict with macaques has resulted

---

[1] Official figures of monkey population estimation by Forest Department, Government of Himachal Pradesh, during the year 2003–04

[2] Official data of Himachal Pradesh Government (http://hpplanning.nic.in/Statistical%20 data%20of%20Himachal%20Pradesh%20upto%202009–10.pdf)

in farmlands in many places being completely abandoned. By various estimates, the annual economic loss due to conflict is more than $50 million
USD[3].

In the past, a large proportion of Himachal macaques were exported for
various purposes and this controlled a major part of the conflict, but in April
1978, this export was completely banned. Later during 2004–6[4], the Forest
Department of Himachal started a translocation program that carried partially captured groups to nonconflict areas or forests. Instead of taking care
of the problem, the problem multiplied. Reeling under the stress of addressing this issue, a multi-pronged strategy was adopted by Himachal, which
included mass education of the tourists and the general public to restrict the
feeding of macaques, legislation to fine feeding offenders, garbage management, natural habitat enrichment by plantation of natural fruit-bearing
trees, maintenance and upkeep of water holes, and mass sterilisation of the
macaques. After initial successful trials of sterilisation using endoscopic
tubectomy in females and CO_2 laser vasectomy in males, a full-fledged
mass sterilisation campaign of macaques was initiated on 17 February
2007 at the Monkey Sterilisation Centre (MSC), in Shimla, the capital of
Himachal state.

During the year 2009, two more MSCs were developed in two different
districts, Hamirpur and Kangra. With a sterilisation capacity of 5,000–10,000
macaques per year in one MSC, the objective of the sterilisation campaign
is to sterilise about 50,000[5] macaques over the next three years. To date,
more than 15,000 macaques have been sterilised. As a measure of conflict
management, 123 of the sterilized macaques were released into a Primate
Protection Park that was established on 11 March 2008 at Taradevi forest
reserve where they appear to be habituating well to their new environment.

Nearly all strategies to reduce human-macaque conflict are ineffective
in the short term. All methods require varying degrees of time before success can be evaluated. Since extreme steps like culling have never been part
of viable solution adopted by the Himachal government, the mitigation of
the conflict can only be achieved by adopting a multi-pronged solution.
Targeting the sterilisation of only female macaques that are in direct conflict

---

[3] Data estimates of Bharat Gyan Vigyan Samiti, Shimla; India's costs stand around 500 crore
Rupees (a crore is ten million).
[4] Official data of Forest Department, Himachal 2004–05, from Shimla town, a total of 3,406
monkeys were captured and translocated and during 2005–06, from Shimla and other parts
of state of Himachal, a total of 996 monkeys were captured and translocated.
[5] Population of macaques in towns and urban areas as per official figures of monkey population estimation by Forest Department, Government of Himachal Pradesh, during the year
2003–04.

> with humans ensures the populations in the forests and nonconflict areas of the state remain untouched. Adoption of other strategies like establishment of Primate Protection Parks, education programs, garbage management, and habitat enrichment will provide a synergistic approach to address the problem in a sustainable manner. The strategy envisages a future with minimal human-macaques conflict using a holistic approach for solving this problem.

Much of the need for this awareness comes from the daily media which emphasize the "monkey menace" in terms of human safety. In addition to problems of agricultural crop losses, for several years serious, even tragic, accidents have been caused by monkeys throughout northern India. In April of 2004 a newlywed young woman fleeing a group of threatening rhesus on the rooftop of her home fell to her death in Patna, the capital of Bihar, eastern India. An even more newsworthy case occurred in October 1997 when the Deputy Mayor of Delhi fell to his death from his rooftop. Shri S. S. Bajwa was reading a newspaper when four rhesus appeared on his rooftop. He waved a stick and flashed the newspaper at them, one or two males lunged at him in a threatening mode, he stepped back and fell off the rooftop. He died in the hospital of head injuries.

Delhi has since had numerous accounts of monkey attacks and bites, including an attack in her home on Priyanka Gandhi, the daughter of Sonia Gandhi, Chief of the Congress Party in India, and widow of Rajiv Gandhi, former Prime Minister of India. Rhesus in the cities of India often enter homes and offices through open windows and doors. Delhi has reported 40,000 monkey and dog bites annually, a serious public health threat considering the potential of rabies in dogs and Monkey B virus (i.e., a herpes virus) in rhesus. Monkey B virus, transmitted through monkey saliva, causes no serious or obvious pathology in rhesus, but is usually fatal in the rare cases it has infected human beings. Fortunately, the virus in rhesus has a short ephemeral viremia with rare transmission to humans. Nonetheless, B virus is a generally unknown public health danger from rhesus bites. A common cause of death in India as indicated in coroner's reports is "FUO", "Fever of Unknown Origin."

Delhi has recently had a 10 million Rupee annual budget (US $253,000) to control rhesus by capture and removal. The problem has always been "removal to where?" No other states in India have been willing to receive them. Some states have insisted on substantial grants to accept Delhi's monkeys. Holding facilities have been developed outside Delhi and New Delhi, but at best they have been able to accept only a few hundred monkeys, whereas the rhesus population of Delhi has been estimated at several thousand. No completely satisfactory solution to the problem of Delhi's urban "monkey menace" has yet been found.

In northwest India, in the state of Himachal Pradesh where rhesus are both a serious agricultural pest, and an urban pest in the city of Shimla, a plan has been developed to sterilize 65,000 male monkeys. Dr. S. K. Sahoo, who has studied rhesus behavior and ecology in HP for many years, feels the program will be expensive, and is not likely to be successful. The country of Tajikistan has offered to accept some of HP's monkeys, probably with the idea of creating a profitable trade in rhesus monkeys for biomedical research and commercial pharmaceutical use. The international supply of rhesus now comes primarily from China, and healthy rhesus (known as SPF or Specific Pathogen Free) have become very expensive, in the order of several thousand dollars per individual.

As a primate species of both cultural and commercial importance, rhesus have been victims of two unfortunate management practices. During the 1950s and '60s, rhesus were certainly subject to excessive trapping and export which led to serious population decline and a threatened status. However, India's decision in 1978 to halt all export for legitimate biomedical and pharmaceutical use, went too far. It failed to consider the high reproductive rates, potential for rapid population growth, commensal nature, and destruction of natural habitat for rhesus. As a result, rhesus entered agricultural, village, and urban environments in increasing numbers, becoming serious pests. Rhesus also have a potential impact on public health as reservoirs of many human pathogens, including those of measles, influenza, encephalitis, hepatitis, tuberculosis, dysentery, and amoebiasis (Jones-Engel et al., 2006; Shah and Southwick, 1965).

In retrospect, it would have been more reasonable to recognize that moderate harvest and humane utilization of rhesus would be beneficial for both human and rhesus populations. Rhesus numbers could have been kept within reasonable levels from the standpoint of crop losses, and both human and monkey health and safety. A comparable situation in the United States to the total cessation of rhesus export from India would be to stop all hunting harvest of white-tailed deer. White-tailed deer are a commensal species with a high reproductive rate, and potentials for rapid population growth. Excessive populations have an important ecological impact through serious forest damage. They are also a factor in human health in harboring the bacterium, *Borrelia*, which causes Lyme disease, the most common tick-borne illness in the northern US. In humans, it results in high fevers and may lead to long-term chronic illness.

Scientists have recognized for a long time the value of rhesus as models in biomedical and behavioral research because of their basic biological analogies to us. Their biological analogies are true in their disease spectrum, immunology, hematology, and general physiology. In these areas, rhesus are accepted as valuable surrogate models for human beings.

It is less accepted, and rarely recognized, that rhesus may also serve as a model for human population ecology. They share several population qualities with us, including potential high rates of reproduction, adaptations to a wide range of habitats, and tendencies to urbanize. Behaviorally and socially, similarities are also apparent: definite group structures, prominent dominance hierarchies, high intelligence and adaptability, strong patterns of both affection and aggression, and overall ecological success in the world.

Much of the rapid population growth of rhesus macaques depends on its high birthrates in a variety of habitats, especially commensal habitats. Birthrate analyses of India data show that rhesus populations in commensal habitats (i.e., urban, temple, and village habitats) have significantly higher annual birth rates (85 percent to 91 percent) than those in forest habitats (68 percent) (Southwick and Siddiqi, 2008).

Birthrate data for other species of macaques are difficult to compare, especially for tropical species where no clear birth season is evident. In *M. fascicularis*, for example, births occur throughout the year, so the best that can be done is to compare infant/adult female ratios as indicators of birth rates. In commensal groups in a small forest in Ubud, Bali, four years of study up to 1992 showed that infant/adult female ratios averaged 41.2 percent (Wheatley and Harya Putra, 1994), whereas comparable data for rhesus are 75 percent to 90 percent. Despite the relatively low natality of *M. fascicularis*, also accompanied by high mortality, the *M. fascicularis* population of Ubud increased 200 percent in fifteen years prior to 1992. This Ubud population has maintained an annual population increase of 6 percent since 1998 (Fuentes, Southern and Suaryana, 2005; Fuentes *et al.*, Chapter 6) Other species of macaques, such as the Bonnet macaque, *M. radiata*, in south India (Chakravarthy and Thyagaraj, 2005), and the Japanese macaque, *M. fuscata* (Watanabe and Muroyama, 2005), have also shown remarkable population growth, including range and habitat expansion.

The ecology and behavior of rhesus monkeys, along with their "extremely advanced"… social structure and impressive geographical distribution prompted the well-known anthropologist, John Napier, to comment: "I am convinced that had not man arrived on the primate evolutionary scene, these ubiquitous creatures would have ruled the world" (Napier, 1976). Philip Lutgendorf, a leading scholar of Hindi and modern India, adds to this in a major work on Hanuman by noting that the "monkey problem" in India, and the popular deity of Hanuman "blur the boundary between human and animal…." This is certainly true for both traditional Hinduism and modern-day India, creating an ethical dilemma in classical Hindu beliefs.

Summary

A 50-year population study of rhesus in Aligarh District of north central India beginning in 1959 showed a substantial decline throughout the 1950s and 1960s, followed by a stabilization and slow recovery of rhesus population numbers in the 1970s and '80s, as rhesus trapping and export declined.

Since the export ban, the rhesus population of Aligarh District has increased markedly, due to high birth rates and low mortality. Overall, the original Aligarh sample of 403 monkeys increased from a low of 175 in 1969 to a high of 744, an increase of 325 percent. Unfortunately, this increase has occurred primarily in commensal habitats; that is, human dominated habitats, such as roadsides (Figure 11.4), agricultural areas, canal banks, villages, towns and cities. Similar increases in rhesus populations have occurred throughout northern India, creating problems of agricultural losses, and human health and safety.

The habitat distribution of rhesus from forests to human habitats is due to both the loss of natural forests to agriculture and single-species timber production, and the commensal nature of rhesus to adapt readily to human environments. In contrast, almost half the non-human primate species of India which are obligate forest dwellers are suffering from severe habitat loss, and are in a threatened or endangered state. The shift of rhesus populations from respected and worshipped representatives of Hanuman, the Monkey God, to economic and public health pests has created not only cultural and philosophical conflicts for many people of India, but a set of serious and expensive management problems as well.

Unfortunately, rhesus have suffered from two management mistakes: excessive trapping and export in the mid-twentieth century, followed by a virtually complete ban on moderate utilization. With very high reproductive capacities, rapid population growth, and commensal habits, rhesus should have a reasonable degree of sustainable harvest similar to many wildlife populations, such as white-tailed deer in the United States. Population evidence indicates that a sustainable harvest rate of 20,000 per year could be of benefit to the welfare of both the rhesus population and human population of India. This suggestion, however, while based on scientific realities, represents a difficult and almost insurmountable cultural and religious conflict for many people of India.

Acknowledgements

This field project was begun when CHS was a Fulbright Research Fellow at Aligarh Muslim University in India with support from the US Educational Foundation in India, and the USPHS-NIH in Bethesda. At that time, we are

Figure 11.4. Roadside rhesus macaques in an agricultural area in Aligarh District in the 1960s (a) and the 1980s (b).

particularly indebted to the late Drs. M. B. Mirza of AMU, Olive Reddick of USEFI in New Delhi, Carl Frey of Ohio University, and Drs. M. Rafiq Siddiqi and M. A. Beg of AMU. Over subsequent years, our field work was supported by grants from the NIH and National Geographic Society to Johns Hopkins University and the University of Colorado. More recently, we have been supported by the Indo-US Primate Project, directed by Prof. S. M. Mohnot of Jodhpur, and financed by the Division of International Conservation of the US Fish and Wildlife Service through the office of David Ferguson, and the Ministry of Environment and Forests of the Government of India. For cooperation and support, we are indebted to the late Drs. F. B. Bang, Director of the Johns Hopkins International Center for Medical Research and Training in Calcutta, the late Drs. M. L. Roonwal and K.K. Tiwari, Directors of the Zoological Survey of India, and R. K. Lahiri, Director of the Calcutta Zoological Gardens, and Dr. R. P. Mukherjee and Dr. J. R. B. Alfred, former Directors of the ZSI. For additional advice and assistance, we are grateful to Drs. T. W. Simpson, K. V. Shah, and C. Wallace of the Johns Hopkins ICMRT in Calcutta, and Dr. I. Malik, founder of the conservation organization, Vatavaran, in New Delhi. In New Delhi, we also appreciate the support and consultative advice of many science officers and administrative staff in the US Embassy, including D. Johnson, R. Whitney, and A. Saxena of the India International Centre, and innumerable local, District, and Regional Forest Officers in the State and National Governments of India. For field assistance, we are further indebted to Drs. D. G. Lindburg, C. Louch, Phyllis Jay, R. Johnson, M. Bertrand, M. S. Rai, B. C. Pal, M. Y. Farooqui, J. R. Oppenheimer, S. W. Ashraf, Ajoy Ghosh, J. A. Cohen, and Ramesh De. Many recent colleagues in the Indo-US Primate Project have also given advice and support: Drs. P. C. Bhattacharjee, J. Bose, M. Singh, D. Chetry, R. M. Chetry, L. Rajpurohit, A. Changani, S. K. Sahoo, A. Srivastava, R. Mathur, S. Nandi, J. Das, V. Menon, J. Biswas, and P. Sarkar. Several Americans have helped with advice and support the IUSPP, including Drs. I. Bernstein, M. Cooper, R. Horwich, R. Mittermeier, J. Oates, R. Kyes, A. Rylands, D. Ferguson, and T. Struhsaker. Finally, we thank the people of India for their interest and cordial hospitality. Villagers and Indian people in all walks of life have made these field studies a pleasure in the sometimes difficult climates and terrain of India.

References

Bennett, B. T., Abee, C. R., and Henderson, R. (eds.). 1995 *Nonhuman Primates in Biomedical Research*. New York: Academic Press. 428 pp.

Blum, D. 1994. *The Monkey Wars*. New York: Oxford University Press. 294 pp.

Chakravarthy, A. K. and Thyagaraj, N. E. 2005. Coexistence of Bonnet macaques with planters in the cardamom and coffee plantations of Karnataka, South India: Hospitable or hostile?. In *Commensalism and Conflict: the Human-Primate Interface*. J. D. Paterson and J. Wallis (eds.). Norman, OK: American Society of Primatologists. pp. 270–293.

Corbett, J. 1953. *Jungle Lore*. Bombay: Oxford University Press. 168 pp.

Fuentes, A., Southern, M., and Suaryana, K. G. 2005. Monkey forests and human landscapes: Is extensive sympatry sustainable for *Homo sapiens* and *Macaca fascicularis* on Bali? In *Commensalism and Conflict: the Human-Primate Interface*. J. D. Paterson and J. Wallis (eds.). Norman, OK: American Society of Primatologists. pp. 168–195.

Imam, E. 2000. A study on some behavioural aspects of rhesus monkey in Aligarh and adjoining districts. Ph.D.thesis. Aligarh Muslim University. 216 pp.

Jones-Engle, L., Engel, G., Schillaci, M. *et al.* 2006. Considering human-primate transmission of measles virus through the prism of risk analysis. *American Journal of Primatology* **68**: 868–879.

Lutgendorf, P. 2007. *Hanuman's Tale: The Messages of a Divine Monkey*. New York: Oxford University Press. 434 pp.

Mohnot, S.M. (ed.) 2005. *Indo-US Workshop for Indian Primates: National Action Plan*. New Delhi and Jodhpur. Primate Research Centre. 66 pp.

Napier, J. 1976. *Monkeys Without Tails*. New York: Taplinger Publ. pp. 21, 96.

Richard, A., Goldstein, S. J., and Dewar, R. E. 1989. Weed macaques: the evolutionary implications of macaques feeding ecology. *International Journal of Primatology* **10**: 569–594.

Sahoo, S. K. 2005. Rhesus macaques in Himachal Pradesh, north India. In *Indo-US Workshop for Indian Primates*, S. M. Mohnot (ed). Primate Research Centre, Jodhpur, and Indo-US Science and Technology Forum. New Delhi. pp. 13–18.

Sanderson, I.T. 1957. *The Monkey Kingdom: An Introduction to the Primates*. New York: Hanover House. 200 pp.

Shah, K. V. and Southwick, C. 1965. Prevalence of antibodies to certain viruses in sera of free-living rhesus and of captive monkeys. *Indian Journal of Medical Research* **53**: 488–500.

Siddiqi, M. F. and Southwick, C. H. 1988. Food habits of rhesus monkeys (*Macaca mulatta*) in the northern Indian plains. In *Ecology and Behavior of Food Enhanced Primate Groups*, J. E. Fa and C. H. Southwick (eds.). New York: Alan Liss, Inc. pp. 113–124.

Southwick, C., Beg, M., and Siddiqi, M. R. 1961a. A population survey of rhesus monkeys in villages, towns and temples of northern India. *Ecology* **42**: 538–547.

1961b. Transportation routes and forest areas. *Ecology* **42**: 698–710.

1965. Rhesus monkeys in north India. In *Primate Behavior: Field Studies of Monkeys and Apes*, I. DeVore (ed.). New York: Holt, Rinehart and Winston. pp. 110–159

Southwick, C.H., Richie, T.L., Taylor, H., Teas, J., and Siddiqi, M.F. 1980. Rhesus monkey populations in India and Nepal: Patterns of growth, decline and natural regulation. In *Biosocial Mechanisms of Population Regulation,* M. N. Cohen, R. S. Malpass, H. G. Klein (eds.). New Haven, CT: Yale University Press. pp. 151–170.

Southwick, C., Siddiqi, M. R., Cohen, J., Oppenheimer, J., Khan, J., and Ashraf, S. (1982) Further declines in rhesus populations in India. In A. Chiarelli and R. Corruccini (eds.). *Advanced Views in Primate Biology.* Berlin: Springer-Verlag. pp. 128–137.

Southwick, C. H. and Siddiqi, M. F. 1988. Partial recovery and a new population estimate of rhesus monkey populations in India. *American Journal of Primatology* **16**: 187–197.

2001. Status, conservation and management of primates in India. *ENVIS Bulletin (India)* **1**: 81–91.

2008. Socio-ecologic influences on birth rates in rhesus macaques in India. *XXII Congress of the International Primatological Society,* Abstracts No. 528.

Srivastava, A. and F. Begum. 2005. City monkeys: A study of human attitudes in J. Paterson and J. Wallis. *Commensalism and conflict: The Human Primate Interface.* Norman, OK: American Society of Primatologists. pp. 259–269.

Tiwari, K.K. and R.P. Mukherjee. 1992. Population census of rhesus macaque and Hanuman langur in India – a status survey report. *Rec. Zool. Surv. of India* **92**: 349–369.

Watanabe, K. and Muroyama, Y. 2005. Recent expansion of the range of Japanese macaques, and associated management problems. In *Commensalism and conflict: The Human-Primate Interface,* J. D. Paterson and J. Wallis (eds.). Norman, OK: American Society of Primatologists. pp. 400–419.

Wheatley, B. and Harya Putra, D. K. 1994. The effects of tourism on conservation at the monkey forest in Ubud, Bali. *Revue d'Ecologie: La Terre et la Vie* **49**: 245–258.

Part V

Understanding and managing the human–macaque interface

12 *Developing sustainable human–macaque communities*

LISA JONES-ENGEL, GREGORY
ENGEL, MICHAEL D. GUMERT AND
AGUSTÍN FUENTES

Introduction

A variety of factors make the management of free-ranging macaque populations challenging, both for communities and policy makers. The objective of this chapter is to discuss management options that promote sustainable coexistence of species. Well-conceived strategies will support the economic and social goals of humans, the conservation of macaque populations, and the health of both human and macaque populations.

Potential goals for managing the human-macaque interface

In any place where humans and macaques exist in proximity, individuals or groups will have specific interests, goals or objectives concerning the management of the human–macaque interface. Frequently, the interests of one group or individual will be in conflict with others, making it difficult to reach consensus on implementing any specific management practice. For example, in communities where macaques are a tourist attraction, tour guides and food vendors may perceive the recreational provisioning of food to macaques as desirable, as it is popular with tourists and often economically rewarding for vendors. Conservationists and public health officials, however, may seek to prevent tourists from feeding monkeys in order to reduce the human impact on macaque populations and reduce the risk of bi-directional transmission of infectious agents. Primatologists are often engaged in studying synanthropic macaque populations, and these

Monkeys on the Edge: Ecology and Management of Long-Tailed Macaques and their Interface with Humans, eds. Michael D. Gumert, Agustín Fuentes and Lisa Jones-Engel. Published by Cambridge University Press. © Cambridge University Press 2011.

researchers also have specific interests and goals regarding management of conflict. In general, we place a high value on supporting the endurance and health of macaque populations, increasing the public appreciation of macaques, and facilitating a sustainable intra-specific relationship by reducing interspecies conflict.

Common contexts for human interaction with long-tailed macaques include urban landscapes, village and gardens, temples, and parks, all of which provide "edge" niches readily exploited by long-tailed macaques (see Chapter 1, Table 1.2). In most human macaque interface zones, conflict arises principally from (1) overlap in use of space, (2) the raiding of food sources by macaques, and (3) damage to property. Core management goals include the creation of landscapes and infrastructure that minimize overlap between humans and macaques, effective waste management, and measured public awareness of appropriate behavior around macaques.

Reducing human/macaque conflict at the interspecies interface

There are three approaches to managing conflicts at the human-macaque interface: (1) modifying the environment in which humans and macaques come into contact; (2) attempting to change the composition and/or behavior of the macaque population; and (3) attempting to change the composition and/or behavior of the human population. Specific situations may call for interventions in one or more of these areas. Box 12.1 highlights nearly two decades of management strategies in Hong Kong, where macaques on Kowloon Hills have been thriving in a small, human altered-environment. In Singapore, where thriving populations of *M. fascicularis* exist along the edges of highways and urban structures next to forested areas, multiple management strategies are employed to varying effect (Box 12.2).

The environmental dimension

Managing the environment: Landscaping the human-macaque interface

Developing innovative structures and landscapes is an important task for wildlife managers, developers, and other stakeholders charged with constructing and managing communities where humans and macaques interface. Landscaping and infrastructure should keep in mind the capabilities and behaviors of macaques. Table 12.1 summarizes several aspects of the human

Box 12.1   Management of nuisance macaques in Hong Kong

Chung-Tong Shek
Introduction
Hong Kong is a physically small city with a total land area of 1,076 square kilometers. While it is renowned for its urban landscape and dense human population, it surprisingly enjoys a rich biodiversity of flora and fauna. Nearly three-quarters of Hong Kong's land is countryside, and about 40 percent of the total land area has been designated as country parks, special areas or restricted areas for protection of its biodiversity.

Hong Kong macaques
Hong Kong falls within the range of natural distribution of the rhesus macaque. However, the original wild stock is believed to have been extirpated. The existing macaque populations are considered to be the descendents of the individuals which have been introduced to the Kowloon Hills, i.e., Kam Shan and Lion Rock Country Parks in the 1910s (Herlots, 1951; Wong and Ni, 2000). It is believed that rhesus macaques were reintroduced to Hong Kong to control the spread of a local poisonous plant, the strychnos. This plant contains alkaloids which are poisonous to humans but it is a favored food of macaques. In addition to the rhesus macaques, long-tailed macaques were released in the same area in the 1950s, leading to cross-breeding between these two *Macaca* species. According to the result of a survey conducted 1991–1992, over 30 percent of the total population of macaques found in Hong Kong were considered hybrids, see Figure 12.1 (Wong and Ni, 2000). These macaques have adapted to the environment and have formed a well-known macaque population in the Kowloon Hills, locally known as "Monkey Hill".

Macaque nuisance
In Hong Kong, some people enjoy feeding wild macaques in the Kowloon Hills, as they worry that macaques are starving in the wild and they need to be fed regularly like pets. Due to heavy human provisioning, the population of macaques in Hong Kong has increased dramatically from about 100 to over 2,000 individuals in the past 30 years. Through frequent contact with humans over the years, some macaques have become habituated to humans. Because people who feed macaques usually carry plastic bags containing food, the macaques have learned to snatch plastic bags from visitors. The macaques' sometimes aggressive behavior has led to conflict with local

Figure 12.1. A hybrid of the rhesus macaque and the longtailed macaque.

human inhabitants. Some macaque troops even stray into nearby suburban residential areas searching for food, spreading the conflict zone from parks to residential areas.

Measures to deal with macaque nuisance

Feeding ban
In view of growing complaints about macaques, in 1998 the Hong Kong Agriculture, Fisheries, and Conservation Department (AFCD) implemented a management plan for wild macaques in Hong Kong in 1999 which included various measures. In order to help wild macaques revert from reliance on human provisioning to natural food foraging in the countryside, a ban on feeding wild macaques, known as the Wild Animals Protection Ordinance (Cap 170) was introduced in 1999. This law has helped reduce

Figure 12.2. Giant remote controlled trapping cage is found effective to capture as many as 130 macaques at one time.

the uncontrolled population growth. To provide macaques with sufficient natural food in their habitat, over 300,000 macaque food trees were planted in the last ten years in the country parks concerned so as to attract wild macaques to stay in countryside areas for feeding.

Managing 'nuisance macaques'

In order to better control nuisance macaques in affected residential areas, a special team was established to handle macaque-nuisance complaints promptly. Capturing macaques in residential areas was found to be the most difficult and time consuming part of the nuisance management. Various capture methods were tested, including cage traps, snares, net guns, dart guns, and placing sedatives in food, but most were unsuccessful. Fortunately, a collaboration between contractors and the Hong Kong Productivity Council produced two effective trapping devices, giant remote controlled trapping cages (2003, see Figure 12.2), and plastic jar traps (2007) that are now used in residential areas where macaques have become a nuisance. In addition, trained security guards have been deployed at "nuisance hot spots" in order to drive the macaques back to the countryside. Advice has been given to property management companies and the concerned government departments, instructing them to replace rubbish bins with wild animal-proof litter bins or to manage the rubbish bins and refuse collection points properly so as to remove food sources for stray macaques in residential areas.

Contraceptive/neutering program

In order to control the population growth of macaques in Hong Kong, a large-scale macaque contraceptive/neutering program has been in effect since 2007 which involves treatments of 30–130 macaques in each operation (Shek and Cheng, 2010). The macaques are trapped using a giant trapping cage, females are injected with an immuno-contraceptive vaccine named SpayVac™ and males are vasectomized by injecting scarring chemicals in the epipidymis, resulting in blockage of the vas deferens. The program has been running for three years and up to March 2010, over 1,400 macaques have been treated, which represents over 50–60 percent of the total population in Hong Kong. According to a recent population survey in 2010, the total population of wild macaques in Hong Kong has dropped by around 5 percent in two years and the average annual birth rates of the troops, which received treatments before the mating seasons, dropped by about 30 percent. Thus, the program has effectively controlled the macaque population. As the program continues, we expect to see a further decline in the total population in the coming few years. In late 2009, a new technique for permanent sterilization of females by endoscopic tubectomy was introduced to the program by the Ocean Park Conservation Foundation Hong Kong (Martelli, 2009) which will further control the population growth of local macaques in the long run.

Public education

Public education is one of the important aspects of managing the wild macaques in Hong Kong. In the affected country parks, large notice boards have been installed at the main entrances with information advising visitors on safety precautions and techniques to avoid nuisance macaques. Moreover, large banners have been erected at recreational sites to remind visitors that it is against the law to feed macaques. A macaque display corner has been established inside Shing Mun Country Park Visitor Center providing information on wild macaques and providing advice on what to do and what not to do when encountering macaques. Seminars and on-site field

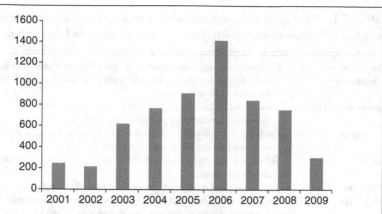

Figure 12.3. Reported cases of macaque nuisance in Hong Kong from 2001–9.

demonstrations on macaque nuisance handling techniques have been given to property management companies and other relevant government departments in the residential areas where macaque nuisance cases are reported. Relevant posters and leaflets providing tips to handle nuisance macaques have also been made available to concerned organizations and government departments.

Conclusion

Hong Kong has adopted an integrated approach in dealing with macaque nuisance which includes feeding bans, nuisance handling and a contraceptive/neutering program, and public education. The decreasing trend in the number of reported macaque nuisance cases since 2007 proves the initial success of the above measures (see Figure 12.3). Through these continued measures we believe macaque nuisance will be effectively managed.

References

Herklots, G. A. C. 1951. The Hong Kong Countryside. *South China Morning Post*, Hong Kong.

Martelli, P. 2009. *Endoscopic Tubectomy in Macaques (DVD)*. Karl Storz – Endoskope.

Shek, C. T. and Cheng, W. W. 2010. Population survey and contraceptive/neutering programme of Macaques in Hong Kong. *Hong Kong Biodiversity* **19**: 4–7.

Wong, C. L. and Ni, I. H. 2000. Population dynamics of the feral macaques in the Kowloon Hills of Hong Kong. *American Journal of Primatology* **50**: 53–66.

Table 12.1. *Types of landscaping and structures that could be strategically modified in human–macaque interface zones in an effort to lessen conflict.*

Trash bins and containers	*Enclosed in caging, automated or complex locks and opening/closing mechanisms, repellants and deterrents for entering, all litter must be removed from community*
Fences and walls	*Smooth and difficult to scale, electrification, motion-triggered water spurts, no handholds, top cannot be traversable*
Car parks	*Enclosed, enter directly into homes or apartments, encourage owners to cover cars if concerned with damage*
Doors	*Self closing and reinforced, use of barred-door as barrier to main door, no pet entries*
Windows	*Self-closing and reinforced, ornamental bars over window*
Walkways	*Underground or enclosed, make wide enough for people to pass a macaque at a >1m distance, shuttle buses or bicycles to carry people, walkways on both sides of road, employ guards*
Bridges	*Enclosed or do not make too narrow, provide two separate pedestrian bridges where possible*
Gardens	*Enclosed, electrified fencing, motion-triggered water spurts, alarms*
Homes and apartments	*Roofing that cannot be pulled off, structures that prevent or do not allow easy scaling up side, enclosed in fencing*
Streets and roads	*Speed bumps in areas of high interface, provide signs and warning in areas where macaque may be in streets*
Signage and communication	*Development of well-researched and clear signage, public announcements indicating proper behavior and ways to avoid conflict with macaques*

landscape that merit special attention when attempting to manage conflict at the human-macaque interface.

Disposal of refuse

Refuse represents an easily accessible, high-yield, and reliable food source for macaques. Thus, refuse bins and waste containment are among the most important anthropogenic aspects of human-macaque interface zones. Macaques that can access refuse will stay in the vicinity to rest and socialize before moving to a new location (Gumert, unpublished data)., Therefore, removing access to trash is one of the most effective methods for limiting overlap between humans and macaques and reducing conflict. Refuse bins and waste containers must be macaque-proofed (see Figures 12.4 and 12.5). For example, trash bins should self-close and have mechanisms that are operable by humans, but not macaques. Similarly, litter draws macaques, so attempts at reducing litter must be vigorous, and anti-litter enforcement is particularly important.

Figure 12.4. This macaque outside Bukit Timah Nature Reserve in Singapore helps herself to a non-macaque-proofed trash bin at a construction site bordering the reserve, showing the importance of proper trash containment in areas of interface.

Figure 12.5. This road sign in Petchaburi, Thailand appears to clearly warn local citizens and foreign guests that caution is needed around macaques.

Considerations regarding the human infrastructure

After access to refuse by macaques is eliminated, other aspects of the land-scape can be considered. For example, fences and walls should be built to make it difficult for macaques to utilize them for locomotion. Fences and walls should be smooth, without possible handholds or tops that can be easily walked on and, where possible, electrification or motion-triggered water spurts can be installed to startle and repel macaques. Doors and windows can be specifically monkey-proofed to prevent macaques from entering dwellings. People who live within macaque ranges should be advised to keep windows and outside doors monkey-proofed or closed. Without such precautions, once macaques learn they can enter a home and find food, they will continually return to the home to forage, causing conflict with occupants. Doors should self-close, so people do not accidently leave them open. Buildings should be difficult to scale, and projections that could be used as handholds for climbing macaques should be spaced greater than 1.5 meters apart (i.e., larger than the arm span of an adult male long-tailed macaque). On buildings where the roof cannot be adequately protected, shingles or roofing material should be reinforced so macaques cannot pull off panels, creating leaks. Lastly, homes, yards, and gardens should be enclosed to prevent access by macaques. Walkways and bridges can be designed with both humans and macaques in mind. Passing within one meter of a macaque, in a confined space, may be perceived as threatening to the animal. Therefore, paths and passages should allow people to give adequate berth to macaques. Walkways should be at least 2.5 meters wide so that a human can make at least a 1-meter detour around a macaque. In general, alternative options (i.e., redundant paths or enclosed walkways) should be available for foot traffic, especially in areas where people are likely to be carrying food or bags. In high-interface areas specially trained guards or rangers can be extremely valuable in assisting people moving around macaques and encouraging people to behave responsibly (see Box 6.1 and Box 12.1).

Food and refuse left in open cars attract macaques. Enclosed car parks are a solution in some areas. In other areas, car owners should be careful to close car windows and deposit all refuse in macaque-proof bins. In regions where there are no enclosed car parks and people are concerned with damage to the appearance of their vehicle by macaques, owners should use a car cover to protect their vehicle.

The macaque dimension

At the human-macaque interface, human behavior is a major determinant of macaque behavior. Yet there is often a lack of consensus regarding which types of macaque behavior should be promoted and which should be discouraged. As mentioned, at sites where macaques are tourist attractions, feeding of the animals is often encouraged, as the sale of monkey food generates revenue for the community and the feeding attracts more people to visit. Promotion of feeding can cause problems as, when uncontrolled, it can potentially lead to the development of more nuisance macaques and population growth (see Box 6.1). Modifying macaque behaviors requires a commitment to change among stakeholders. In this section we offer some options for dealing with nuisance animals as well as population-level management considerations.

Particularly in urban areas or in areas with ample financial resources, macaques can be tagged with sensors for tracking. Satellite or radio tagging could be used on some individuals in groups that frequently interface with people. After tagging, these groups can be tracked by wildlife authorities allowing for rapid and appropriate responses. Additionally, placing transponder receivers in trashcans, homes, and other places where macaques are not wanted allows a record of which macaques are most frequently accessing these points. Such information will be useful to wildlife authorities when making management decisions. These transponders might also be useful for triggering alarms, water spurts, or other deterrents when sensed by receivers.

Controlling macaque population size

Another management option for reducing human-macaque conflict is to control the size and demographics of the macaque population. When considering the management of synanthropic macaque populations, a census of the macaque population and surrounding nonsympatric populations can help determine what percentage of the regional population will be affected by management decisions. After the population is well assessed, population control measures may be considered if indeed the conflict can be related to a high number of macaques. This is important to determine, as often high numbers are not necessary for human-macaque conflict. For example, in Singapore human-macaque conflict occurs despite having a small population because nearly all of the macaques in the population interface to some degree with people (Sha *et al.*, 2009b). Similarly, on Karimunjawa in Indonesia, a population of less than 300 macaques poses significant challenges to the local community (Box 1.1). In these cases, distribution is the main issue, and population control would not alleviate the conflict, as the remaining macaques would still be distributed in the same location. If the population is determined to be excessively large and/or growing, then population control should be considered, but should not be thought of as the sole solution in every context.

Methods of population control have included trapping, relocation, culling, and sterilization programs. Singapore is one of the few countries that is transparent in population control efforts, and thus we are able to learn from their management efforts (see Box 12.2). They regularly use trapping to remove nuisance macaques from the population (Sha *et al.*, 2009a), and have culled local populations, such as in the 1970s in the Botanical Gardens. Culling has also been widely used in attempts to entirely remove macaques from Ngeaur Island in Palau (Wheatley, Chapter 10). In Malaysia, recent efforts to ameliorate conflict in the Selangor province around Kuala Lumpur included trapping and culling of a large proportion of the estimated 250,000 macaques in the region (Md-Zain *et al.*, Chapter 4).

In an effort to reduce direct killing of macaques in Southeast Asia, sterilization has been proposed as a means to stop or lessen population growth and reduce human-macaque conflict. Authorities in Hong Kong have put the most effort into sterilization programs (Box 12.1; Wong and Chow, 2004) and experimental sterilization programs have also been implemented in Singapore (Box 12.2) and in Palau (Wheatley, Chapter 10). Currently, Thailand is working on a sterilization program in Lopburi (Bunluesilp, 2009; Malavijitnond *et al.*, Chapter 5), and there are efforts to use sterilization to manage the large macaque problem in Selangor in Malaysia (Md-Zain *et al.*, Chapter 4). Although an attractive management approach because it does not result in the direct killing of animals, there are no systematic long-term studies on sterilization programs and their effects on human-macaque conflict. Without objective and rigorous monitoring, the efficacy of sterilization as a management tool is undetermined. Data on the effects of sterilization programs on populations needs to be accessible so that stakeholders can access information to inform strategy-making and decisions.

Macaque sterilization programs must consider numerous factors, most notably that macaque *distribution*, not population size, is the main driver of human-macaque conflict. Macaques have long life spans, and after sterilization they will stay in their location and continue to be in conflict with humans. Reduction in numbers of individuals may only minimally reduce interspecies conflict because a group of 20 macaques has the potential to be as damaging as a group of 40 macaques living in the same space. Other factors to consider are that the goals and outcomes of the sterilization program must be clearly defined prior to the initiation of the program. For example, if the goal is to maintain a certain population size, then that must be clearly articulated so that there are no expectations that sterilization will completely remove all conflict. Sterilization programs can be successful at maintaining set population sizes,

Box 12.2   Lessons and challenges in the management of long-tailed macaques in urban Singapore

Benjamin P. Y-H. Lee and Sharon Chan

Introduction

It is estimated that more than 50 percent of the world's human population now reside in large urban centers (United Nations, 2008) and Singapore is no exception, a compact island city-state with an area of about 700 sq. km but housing some 5 million people. Such a phenomenal population growth since its founding as a shipping port by the British in 1819 has taken a heavy toll on its natural environment and biodiversity (Corlett, 1992; Brook *et al.*, 2003). Urbanization represents a major threat to wildlife, as this process both reduces natural habitat and causes fragmentation (McKinney, 2006) and also brings people in much closer contact with wildlife.

There are documented human-wildlife conflicts in Singapore involving nonprimate wildlife such as pythons (Lee and Chong, 2006) and birds (Sodhi and Sharp, 2006) but the single most common human-wildlife conflict that invariably makes the headlines involves the long-tailed macaque (*Macaca fascicularis*) (see Murdoch, 2007; Sua, 2007; Neo, 2010). This species of macaque, first described and named by the founder of modern Singapore, Sir Stamford Raffles (1821), has persisted till today, and is the largest and most common wild mammal in Singapore.

Before 2000, the wild population of long-tailed macaques, particularly those at Bukit Timah Nature Reserve (BTNR), contributed much to our understanding of the ecology of the tropical rain forest with respect to seed dispersal and forest phenology (see Corlett and Lucas, 1990; Lucas and Corlett, 1991; Lucas and Corlett, 1992; Lucas and Corlett, 1998). Recent research on Singapore's long-tailed macaques has focused on disease transmission risk (Jones-Engel *et al.*, 2006, Jones-Engel *et al.*, 2007), human-macaque interactions (Fuentes *et al.*, 2008; Sha *et al.*, 2009a, 2009b) and morphological variation (Schillaci *et al.*, 2007). All these studies contributed immensely to understanding the ecology and taxonomy of this species.

In modern Singapore, the conflict arising from the human-macaque interface usually stems from food provisioning, both intentional and unintentional, and also intolerance, and an irrational fear of macaques. The areas with the most severe conflicts usually involve residential areas in the vicinity of the forested nature reserves, as the long-tailed macaques favor forest edges (Sha *et al.*, 2009a). It is worth noting that human-macaque conflict in Singapore is not a recent phenomenon but started sometime in the early

1970s when the Singapore Botanic Gardens removed groups of "nuisance monkeys" within its premises through shooting them despite public outcry (Tan *et al.*, 2007; Anon, 2010). Since that episode, the Agri-Food and Veterinary Authority (AVA) has publicly stated that culling will only be a "last resort and only if the monkeys become aggressive and pose a danger to the public." (Goh, 2008).

This is a brief review of the main management practices that NParks has adopted to deal with the human-macaque conflict situation in Singapore, often in cooperation with other agencies such as the AVA. Translocation and sterilization (Martelli, B., *unpublished data*) were also explored and implemented in the past but the former technique is not feasible due to rehabilitation needs and the lack of forested areas for release of macaques without causing further human-macaque conflict. Sterilization, as a population control measure, is worth considering as it has shown to be partially effective in places with commensal macaques such as Gibraltar and Hong Kong (Cortes and Shaw, 2006, Wong and Chow, 2004) although Cortes and Shaw (2006) have questioned the implications of sterilization on macaque behavior in the long-term.

Management practices

Total feeding ban and elimination of artificial food sources
Lucas (1985) reported that visitors and staff alike at Bukit Timah Nature Reserve (BTNR) used to feed the moneys in the late 1980s despite notices asking them to refrain from food provisioning, but this situation is no longer observed in the reserve. There is a feeding ban under the Parks and Trees Regulations, which prohibits the feeding of animals in our parks and nature reserves, and this rule is regularly enforced. In fact, enforcement efforts have been increased. From January to May 2010, there were 146 fines issued for monkey feeding alone, compared to the same five-month period in 2009 where only thirty-four fines were issued. The penalty for feeding monkeys have also risen in the last few years from SGD$200 to $250 in May 2007, and from SGD$250 to a hefty SGD$500 in Feb 2008 (National Parks Board, 2008). An example was also made of an offender in national newspapers who had to pay a fine of SGD$4,000, the highest on record for feeding monkeys, after he refused to pay an initial fine in January 2008 (Chong, 2008). For effective enforcement, auxiliary police as well as surveillance cameras are used to deter and apprehend monkey feeders. The increase in enforcement efforts and the stiffer penalties have discouraged most people from going to Old Upper Thomson Road (a popular location

Figure 12.6. Signs along the streets and park trails warn guest to parks in Singapore that feeding macaques is punishable by fines.

for feeding monkeys) to feed the monkeys, and for a while, the monkeys were nowhere to be seen near the road (*personal observations*) (Figure 12.6). The number of troops found along Old Upper Thomson Road has been reduced from six troops to three due to the lower frequency of food provisioning by humans, and some troops spend more time foraging in the forests.

For refuse collection points next to the nature reserves such as BTNR, the trash is contained in a cage or secured trough until the garbage truck arrives to take it away. Monkey-proof dustbins have also been deployed at the visitor centers of nature reserves as well as within the reserves to prevent monkeys from raiding the bins for food scraps. These bins have worked well and some residents living near the reserves have also modified their bins to keep monkeys out. Despite this, many residents still keep trash bins that are easily opened by macaques, and thus the problem of trash accessibility is not fully solved. Similarly, fruit trees formerly planted along the streets in residential areas experiencing monkey nuisance have been replaced by street trees that are less attractive to monkeys as a food source.

Public education and outreach

Educating the public remains a cornerstone of our strategy in dealing with human-macaque conflict over the long term. Public events, such as the "Monkey Blitz" are conducted, where groups of volunteers go to the parks and advise and educate people about not feeding monkeys. NParks has also organized talks, seminars and guided walks in an effort to enlighten the public on macaque social behavior, reduce human-macaque conflict through the promotion of coexistence, and point out that artificial feeding contributes to an unnatural increase in the macaque population.

Trapping

Public complaints regarding conflict with macaques between 2001 and 2007 (N = 2115, mean 302 per year) have more than doubled from the annual average of complaints recorded between 1996 and 2000 (N = 590, mean 118 per year) (Agri-Food and Veterinary Authority, *in litt.*). In cases where monkeys are reported to be causing a nuisance to residents or members of the public, each case is investigated and monitored either by an NParks or AVA officer before action is taken. This is because not all monkey nuisance cases warrant trapping. In some cases, the nuisance arises due to a lapse in human housekeeping such as not securing a refuse chute, or a domestic helper or family member who is feeding a monkey in the backyard of the house. These cases are easily avoided with advice and constant reminders to residential areas. If a monkey is found to be aggressive and coming back regularly to a house or park to cause a nuisance, a trapping cage is set up to capture and cull the individual.

Island-wide population assessments

Surveys for macaque troops and understanding their distribution in Singapore are crucial for macaque population control, predicting likely areas of human–macaque conflict in relation to land-use planning (Sha *et al.*, 2009a). Several attempts were made to census the macaque population in Singapore in the past and numbers ranged from less than 1,000 individuals counted by the members of the then Malayan Nature Society (Singapore Branch) in August 1986 (Lucas, 1995) to a more definite 635 individuals in forested areas of both the Bukit Timah and Central Catchment Nature Reserves (Agoramoorthy and Hsu, 2006). One must note that both of these surveys did not include population assessments of macaques in the offshore islands and urban areas. To date, the most thorough census of *M. fascicularis* in Singapore estimated the total population to be between 1,218 and 1,454 individuals on both mainland and her offshore islands in a study commissioned by NParks and AVA (Sha *et al.*, 2009a). In comparison with two other places with human-macaque interface, Singapore experiences the lowest human-macaque interface with 28.2 individuals per km² in the nature reserves and its vicinity, compared with Hong Kong (326 indiv per km²) (Wong and Ni, 2000) and Bali (1111 indiv per km²) (Wheatley, 1999). These numbers should put things into perspective for the public and suggest that population density of macaques may not be the main driver of perceived human-macaque conflicts (Sha *et al.*, 2009a).

Conclusion

Human-macaque conflict does exist in Singapore but it is much more benign compared to human-macaque interfaces in other parts of the world (Fuentes *et al.*, 2008, Jones-Engel *et al.*, 2006, Sha *et al.*, 2009b). Research has show that human behavior is the main cause of macaque-to-human interaction, and the challenge would be to change human behavior and habits to ameliorate conflicts in interface zones (Sha *et al.*, 2009b). In this same study, the authors have shown that majority (90 percent) of the interviewees argued for the conservation of macaques rather than their entire eradication. Coupled with the research conducted on Singapore macaques to date, this attests to the conservation, scientific and intrinsic values of Singapore's long-tailed macaques, and that the preservation of the species has an important place in safeguarding Singapore's natural heritage.

References

Agoramoorthy, G. and Hsu, M. 2006. Population status of long-tailed macaques (*Macaca fascicularis*) in Singapore. *Mammalia* **70**: 300–302.

Anon. 2010. Monkey shootings draw celebrity protests. P 109. *Chronicle of Singapore – Fifty Years of Headline News*. Peter Lim (ed.).

Brook, B. W., Sodhi, N. S., and Ng, P. K. L. 2003. Catastrophic extinctions follow deforestation in Singapore. *Nature* **424**: 420–423.

Chong, E. 2008. $4,000 fine for feeding monkeys. *The Straits Times*. January 24, 2008.

Corlett, R.T. 1992. The ecological transformation of Singapore: 1819–1990. *Journal of Biogeography* **19**: 411–420.

Corlett, R.T. and Lucas, P.W. 1990. Alternative seed-handling strategies in primates – seed-splitting by long-tailed macaques (*Macaca fascicularis*). *Oecologia* **82**: 166–171.

Cortes, J. and Shaw, E. 2006. The Gibraltar macaques: Management and future. In *The Barbary Macaque: Biology, Management and Conservation*, J. K. Hodges and J. Cortes (eds.). Nottingham Uniersity Press. pp. 199–210.

Fuentes, A., Kalchik, S., Gettler, L., Kwiatt, A., Konecki, M., and Jones-Engel, L. 2008. Characterizing human-macaque interaction in Singapore. *American Journal of Primatology* **70**: 1–5.

Goh, S. Y. 2008. Management of monkey nuisance. www.ava.gov.sg/NR/ rdonlyres/5780829E-23A9-4A74-979F-7B1B9495F685/22399/ STMonkeynuisance_4Nov08.pdf, accessed 13 June 2010.

Jones-Engel, L., Steinkraus, K. A., Murray, S. M., *et al.* 2007. Sensitive assays for simian foamy viruses reveal a high prevalence of infection in commensal, free-ranging Asian monkeys. *Journal of Virology* **81**: 7330–7337.

Jones-Engel, L., Engel, G. A., Schillaci, M. A., *et al.* 2006. Considering human-primate transmission of measles virus through the prism of risk analysis. *American Journal of Primatology* **68**: 868–879.

Lee, H. C. and Chong, C. K. 2006. Python squeezes pet dog to death at River Valley condo. *The Straits Times*. 19 November 2006.

Lucas, P. W. 1995. Long-tailed macaques. *The Gardens' Bulletin (Singapore)* **Suppl. 3**: 105–119.

Lucas, P. W. and Corlett, R. T. 1991. Relationship between the diet of *Macaca fascicularis* and forest phenology. *Folia Primatologica* **57**: 201–215.

   1992. Notes on the treatment of palm fruits by long-tailed macaques (*Macaca fascicularis*). *Principes* **36**(1): 45–48.

   1998. Seed dispersal by long-tailed macaques. *American Journal of Primatology* **45**: 29–44.

McKinney, M. L. 2006. Urbanization as a major cause of biotic homogenization. *Biological Conservation* **127**: 247–260.

Murdoch, G. 2007. Stressed Singaporeans crack down on thieving monkeys. *The Star*. 19 February 2007.

National Parks Board (Singapore). 2008. Tougher measure against monkey feeding. www.nparks.gov.sg/cms/index.php?option=com_news&task=view&id= 55&Itemid=50, accessed 13 June 2010.

Neo, C. C. 2010. Tiresome monkey business. *Weekend Today*, 16–17 January 2010, p. 12.

Raffles, T. 1821. Descriptive catalogue of a zoological collection, made on account of the Honourable East India Company, in the Island of Sumatra and its vicinity under the direction of Sir Thomas Stamford Raffles, Lieutenant-Governor of Tort Marlborough; with additional notices illustrative of the natural history of those countries. *Transactions of the Linnean Society* **13: 239–340.**

Schillaci, M. A., Jones-Engel, L., Lee, B. P. Y-H., *et al.* 2007. Morphology and somatometric growth of long-tailed macaques (*Macaca fascicularis fascicularis*) in Singapore. *Biological Journal of the Linnean Society* **92**: 675–694.

Sha, J. C. M., Gumert, M. D., Lee, B. P. Y-H., *et al.* 2009a. Status of the long-tailed macaque *Macaca fascicularis* in Singapore and implications for management. *Biodiversity and Conservation* **18**: 2909–2906.

Sha, J. C. M., Gumert, M. D., Lee, B. P. Y-H., Jones-Engel, L., Chan, S., and Fuentes, A. 2009b. Macaque–human interactions and the societal perceptions of macaques in Singapore. *American Journal of Primatology* **71**: 825–839.

Sodhi, N. S. and Sharp, I. 2006. Winged Invaders: *Pest Birds of the Asia Pacific with Information on Bird Flu and Other Diseases.* Singapore: SNP International Publishing, 184 pp.

Sua, T. 2007. Monkey mayhem at MacRitchie reservoir. *The Straits Times.* 29 December 2007.

Tan, H. T. W., Chou, L. M., Yeo, D. C. J., and Ng, P. K. L. 2007. *The Natural Heritage of Singapore.* Prentice Hall. 271 pp.

United Nations. 2008. World urbanization prospects: The 2007 revision. *Population Newsletter* **85**: June 3–7. Wheatley, B. P. 1999. *The Sacred Monkeys of Bali.* Waveland Press. 189 pp.

Wong, C. L. and Chow, K. L. 2004. Preliminary results of trial contraceptive treatment with SpayVac™ on wild macaques in Hong Kong. *Hong Kong Biodiversity* **6**: 13–16.

Wong, C. L. and Ni, I. H. 2000. Population dynamics of the feral macaques in the Kowloon Hills of Hong Kong. *American Journal of Primatology* **50**: 53–66.

but further research is needed to fully understand how sterilization affects human-macaque conflict.

Additionally, it is important to document behavioral repercussions of reproduction control, as well as possible selective effects. For example, animals that are easily captured are often those that are the most habituated and the least aggressive to humans. Therefore, we need to consider how the removal of these individuals might affect the genetic composition of subsequent generations of synanthropic macaques (see Chapter 13). Another consideration is that sterilization may make females more assertive around people as they will be less fearful without infants to carry and defend.

The human dimension

Managing human behavior at the human–macaque interface is critical. Human behavior can unwittingly reinforce undesirable macaque behavior. For example, feeding macaques reinforces their association that humans provide easy access to food. Governmental support is vital to the regulation of no-feeding and/or no-littering laws (Fuentes *et al.*, 2008; Sha *et al.*, 2009a, 2009b). While the presence of food significantly promotes macaque nuisance behaviors, reckless human behaviors (e.g., chasing, taunting, and throwing things) can result in serious injury for humans and macaques (Mccarthy *et al.*, 2009; Sha *et al.*, 2009b). Macaques frequently react aggressively to people who taunt, attempt to repel, or throw objects at them. It is therefore important that we propose alternative repellant tactics for people at the human–macaque interface. This can include repelling macaques with water from hoses from a distance, a safe and effective measure unlikely to provoke an aggressive response, as they don't appear to associate being struck with water as an attack from humans. Another important consideration is that counter-aggression displayed by macaques may not be directed at the individual performing the irresponsible behavior, but to innocent bystanders.

Education and persuasive communication efforts are cited as preferred mitigation strategies for addressing undesirable conservation-related and public health behaviors (Box 12.1 and Box 12.2). Education may be preferred for a number of reasons: (1) it is an enduring solution that transcends many contexts; (2) it retains freedom of choice and is typically less intrusive; and (3) it is thought to be less expensive than other alternatives.

Educational activities can teach how to safely interact with macaques and instill a commitment on the part of the community to a more peaceful interface. Educational campaigns can focus on children, as well as adults, and should be particularly focused on people at the interface. Campaigns can be tailored by utilizing research from macaque programs in that area. Teaching communities about their macaques, and about the individuality of these macaques, can establish familiarity, which can foster positive attitudes towards macaques. Educational campaigns can also disseminate information about regulations established to control interface zones, and delineate norms for moving and living around macaques. Overall, educational campaigns alone will not remove human-macaque conflict, but they can alter a community's perception, establish greater tolerance, and direct people to act safely around macaques. Finally, education can convince community members to value the human-macaque community.

It is important to note that people who implement educational campaigns often embark on them rather naively, assuming that by simply making information available, behavior change will follow. In reality, effective programs based on communication are notoriously difficult to develop. Various factors confound the ability to persuade someone with informational messages (Wood, 2000). For instance, the extent of attitude and behavior change may depend upon the channel of communication, the source of the message, the strength of the arguments presented, the subject's prior knowledge, and the strength and function of his or her existing attitude(s) (Eagly and Chaiken, 1993; Petty and Cacioppo, 1996; Wood, 2000). While research has been unable to identify broadly general conclusions about persuasion, important conceptual advancements have been made in recent decades that can effectively guide communication programs (Crano and Prislin, 2006). The long-term goal of the education campaigns should be to alter conflicts at the human-primate interface. While there have not been any studies to date that specifically address these conflicts, there has been significant work done in other areas of human-animal conflict that utilize attitude-behavior theory and attitude change theory (Bright and Manfredo, 1997; Manfredo, Teel, and Henry 2009; Needham *et al.*, 2007; Teel and Manfredo, 2010; Whittaker, Vaske, and Manfredo, 2006). Approaching conservation communication and educational campaigns using the available scientific frameworks is strongly encouraged.

Management of specific contexts

Options to control crop-raiding macaques

Long-tailed macaques are not always the most intensive crop-raiders (relative to other mammalian raiders), but they can be significant crop-raiders in some areas (Marchal and Hill, 2009) (Gumert, Chapter 1). Crop-raiding macaques present difficultlies for farmers and strategies to alleviate the damage are needed. There is no fully effective single strategy for preventing primate crop raiding, although studies on other primate species offer a framework for mitigation. One effective technique is vigilant guarding, and this is a common practice in several localities (Chhangani and Mohnot, 2004; Hill, 2000; Naughton-Treves, 1997; Sprague and Iwasaki, 2006; Tweheyo, Hill, and Obua, 2005; Wang, Curtis, and Lassoie, 2010). There are few alternatives more effective than having people chasing animals as they come into crop areas. Guarding, however, is manpower intensive, and in order to lessen the need for continual guarding, some farmers have constructed fencing systems (Chhangani and Mohnot, 2004; Wang, Curtis, and Lassoie, 2010). Unfortunately, fencing must cover large areas and the primate raiders are often able to navigate around fences, even if electrified (Strum, 1994). Details on the shortcomings of constructed fences are scarce and future studies of fencing and barricades are needed to develop improvements.

Deterrent methods include chemicals such as the scent of predators, or the sounds of predators, gunshots, firecrackers, water hoses or guns, sling shots and/or primate alarm calls to inhibit entering cropland. These can all be useful strategies in the short term (Strum, 1994), but the effectiveness of some techniques can wear off if they are used frequently enough to habituate animals to the aversive stimulus. It is best to use a variety of detterents to keep the animals more vigilant and apprehensive. The use of chemical repellants, such as oleo capsicum and HATE 4C to modify the taste of food has not been found to be significantly effective at stopping baboon raiders, and thus may also not work with macaques. In contrast, Strum (1994) has shown that "conditioned taste aversion systems" can produce positive effects in reducing crop raiding (Strum, 1994). This technique utilizes lithium chloride, or other chemical emetics, to induce vomiting after feeding. Experimental work on taste aversion has shown that only one exposure is necessary for an animal to acquire and learn a taste aversion (Garcia, Ervin, and Koelling, 1966).

Other programs to stop crop raiding are focused on removal of primates. This tactic has been demonstrated to be effective against both baboon troops and rhesus monkeys (Strum 1994; Southwick, 1986). Other authors have criticised translocation as an anti-crop-raiding strategy on the grounds that it is often difficult to find suitable areas remote from human settlement in which to relocate macaques. Translocated animals may come into conflict with humans in the new location while new macaques may move into the original conflict area, replacing translocated individuals (Pirta, Gadgil, and Kharshikar, 1997). Also, little is known about the negative effects of translocation on regional population dynamics. Another proposed option has been simply to kill raiding macaques, which is difficult to do. Total eradication is often impossible (Strum, 1994; Wheatley, Chapter 10).

The solution to crop raiding does not lie simply in changing the behavior and location of the animals. The modification of human land-use methods is also a critical component. Land within half km of the forest edge is where the majority of crop raiding occurs (Hill, 2000; Naughton-Treves, 1997). Developing strategies to reduce use of these border regions, as well as planting less palatable crops, or placing barrier plants close to the forest edge should also be considered (Hill, 2000; Naughton-Treves, 1998). Another option is crop insurance schemes, which farmers could buy into or which are government subsidized. Insured farmers would have assistance in deflecting the costs of crop loss, which would also make them more tolerant of their wildlife neighbors (Sinha *et al.*, 2006). The main problem with insurance programs is that they are costly to unaffected parties if the local people cannot afford or are unwilling to purchase insurance themselves, which is often the case in developing countries. Perhaps in such communities, rather than currency, shared subsistence programs could be devised where local farmers provide food assistance to farmers with damaged crops in exchange for services (e.g., assist in harvesting and distributing crops).

Island colonies

Over the past decades, relocation of primates to islands has been put forward as a way to address the issue of animals that have been removed from their native group and basically have no other place to go. Proposals to establish island colonies have been considered for confiscated pet macaques and other "surplus" animals in places like Singapore and Indonesia. The motivation behind the establishment of these islands has been both practical and humanitarian. The appeal of this approach is readily appreciated: the animals are winsome, their "advocates" often fervent and well-meaning. However, this approach to the issue of unwanted animals is problematic for several reasons. First, investing in "macaque islands" may divert attention and resources from the issues that produced the unwanted macaques in the first place. Rather than "saving" individual macaques, resources, including charitable funding, might more effectively be used to promote sustainable population-level relationships between humans and macaques at existing interfaces. For example, resources are required for more rigorous enforcement of laws against trapping and/or trading primates. A related concern is that establishing special islands for the release of macaques removed from human-macaque conflict zones might create the perspective in local communities that macaques belong on these types of islands and nowhere else. This perception might erode support for investments in long-term coexistence and conservation of synanthropic macaques. There is even concern that relocation could actually create a justification to exterminate populations of synanthropic macaques.

The "monkey islands" themselves present concerning challenges. Merely relocating macaques away from areas of conflict does not ensure that the macaques' new situation is desirable or adequate. In some instances, "rescued" macaques have been housed in inadequate cages, where quality of life is debatable, negating the ethical justification for relocation. Additionally, releasing macaques to range freely on islands raises other questions. Depending on how animals are selected for release, abnormal age-sex ratios are likely to result, leading to skewed social functioning in the groups. Provisioned with food, population growth may lead to overcrowding and the need to expend resources to manage these populations. The impact of relocated macaque populations on island environments is another potentially significant problem. There is concern that these islands might essentially become poorly funded monkey zoos, reliant on governmental and NGO support. These resources might be better employed in developing sustainable human-macaque ecosystems at the existing interface.

Managing introduced populations

Dealing with ethnophoresy, the introduction of long-tailed macaques to new locations, is a very important aspect of managing them. The management strategies outlined above also apply to introduced populations, with a crucial caveat. In areas where introduced macaques threaten native flora and fauna, the option of complete removal of all macaques may be considered. Such a dramatic and irrevocable intervention should be informed to the extent possible by acquisition of relevant data on the impact of the macaques and after a review of the multiple management options available, as proposed by a variety of stakeholders. In areas where removal is not feasible, management strategies will need to be implemented that reduce the impact of macaques and raise the value of macaques to the local communities living with them.

Promoting human and macaque health at the interspecies interface

The literature on health in primate populations is dominated by interspecies disease transmission, both human-to-primate as well as primate-to-human. Chapter 7 outlines some of the infectious diseases that pose a risk to humans and primates at the interface. From the standpoint of disease transmission, all contact, direct and indirect, between macaques and humans presents some risk of disease transmission, in both directions. However, as coexistence between humans and long-tailed macaques is desirable, a reasonable approach is sought that reduces health risks to both while maintaining the benefits of human-macaque interaction. Health promotion, among other goods, should be considered desirable for both human and primates. In most situations, the infectious danger that should be emphasized is the danger posed by humans *to* macaques. Conversely, focusing on the risks of macaque-to-human transmission can increase the risk that macaques become viewed as a potential danger to public health, which can lead to actions that harm them.

Human-to-macaque transmission

As discussed in Chapter 7, surprisingly little is known about actual causes of mortality and morbidity in free-ranging populations of long-tailed macaques. This lack of data is a significant barrier to understanding how endemic human pathogens affect macaques. The recommendations put forward here are informed by available data describing evidence for exposure of free-ranging macaques to human pathogens, data on infection of laboratory macaques (i.e., mostly rhesus and long-tailed and pig-tail macaques) and epidemiological data on pathogen spread in human populations.

Respiratory pathogens are transmitted through inhalation of aerosolized particles produced by speech, coughing, and sneezing, and through contact of pathogens with mucosal surfaces (i.e., conjunctivae, nasal mucosa, and oral mucosa) (see Chapter 7). Contexts in which close proximity and/or physical contact occur can lead to respiratory pathogen transmission. Not all human demographic groups are equal with respect to their capacity to spread disease. Targeting interventions at higher-risk groups may be a useful strategy for reducing risk of disease transmission. For example, children typically suffer from upper respiratory infections at rates that far exceed adults, and must therefore be considered to pose a higher risk of transmitting infection. Obviously, people who display symptoms of infection (e.g., coughing, sneezing, runny nose, etc) are likely to be infectious. Focusing interventions on specific high-risk groups can facilitate implementation, decrease the amount of resources required and make interventions more acceptable to the public.

Specific management options include screening for illness (for example, asking visitors to a monkey temple whether they have cold symptoms), providing face masks to symptomatic individuals, and requiring people to wash their hands before they enter an area where they might have direct or indirect contact with macaques. Certainly, providing well researched and tested education materials via pamphlets, signage, and having informed guides available can increase awareness of the dangers (especially to macaques) of pathogen transmission, and can potentially lead people to behave in ways that decrease infectious risk for macaques. This approach is used by ecotourism groups (e.g., in Africa) to reduce the risk of transmission of endemic human pathogens to great apes (Köndgen et al., 2008). Specific activities that are likely to bring people and macaques into contact or close proximity, such as feeding or encouraging macaques to climb onto a person's shoulders, present a high risk for pathogen transmission and should be curtailed to the extent possible.

Oral-fecal pathogens (e.g., enteroviruses, bacteria, and intestinal parasites like giardia and amoebae) are transmitted through the ingestion of contaminated food, water, and fomites (i.e., inanimate objects contaminated with an infectious agents). A person handing a piece of food to a macaque can thus transmit a pathogen present on his hand to the macaque, which becomes infected when it eats the contaminated food. Strategies to reduce the risks of human-to-macaque transmission of these agents should prioritize proper containment of refuse (see section on refuse disposal, above) and discourage direct and indirect (i.e., though food and fomites) physical contact between people and macaques. Disposal of human waste should take into account the need to prevent the contamination of water supplies likely to be used by macaques. Thus, effective and safe sewage disposal and clean water sources, important safeguards of public health, can also have "spin-off" benefits for the health of macaque populations.

Macaque-to-human transmission

Transmission of respiratory and oral/fecal pathogens can occur both from humans to macaques and from macaques to humans. As a result, implementation of the above recommendations for reducing the likelihood of human-to-macaque disease transmission will also decrease the chance that humans become infected with respiratory and oral/fecal pathogens present in the macaque population. Transmission of a *parenterally* (i.e., taken into the body in a manner other than through the digestive or respiratory tract) transmitted simian virus to a human has been modeled mathematically (Engel *et al.*, 2006) for visitors to a monkey temple in Bali, employing the assumption that macaque bites were the mode of viral transmission. The analysis suggested that wound washing can significantly reduce the likelihood of a person becoming infected as a result of a macaque bite. Therefore, increasing the awareness of the importance of, and certainly the availability of, prompt and thorough wound cleansing following a macaque bite would likely decrease the risk of an individual becoming infected with parenterally transmitted viruses.

Management considerations of health risks in specific contexts

The frequency and dynamics of interspecies contact vary by context and population. Specific sites have unique features or combinations of features, making it difficult to generalize about interspecies contact and disease risk. Local culture, history, socioeconomic, and other factors contribute to patterns of interspecies interaction and, consequently, disease transmission. However, we will venture a few generalizations.

Urban macaques typically are restricted to specific ranges within cities, often centered on parks, temples, or green belts. People who live, or work, or travel within these ranges can come into contact with macaques, and, over a lifetime, the likelihood of some kind of contact is additive. Some individuals deliberately seek out contact with macaques, for religious reasons, or personal inclination. Targeting these individuals for education about the risks of interspecies transmission (i.e., both *from* and *to* macaques) is one strategy for decreasing risk of macaque-to-human transmission of infection.

Monkey temples are religious sites that, over time, have become associated with groups of free-ranging monkeys, most often macaques. They are prevalent in Southeast Asia, where some have become major tourist destinations, drawing a minimum of 100,000 visitors per year (Fuentes and Gamerl, 2005). From the standpoint of human health, temples are significant because they bring together macaques with people who are "macaque naïve," and thus more likely than local people to behave in a manner that provokes macaque aggression, with concomitant risk of injury and transmission of infectious agents. From the standpoint of the macaques' health, temples with macaques constitute a location where humans from all corners of the globe converge, bearing pathogens, some of which may be novel to the macaques. Introducing novel pathogens to immunologically naïve macaque populations poses a significant health risk for them and makes advisable the implementation of measures to decrease physical contact and proximity between macaques and visitors. Educational signboards are an important feature of this effort, but there is no substitute for the enforcement of common-sense proscriptions against feeding and teasing macaques.

Live animal markets, as they tend to bring multiple animal species together in close proximity and in confined space, provide pathogens with unique opportunities. Animals may be ill and immunocompromised, increasing both the "availability" of pathogens in the environment and the likelihood that individuals are susceptible to disease. Markets may act as "mixing pots" where infectious agents from one animal reservoir can come into contact with a host species it would not normally find in a "natural" setting. Such a scenario is thought to have initiated the SARS epidemic in 2002. Therefore, regular health monitoring of animal populations, including non-human primates, as well as people who work at pet markets, is a logical strategy for detecting new patterns of disease, including the emergence of new diseases.

Pet owners, and those who own "performing monkeys," are unique in the long-term, close contact they typically have with macaques (Jones-Engel *et al.*, 2006; Schillaci *et al.*, 2006; Schillaci *et al.*, 2008). It is the rule, rather than the exception, for a pet owner to have been bitten at some point by his/her pet. Pets and their owners likely "share" a broad range of infections, including pathogens that are transmitted by respiratory route, oral/fecal and parenteral transmission.

Pets may have additional significance as a potential mechanism for transmitting pathogens between human populations and free-ranging, including unhabituated, groups of macaques. Pet macaques are typically acquired when they are very young, at a stage when they are cute and unthreatening. As the macaque matures, developing strength and large teeth, pet owners often are no longer interested in keeping the pet and may sell, kill, or release it. In the latter case, a pet harboring infectious agents acquired from the people with whom it has lived may come into contact with and transmit pathogens to free-ranging macaques. Free-ranging, especially unhabituated animals may be immunologically "naïve" to human pathogens, and therefore vulnerable to these pathogens. Seen in this light, pet ownership is a potentially problematic practice, as pets may act as "pathogen conduits" from humans to free-ranging macaques.

Role of Public Health Ministries

Chapter 7 referred to research on measles in free-ranging macaques, contrasting the seroprevalence of measles antibody among rhesus macaques in Nepal, a country with relatively low immunization rates, compared to Singapore, a country with high rates of immunization. An effective public health system that maintains high immunization levels among human populations is likely to protect commensal macaque populations from infections, such as measles. Good nutrition and healthcare are also effective barriers to diseases such as tuberculosis in the human population, and by extension, in primate populations. However, improvements in vaccination and nutrition do not occur overnight, or at low cost. From the perspective of preventing the spread of disease from humans to macaque populations, prioritizing those individuals most likely to have contact with macaques for immunization should be considered. Investments in public health infrastructure may have positive "spin off" effects for populations of macaques.

Communication between health professionals (i.e., public health officials, physicians, nurses, veterinarians) and those responsible for managing macaque populations (i.e., government ministries including forestry, wildlife, and environment) and others (e.g., ministries of tourism) is a potentially powerful tool to promote health. The detection of emerging patterns of disease (e.g., a flu epidemic) in the human population may have significant implications for the health of macaque populations and vice versa. An interlinked system of health information has the capacity to inform and coordinate efforts to both conserve macaque populations and protect public health.

The contribution of public health agencies extends beyond prevention of infectious diseases. For example, environmental pollution is a widely acknowledged threat to human populations. By extension, toxic substances in the environment (e.g., lead, mercury, and insecticides) may also harm synanthropic macaques, though there is scant research in this area. As synanthropic macaques show promise as possible sentinels for human exposure to environmental toxins, this is another potential area of linkage between primate conservationists, public health organizations, and the medical and veterinary fields (Engel *et al.*, 2009).

Managing interface zones: The role of governance

Among stakeholders at the human-macaque interface, governments are key players, capable of enacting and enforcing policies, as well as creating infrastructure and altering the landscape. The government's role in managing the human-macaque interface is complex, with multiple mandates that may at times conflict. Governmental responsibilities include supporting economic growth, promoting public health, and conserving natural resources, including wildlife. Moreover, issues arising at the human-macaque interface can fall under the jurisdiction of more than one government agency, such as wildlife, housing, parks, forestry, health, trade, tourism, and others. The government may also be called upon to mediate between nongovernmental groups.

At the regional level, governments can coordinate with each other and perhaps with nongovernmental organizations (NGOs) to define priorities for the conservation and management of macaques. Optimally, this should include acquiring and harnessing data that inform cohesive, overarching, long-term strategies and policies that support sustainable human-macaque coexistence at the interface. These policies should set the stage for individual communities' policies regarding specific populations of macaques. NGOs can assist communities by providing resources that support these policies. Governments also have a role in marshalling and distributing resources that further conservation goals. As governments benefit from tax revenue generated from the lucrative sale of macaques to the biomedical industry, it can be argued that these monies should be used by governments to promote research and conservation of free-ranging macaque populations. In Mauritius, some revenue from macaque sales has been invested in general conservation (see Padayatchy, Chapter 9), but we have yet to see a program where the funds generated from macaque trade actually go towards the many needs of macaque populations. Such funding is particularly important for a species such as the long-tailed macaque that is relatively common, as NGOs and governmental organizations are generally most disposed to fund programs aimed at endangered species.

As long-tailed macaques thrive in "edge" environments, governments can use this ecological predisposition to attempt to control ranging patterns. The government can use land development policy to create "buffer zones" around forest edges and farms, prioritizing the planting of crops unlikely to attract long-tailed macaques to these areas. Similarly, government can restrict development of residential properties in areas near macaque ranges. In residential areas already established near macaque habitat, governmental regulations can mandate building standards (see above) that take into account the presence of macaques. Government can also promote awareness of macaque conservation/ management issues in areas that abut macaque habitats.

Particularly in densely settled areas at the human-macaque interface, government agencies are frequently called on to address complaints of citizens bothered by macaques (Sha *et al.*, 2009b). Such complaints can be filed directly with government agencies, or can be directed through the media. Pressure to act on these complaints may be significant. Policy solutions that take into account the interests of multiple groups, using rigorously collected data when possible, are desirable. The formation of "Macaque Boards" (i.e., groups of individuals representing different interest groups), informed by professionals with expertise and experience, may be a way to find acceptable solutions to problems that occur at the interface.

Government agencies are likely to play an important role in enacting and enforcing laws that regulate human behavior at the interface. Collaboration between governmental agencies and communities (e.g., coordination of efforts to patrol neighborhoods) has the potential to increase buy-in by communities and may decrease the amount of resources the government needs to expend to enforce regulation and manage macaque populations. Monitoring and regulation of trade in primates is another important government function, as non-native primates may introduce pathogens to an existing primate population. Similarly, governments are the entities most likely to be able to monitor the health of macaque populations, including free-ranging populations and other contexts, such as pet markets.

Interagency conflict

Macaques do not respect geopolitical boundaries. They may range from an area under the jurisdiction of one government agency (e.g., a park) into a neighboring area under the jurisdiction of a separate agency (e.g., a residential area). Thus, a regulatory response (e.g., enforcing fines for people who feed macaques) initiated in one area may be ineffective, because it is not uniformly applied in all areas in the macaques' range (Sha *et al.*, 2009b). This phenomenon occurs both at the community level and, in a larger sense, between states and nations. Intra and intergovernmental communication and coordination will be critical in the development of a cohesive response strategy to the management challenges of human-macaque interfaces.

Conclusion

The problems of human-macaque interface in Southeast Asia are complex and extensive. A wide range of issues affects both humans and macaques. Every community has a unique set of challenges and resources; there are no one-size-fits-all solutions. Some communities will elect to strive towards sustainable communities with their neighboring macaques, while other poorer communities may pursue eradication, considering the impact of macaques too costly to support. What we hope to provide in this chapter is not a manual of solutions, but a source of ideas about aspects of management in areas of human-macaque conflict. We propose that communities approach the issues of managing their human-macaque communities, with peripheral support, guidance, and regulation from NGOs and governmental agencies. Overall, it will be the communities that will be living with the macaques. Therefore, the future of hundreds of thousands, if not millions, of long-tailed macaques in Southeast Asia, will depend on the tolerance, good will, and innovative problem solving skills, of their neighboring human communities.

References

Bright, A. D. and Manfredo, M. J. 1997. The influence of balanced information on attitudes toward natural resource issues. *Society and Natural Resources* **10** (5): 469–483.

Bunluesilp, N. 2009. No monkey business: Thailand launches primate birth control. *Reuters Life*, 21 August 2009.

Chhangani, A. K. and Mohnot, S. M. 2004. Crop raid by Hanuman langur (*Semnopithecus entellus*) in and around Aravallis (India) and its managment. *Primate Report* **69**: 35–47.

Crano, W. D. and Prislin, R. 2006. Attitudes and persuasion *Annual Review of Psychology* **57**: 345–374.

Eagly, A. H. and Chaiken, S. 1993. *The Psychology of Attitudes*. London: Harcourt Brace.

Engel, G., Hungerford, L. L., Jones-Engel, L. *et al.* 2006. Risk assessment: A model for predicting cross-species transmission of simian foamy virus from macaques (*M. fascicularis*) to humans at a monkey temple in Bali, Indonesia. *American Journal of Primatology* **68**(9): 934–948.

Engel, G., O' Hara, T. M., Cardona-Marek, T. *et al.* 2009. Synanthropic primates in Asia: Potential sentinels for environmental toxins. *American Journal of PhysicalAnthropology* **142**: 453–460.

Fuentes, A. and Gamerl, S. 2005. Disproportionate participation by age/sex classes in aggressive interactions between long-tailed macaques (*Macaca fascicularis*) and human tourists at Padangtegal monkey forest, Bali, Indonesia. *American Journal of Primatology* **66**(2): 197–204.

Fuentes, A., Kalchik, S., Gettler, L., Kwiatt, A., Konecki, M., and Jones-Engel, L. 2008. Characterizing human-macaque interactions in Singapore. *American Journal of Primatology* **70**(9): 879–883.

Garcia, J., Ervin, F., and Koelling, R. A. 1966. Learning with prolonged delay of reinforcement. *Psychonomic Science* **5**: 121–122.

Hill, C. 2000. Conflict of interest between people and baboons: Crop raiding in Uganda. *International Journal of Primatology* **21**: 299–315.

Jones-Engel, L., Schillaci, M., Engel, G., Paputungan, P., and Froehlich, J. 2006. Characterizing primate pet ownership in Sulawesi: Implications for disease transmission. In *Commensalism and Conflict: The Human-Primate interface*, J. D.

Paterson and J. Wallis (eds.) Norman, OK: American Society of Primatologists. pp. 196–221.

Köndgen, S., Kühl, H., N ' Goran, P. K., *et al.* 2008. Pandemic human viruses cause decline of endangered great apes. *Current Biology* **18**: 260–264.

Manfredo, M. J., Teel, T. L., and Henry, K. L. 2009. Linking society and environment: A multilevel model of shifting wildlife value orientations in the western United States. *Social Science Quarterly* **90**(2): 407–427.

Marchal, V. and Hill, C. 2009. Primate crop-raiding: A study of local perceptions in four villages in North Sumatra, Indonesia. *Primate Conservation* **24**: 107–116.

Mccarthy, M. S., Matheson, M. D., Lester, J. D., Sheeran, L. K., Li, J. H., and Wagner, R. 2009. Sequences of Tibetan macaque (*Macaca thibetana*) and tourist behaviors at Mt. Huangshan, China. *Primate Conservation* **24**: 145–151.

Naughton-Treves, L. 1997. Farming the forest edge: Vulnerable people and places around Kibale National Park, Uganda. *Geography Review* **87**: 465–488.

1998. Predicting patterns of crop damage by wildlife around Kibale National Park, Uganda. *Conservation Biology* **12**: 156–158.

Needham, M. D., Vaske, J. J., Donnelly, M. P., and Manfredo, M. J. 2007. Hunting specialization and its relationship to participation in response to chronic wasting disease. *Journal of Leisure Research* **39**(3): 413–437.

Petty., R. E. and Cacioppo, J. T. 1996. *Attitudes and Persuasion:Classical and Contemporary Approaches*. Boulder, CO: Westview Press.

Pirta, R. S., Gadgil, M., and Kharshikar, A. 1997. Management of the rhesus monkey *Macaca mulatta* and Hanuman langur *Presbytis entellus* in Himachal Pradesh, India. *Biological Conservation* **79**: 97–106.

Schillaci, M. A., Jones-Engel, L., Engel, G. A., and Kyes, R. C. 2006. Exposure to human respiratory viruses among urban performing monkeys in Indonesia. *American Journal of Tropical Medicine and Hygiene* **75**(4): 716–719.

Schillaci, M. A., Jones-Engel, L., Heidrich, J. E., Benamore, R., Pereira, A., and Paul, N. 2008. Thoracic radiography of pet macaques in Sulawesi, Indonesia. *Journal of Medical Primatology* **37**(3): 141–145.

Sha, J. C., Gumert, M. D., Lee, B., Fuentes, A., Chan, S., and Jones-Engel, L. 2009a. Status of the long-tailed macaque (*Macaca fascicularis*) in Singapore and implications for management. *Biodiversity and Conservation* **18**(11): 2909–2926.

Sha, J. C., Gumert, M. D., Lee, B. P., Jones-Engel, L., Chan, S., and Fuentes, A. 2009b. Macaque–human interactions and the societal perceptions of macaques in Singapore. *American Journal of Primatology* **71**(10): 825–839.

Sinha, A., Kumar, R., Gama, N., Madhusudan, M., and Mishra, C. 2006. Distribution and conservation status of the Arunachal macaque, *Macaca munzala*, in Western Arunachal Pradesh, Northeastern India. *Primate Conservation* **21**: 145–148.

Sprague, D. S. and Iwasaki, N. 2006. Coexistence and exclusion between humans and monkeys in Japan: Is either really possible? *Ecological and Environmental Anthropology* **2**: 30–43.

Strum, S. C. 1994. Prospects for management of primate pests. *Revue D'Ecologie (Terre et la Vie)* **49**(3) : 295–306.

Strum, S. C. and Southwick, C. 1986. Translocation of primates. In *Primates: The Road to Self-Sustaining Populations*. K. Bernirschke (ed.). New York: Springer-Verlag.

Teel, T. L. and Manfredo, M. J. 2010. Understanding the diversity of public interests in wildlife conservation. *ConservationBiology* **24**(1): 128–139.

Tweheyo, M., Hill, C., and Obua, J. 2005. Patterns of crop raiding by primates around the Budongo Forest Reserve, Uganda. *Wildlife Biology* **11**: 237–247.

Wang, S., Curtis, P., and Lassoie, J. 2010. Farmer erceptions of crop damage by wildlife in Jigme Singye Wangchuck National Park, Bhutan. *Wildlife Society Bulletin* **34**: 359–365.

Whittaker, D., Vaske, J. J., and Manfredo, M. J. 2006. Specificity and the cognitive hierarchy: Value orientations and the acceptability of urban wildlife management actions. *Society and Natural Resources* **19**(6): 515–530.

Wong, C. L. and Chow, G. 2004. Preliminary results of trial contraceptive treatment with SpayVac™ on wild monkeys in Hong Kong. *Hong Kong Biodiversity: AFCD Newsletter*. **6**: 13–16.

Wood, W. 2000. Attitude change: Persuasion and social influence. *Annual Review of Psychology* **51**: 539–570.

13 Future directions for research and conservation of long-tailed macaque populations

MICHAEL D. GUMERT, AGUSTÍN FUENTES,
GREGORY ENGEL AND LISA JONES-ENGEL

Long-tailed macaques are an edge species, preferring to live along the forest borders of many habitat types (Gumert, Chapter 1). The result of this preference is that long-tailed macaques are adaptable generalists that are frequently found along the edges of human settlements across Southeast Asia. Another consequence is that long-tailed macaques can adjust quickly to living with other species, and thus have commonly expanded beyond the edge to overlap with humans in numerous contexts (see Part II). Due to the close association with humans, macaque populations can be powerfully impacted by human activity. In some cases they have been carried and introduced to areas beyond their normal range (see Part III). The overlap of macaques and humans, and the consequences of this overlap, needs to be better understood. While the basis of our relationship with long-tailed macaques is becoming apparent, much more research will be needed to fully understand their population and the causes and consequences of our interface with them. This chapter is an attempt to focus future research in a few important areas that will be necessary for better understanding the population, ecology, and synanthropic nature of long-tailed macaques. This chapter focuses on three subject areas that warrant special consideration for future scientific research on *M. fascicularis*: population-level research, the issue of ethnophoresy and introduced populations, and the causes and consequences of human-macaque overlap.

Directions for population-level research

Long-tailed macaques perhaps have the greatest amount of intraspecific variation of any primate species (Fooden, 2006). The large variation is not yet well

Monkeys on the Edge: Ecology and Management of Long-Tailed Macaques and their Interface with Humans, eds. Michael D. Gumert, Agustín Fuentes and Lisa Jones-Engel. Published by Cambridge University Press. © Cambridge University Press 2011.

understood, and to date ten subspecies are recognized (Fooden, 1995; Groves, 2001; Gumert, Chapter 1). The taxonomy of long-tailed macaques is based on Fooden's (1995) morphological studies conducted in a museum, and no genetic data has ever been collected to fully validate his classification. Consequently, it will be very important to obtain a good genetic characterization of the entire population.

A genetic database will be important for several reasons. First, understanding the genetic diversity of long-tailed macaques will be important to field biologists seeking to understand the evolutionary origins and diversity of macaques in Southeast Asia. Second, it will also be important for wildlife officials needing genetic information to better understand the history and make-up of the populations they are attempting to manage. Third, a population genetic database will be valuable to investigators of wildlife trade, as it would provide a basis for assessing of the origin of animals being trafficked. Fourth, for this same reason, population genetics should be of great interest to laboratory researchers utilizing long-tailed macaques (Blancher *et al.*, 2008) because an animal's genotype can influence the response to experimental procedures (i.e., drugs, disease, implants, etc.) (Menninger *et al.*, 2002).

There are other aspects of the long-tailed macaque population that are also important to investigate. First, data on actual population levels of this species are scarce, owing to the logistical difficulty in characterizing a population that is patchy in distribution and stretched across a vast and island-laden geographic range. Second, we have limited data characterizing the proportion of the entire long-tailed macaque population that actually interfaces with human beings. As a result, little is known about how human activity impacts synanthropic long-tailed macaque populations. Given the potential increase in synanthropic macaque populations in some parts of Southeast Asia (see Malaivitinond *et al.*, Chapter 5), it will be important to know the degree of interface occurring and what proportion of the population is now synanthropic.

Knowing more about macaque population size and its overlap with humans is necessary, but achieving it will not be easy. Long-tailed macaques are distributed throughout sixteen different nations (see Table 13.1), and thus population level assessments require an international effort. In many regions across their range mass extermination, capture, and/or sterilization programs are being implemented to reduce human-macaque overlap (Di Silva, 2008; Twigg and Nijman, 2008; Bunueslip, 2009; Martelli, 2009; see Malaivijitnond *et al.*, Chapter 5). Moreover, long-tailed macaques are one of the most exploited species found in both legal and illegal trade (see Box 1.2; Shepherd, 2010). Although at present the total population of long-tailed macaques is thought to be large, concerted efforts to exterminate them or control their reproduction may result in rapid population decline. As development continues in

Table 13.1. *Nations with populations of long-tailed macaques*

Bangladesh
Brunei
Cambodia
China (i.e., Hong Kong)
East Timor
India (i.e., Nicobar Islands)
Indonesia
Laos
Malaysia
Mauritius
Myanmar (Burma)
Philippines
Republic of Palau
Singapore
Thailand
Vietnam

Southeast Asia, we need to assess how vulnerable macaques are to human influences (Eudey, 2008).

The question of conserving long-tailed macaques

Should efforts be made to conserve long-tailed macaques? According to the current IUCN Red List, the core long-tailed macaque subspecies *M. f. fascicularis*, is considered to be of least concern, although some species are thought to be declining (Ong and Richardson, 2008). Prior to this classification, *M. fascicularis* was listed as lower risk/near threatened in earlier assessments (i.e., 2000 and 2004), and thus was afforded a greater level of protection in the past. It is unclear why the status has changed and the population is no longer given any serious protection. This is especially curious since the current population is argued to require close monitoring because, even though they are widespread, they are rapidly declining in numbers due to conflict with humans, capture for trade, and habitat loss (Eudey, 2008).

Perhaps related to the demotion in their protected status was the ratification of long-tailed macaques into the IUCN's list of the World's 100 Worst Invasive Species (Lowe *et al.*, 2000). The conservation community has now seriously devalued long-tailed macaques because of the heightened emphasis of human–macaque conflict in Mauritius, Palau, and West Papua (Kemp and Burnett, 2007), while trade has created a greater demand and increased monetary reward for trade with fewer regulations. Another factor could be the increased use of this species in biomedical research in recent years (see Box 1.2), which has led to calls for fewer regulations in the export of animals. The net effect has been to remove conservation protection for the core species of *M. fascicularis* at a time when pressures on their populations are growing. Overall, the core subspecies of long-tailed macaque has essentially been removed from the radar for any strong forms of protection from international governance, are now considered the world's worst primate pest (despite having the same features as several other macaques, baboons, and guenons).

For the other subspecies of macaques, the conservation status differs dramatically from the core subspecies, but because of the status of *M. f. fascicularis* the other subspecies are largely ignored. Most are data deficient, while *M.f umbrosa* and *M.f. condorensis* are categorized as vulnerable and *M.f. philippensis* is near threatened. (Ong and Richardson, 2008). When we assess the subspecies of long-tailed macaques for conservation needs, a very different picture emerges. For the most part, we are entirely ignorant of the status of six of these forms, while the three that we do have any information on all require conservation attention due to small numbers, geographical isolation, and/or pressure on their populations. It is not unlikely that most of the data deficient forms are also in need of some level of conservation support and that we are just unaware of their status and needs. For example, on Karimunjawa less than 300 macaques survive, all of which conflict with humans. The neighboring island of Kemujan, the only other island where this subspecies occurs, has not yet been surveyed, but it is unlikely to have any more macaques than on Karimunjawa (see Box 1.1). As a result, we can safely assume that the *M. f. karimondjawae* situation does not differ much from the situation for *M.f. condorensis*, which is listed as vulnerable by IUCN.

As we better survey the population we may begin to uncover new subspecies of long-tailed macaques, some of which may be threatened and require conservation. There may be forms of long-tailed macaques that will be delineated as new subspecies and may require conservation consideration in the future. For example, it has been suggested that another subspecies exists west of Kalimantan in Karimata (Yanuar *et al.*, 1993). We may find that regional populations distributed on the mainland may also be facing localized threats in various areas. For example, in this volume, San and Hamada., Hamada *et al.*, and Lee all report serious population pressures facing long-tailed macaques in Myanmar (i.e., *M. f. aurea*), Laos (i.e., *M. f. fascicularis*), and Cambodia (i.e., *M. f. fascicularis*) respectively.

Conservation of unique behavioral features of long-tailed macaques also needs to be considered, as there is a large amount of behavioral variability across their range, and some of it may represent unique local traditions. For example, in the Kupang Cave of East Timor, long-tailed macaques were

Figure 13.1. A juvenile long-tailed macaque using a small stone hammer to axe at oysters attached to boulders along an island rocky shore in the Andaman Sea. A tradition found only in this small region of the long-tailed macaque distribution. (Photograph curtsey of M. D. Gumert.)

observed in the past to regularly sleep in caves (Tenaza, 1996). In Sumatra and Kalimantan, three groups of long-tailed macaques have been observed to catch and eat fish (Stewart *et al.*, 2008). In Lopburi, Thailand they use fibers, human hair, and wooden picks to clean and floss their teeth (Watanabe *et al.*, 2007). The behavioral variant of most significance occurs in the Andaman sea region of Thailand and Myanmar, where long-tailed macaques (*M. f. aurea*) customarily use stone tools to open shellfish in intertidal zones along shores and islands (Carpenter, 1887; Malaivijitnond *et al.*, 2007; Gumert *et al.*, 2009a). Here they exhibit a form of hammering not typically seen in other stone using primates (Gumert *et al.*, 2009b) (Figure 13.1). Nothing is known about the extent of this cultural behavior in Myanmar, but in Thailand it occurs in at least five locations close to the Myanmar border (Gumert *et al.*, 2010).

At the time of writing, only two of the five known tool-using sites are protected areas, which are two neighboring islands, Piak Nam Yai and Thao Island, in Laemson National Park. Despite this protection, serious encroachment is occurring on both islands. Congruent with van Schaik's (2002) fragility of traditions hypothesis, human activity is disturbing this behavior. On Piak Nam Yai Island, local people have recently established a settlement (Gumert *et al.*, 2010). They have established a palm oil plantation and also hunt, farm, and harvest other resources from the island. They introduced domestic dogs in 2008, and since then the dog population has grown and the dogs frequently attack the macaques when they are using tools on the open shores and mangroves. The macaque group closest to the human settlement appears to have decreased in size between 2008 and 2009, and now spends less time on the shores (Gumert *et al.*, 2010). The problem is that Laemson National Park may be the only location in Southeast Asia where this behavioral tradition occurs on legally protected land, and even there, protection appears to not be well enforced. An assessment of this tradition, in terms of geographic distribution and number of macaques who follow it is needed in order to adopt an appropriate conservation strategy.

Understanding ethnophoresy and colonization

Their listing as one of the World's Worst 100 Invasive Alien Species (Lowe *et al.*, 2000) has raised major concerns in the conservation community about the presence of long-tailed macaques on islands outside of their original range (Kemp and Burnett, 2007). It also may have contributed to a negative attitude towards conserving these macaques even in their natural range. Further data on the ecological effects of long-tailed macaques can provide a science-based perspective on this issue.

One of the first steps in solving issues related to ethnophoretic long-tailed macaques is to determine what are the major factors involved in allowing this species to become an ecological success after introduction. Long-tailed macaques have been carried to and released onto Mauritius (Sussman and Tattersall, 1981), Ngeaur Island, Palau (Poirier and Smith, 1974), West Papua (Kemp and Burnett, 2007), and Kabaena Island, Sulawesi (Froehlich *et al.*, 2003). The two islands that report the greatest impact are Mauritius (Padayatchy, Chapter 9) and Ngeaur Island (Wheatley, Chapter 10), while West Papua (Kemp and Burnett, 2007) and Kabaena (Froehlich *et al.*, 2003) do not provide any strong evidence that the long-tailed macaques there have become a seriously invasive exotic species. Exploring the differences between these two sets of islands will provide us with a basis for predicting what conditions lead to these so-called long-tailed macaque invasions. Such understanding may help to prevent future translocations, and possibly alleviate some of the problems in areas that have already been colonized.

Mauritius and Ngeuar Island have a few similarities that could account for the long-tailed macaque's success on these two islands. Both islands have experienced significant problems from many exotic species, have had high levels of habitat destruction, and have easily assessable crops and human resources (Porrier and Smith, 1974; Lorence and Sussman, 1986; Wheatley, Chapter 10; Sussman, Chapter 8). The number of exotic species present on these islands makes it difficult to isolate the specific effects of long-tailed macaques. Mauritius has been extensively modified from human activities (Lorence and Sussman, 1986), and habitat destruction has been very intense in Ngeaur, where there was an intensive three-month bombing during the American invasion in World War II, which obliterated the island's forest regions (Poirier and Smith, 1974). Both islands also have reports of extensive crop raiding and dependence by macaques on human farming (Mungroo and Tezoo, 1999; Wheatley *et al.*, 2002), and Dutch settlers may have abandoned Mauritius because of heavy raiding by macaques (Sussman and Tattersall, 1981).

Another clue that specific conditions led to macaque colonization comes from the isolation of macaques on Ngeaur Island in Palau. Although pet macaques have been sold to people on other islands in Palau, there have been no successful colonization events on these islands even though some of these pets have escaped (Wheatley *et al.*, 1999, 2002). This seems to indicate that rather than just being a solely intrinsic ability of long-tailed macaques to easily colonize an island and exploit new niches there are also certain ecological and/or anthropogenic conditions that facilitate these so-called macaque invasions. Perhaps human impact is the key for allowing successful macaque colonization on an island.

It appears possible that a few key ingredients play a significant role in facilitating exploitation of new environments. First, on Mauritius and Ngeaur there was severe environmental alterations and damage from human activity during the development of their macaque population. Given this important factor occurred on both islands, it is possible that anthropogenic habitat disturbance provides niches suitable for macaques. For example, forest edges and new food patches that the macaques can easily exploit with little or no competition from the local fauna may become available following land development by humans. Moreover, long-tailed macaques have been documented to cope well with severe environmental damage (e.g., fire destruction in Kalimantan) (Berenstain, 1986). This may give long-tailed macaques an advantage in stochastic environments compared to local fauna, which are adapted to the predisturbed stable environment.

A high level of sustenance from human agriculture and human-derived food sources may be of equal importance in supporting macaque colonization. Human resources were consistently available on both Mauritius and Palau during the population expansion of their macaques. Therefore, it seems likely that the long-tailed macaque's ability to live in human settlements or near agricultural plots has likely allowed them to survive and flourish while other fauna may not. Analogously, urban populations of Hanuman langurs (*Semnopithecus entellus*) in Jodhpur, India (i.e., another highly synanthropic primate) have survived serious droughts while other local wildlife not as well adjusted to living near humans, succumbed to the severe conditions (Waite *et al.*, 2007). Overall, altered habitat with exposed forest edges combined with stable food resources from human activity may draw long-tailed macaques to human settlements, which then can promote expansion of their population in an exotic environment.

Kabeana and West Papua provide an interesting contrast to Mauritius and Ngeaur, as there the long-tailed macaques have not become as serious a problem. The West Papua macaque population growth is slower than in Mauritius and Palau. The monkeys have been there for 30–100 years, but their population has not expanded much. In fact, the population consists of only six groups of ~10 macaques each, yielding a total of 60–70 monkeys (Kemp and Burnett, 2003, 2007). Moreover, although Kemp and Burnett (2003, 2007) have argued that the macaques on Papua are negatively affecting birds and reptiles in the Jayapura region, flaws in their research design make their conclusions suspect. The researchers compared macaque-inhabited to non-macaque-inhabited forest plots, but the macaque plots were all closer to human settlements, while the non-macaque plots were around two to three times farther from the nearest settlements. Consequently, it is just as likely that the lower native bird and reptile numbers were due to closer proximity to humans and their land alteration, than any direct influence by the macaques. Future studies with better controls will be needed to resolve these ambiguities.

On Kabaena the macaques range over only 25 percent of the island. No serious invasion has been documented, the potential for conflict with native wildlife has only been speculated, and they range over only ~25 percent of the island (Froehlich *et al.*, 2003). Like Papua, Kabaena does not have a highly disturbed ecosystem. It therefore appears possible that on islands with less ecological disturbance by humans, and with less provisioning by people, macaques may persist, but do not extensively proliferate along the human-made edges across the island. Of course, it is possible that a "macaque invasion" could occur if human activity exposes too many forest edges and makes available food resources for these animals to exploit. Moreover, the long history of macaques on Kabaena, suggests an ecological expansion is not inevitable following release and that long-tailed macaques can and do reach an equilibrium with their colonized environment if not given too many advantages by human-altered habitat and food resources.

Papua provides another example of introduced macaques not expanding in an uncontrolled fashion. Macaques have been on Papua as long or longer than they have been on Palau, but have not expanded to the same degree. Why? Papua appears to have greater resistance to colonization by exotic species as evidenced by the fact that Papua has fewer exotic species than many other island habitats. The island is even resistant to highly effective invasive species, such as cane toads (*Bufo marinus*) and Rusa deer (*Cervus timorensis*), which like macaques, have also had limited success in colonizing the island (Heinsohn,

2003). Consistent with our hypothesis, Papua has had lower levels of anthropogenic habitat alteration than has occurred in Mauritius and Palau, partly the result of the mountainous terrain of Papua that is difficult for humans to develop. Moreover, Papua is noted for having a limited crop base in their farmlands, compared to other regions in the Indonesian Archipelago (Diamond, 1997). The combination of decreased habitat disturbance and decreased athropogenic food sources may explain why the macaque population remains small in Papua.

Overall, future investigation will be needed to better understand the process of macaque ethnophoresy and the factors that lead them to be considered invasive. Understanding what allows long-tailed macaques to proliferate in new environments will help prevent future introductions and will give wildlife managers in these island habitats a basis for alleviating their own macaque problems. Where macaques are unwanted aliens, protective actions are imperative, as it is important not to let monkeys overtake sensitive island ecologies with unique flora and fauna. With that in mind, it is also important to consider where exotic macaques have become integrated into their new environments, and where their presence is benefiting humanity.

Research approaches for studying the human–macaque interface

Humans have shared ecosystems with other primates for millennia, and thus represent important biotic components of the ecology that has impacted each other's evolutionary path. Researchers who have focused on understanding human-primate relationships have come to the conclusion that the sympatric relationships between humans and primates are central to the ecology, evolution, and conservation/management of many primate species (Paterson and Wallis, 2005; Riley, 2007, Wolfe and Fuentes, 2007; Fuentes and Hockings, 2010; Riley and Fuentes, 2010). Moreover, it is becoming clearer that it is not fully accurate to consider primates affected by human activity as unnatural or unimportant systems for studying primate evolution. Because of this, it has been suggested that the interface between humans and other primates needs to be more elaborately documented and systematically studied in order to build more complete models of the behavior and evolution of primate societies (Fuentes, 2006a, 2007a; Lane *et al.*, 2010).

Some species of macaques and humans extensively overlap geographically and this is because of similar evolutionary histories and common physiologies between genera. Both genera have ancestors branching out of Africa and the Mediterranean basin around two million years ago that spread into South and Southeast Asia, with equal success (Hart and Sussman, 2008), indicating that during this time macaques and humans were facing similar ecological pressures. Both of these lineages had more success than other primate species lineage over the last two million years, and this has been especially so in Asia, where we find the largest populations of humans and macaques (i.e., particularly rhesus and long-tailed). Some of the similar features we share with macaques include generalized anatomies, flexible social systems, relatively simple guts, and great dietary and ecological plasticity (Hart and Sussman, 2008). These features have all led to the ecological success of humans and macaques and have also allowed for macaques and humans to overlap in many regions of Asia. In India, the species that most excels in this overlap is *Macaca mulatta* and over most of Southeast Asia, it is *Macaca fascicularis*. Long-tailed macaques have likely lived near humans for thousands of years, and evidence shows that from well over 20,000 years ago in Niah Cave, Sarawak, there was an interface between humans and *Macaca fascicularis* (Harrison, 1996). Given this history, it is likely we have had a sufficiently long relationship with syanthropic macaques to have potentially impacted their evolution.

In this section, we provide an overview of research needs for understanding the interface between long-tailed macaques and humans and explore ways to study how humans and macaques impact each other. We present two approaches to studying the relationship. We summarize the *ethnoprimatological approach* and introduce a new perspective we have termed the *evolutionary biological approach*. We highlight important contributions that both approaches could have on future research on human-macaque sympatry, and suggest ways for investigating the proximate and evolutionary impacts of humans and macaques on each other.

The ethnoprimatological approach

There are two major ways primatologists have approached the study of human-macaque overlap. First, traditional socioecological models in primatology have focused on female distribution, patterns of competition, predation, resource availability, and the structure of the local ecology (van Schaik, 1989; Strier, 2006; Sterk *et al.*, 1997). In these models anthropogenic processes affecting human-primate interfaces can potentially impact all of these factors. Therefore, socioecologists typically consider human activity as "unnatural" disturbance that alters the animal's "natural" behavioral patterns. They have argued that investigations of synanthropic primates are not relevant to understanding the behavioral biology of the animal, as the human influence makes them behave in species-atypical ways.

A second approach is the emerging field of ethnoprimatology. This approach starts from the socioecological perspective, but instead of viewing human impact as just "interference" or perturbation of a "natural" state, views human impact as another ecological component to be considered in the natural system (Burton and Carroll, 2005). Therefore, an ethnoprimatological approach seeks to understand the anthropogenic interface affects the behavior of the animal, and considers behavior elicited by anthropogenic factors as a facultative or "normative" response to the new environmental pressures. In other words, the human affect on primate populations should be considered as a significant ecological pressure that shapes the natural behavioral potential of the synanthropic primate.

The similarity across primates, particularly macaques, in morphology and behavior suggests that the interface between humans and other primate species may differ from those between humans and other mammals (Fuentes, 2007b). For example, the high degree of overlap in primate sensory systems and neuropsychology might result in more tightly shared perceptual and emotive experiences between humans and other anthropoid primates. Also, similarities in digestive tracts and dietary requirements can create similarities in the survival needs across species. The high degree of biological overlap suggests the interface between humans and other primates may have some important ecological and physiological differences from similar interfaces between humans and other synanthropes, such as dogs or cattle. Future research will want to focus more on this distinction.

Primatologists usually model the human-primate interface primarily as a competition for space and resources, and thus a contest between humans and the synanthrope. The approach of ethnoprimatology (Sponsel, 1997; Wheatley, 1999; Fuentes and Wolfe, 2002) takes such competition into account, but also brings greater focus to the mutualistic and commensal components of human-primate relationships. By exploring all the potential aspects of the interspecific relationship, an ethnoprimatologist attempts to merge the socioecological approach with an anthropological approach. They study the symbolic interpretations, uses, and perception of the relationship with the non-human primates by the human communities living with them (i.e., socio-cultural anthropology and historical accounts), as well as the direct interactions between humans and non-humans (Fuentes, 2006b). This approach allows us to characterize each human-macaque interface zone studied, and then, by understanding the qualities of the interface, we can potentially begin to discern their ecological affects on the interacting species.

The main foci of an ethnoprimatologist include human predation via hunting, human pet keeping, the role of primates in entertainment, human-other primate overlap, the partitioning of space, the human habitat alteration on primates, bidirectional pathogen exchange, human perceptions, cultural aspects, and the impacts of tourists and researchers on primate populations (Fuentes and Hockings, 2010). Moreover, ethnoprimatologists focus on several aspects of the human-animal relationship. For example, they study how crop raiding, the threat of aggression, and the environmental impact by synanthropic primates can affect the livelihoods and perceptions of humans. They also study how human alteration of the landscape, hunting, religious belief, and pet keeping can affect the behavior and ecology of the synapthropic primate.

Ethnoprimatogists are also interested in human perceptions and behavior towards macaques, as well as macaque behavioral patterns towards humans. Current work has focused on how perceptions and behavior relate to human-macaque conflict (Sha *et al.*, 2009a; Fuentes *et al.*, 2008, Fuentes and Gamerl, 2005; Wheatley and Putra, 1994; Wheatley *et al.*, 2002). On a simple level, human behavior, including smiling, scolding and loud voices can have counterintuitive (for humans) responses from macaques. On the other hand, behavior that macaques see as normative, threats, signs of stress, and norms of spatial proximity are foreign to humans and often misread. Such misinterpretations of signals across species, indicates an evolutionary mismatch in the signals (i.e., the signals have not evolved to communicate to the other species), and as a result, an inability to communicate can lead to overt conflict (i.e., aggression). On a broader scale, human cultural histories, religious perceptions, experience, financial motivations, individual differences, etc. all interact to create very different perspectives on the perception of conflict with macaques. Some people enjoy living near macaques, while other despise it and wish to have all of the animals removed (see Sha *et al.*, 2009a for a discussion of attitude variation in Singapore and Loudon *et al.*, 2006 for variation in Balinese perception of the macaques). Human perceptions alone do not identify the characteristics of the human-macaque relationship. Rather, a communities' set of attitudes and beliefs are just another component of the sympatric relationship, and observational and ecological studies are additionally needed to fully understand the impact of these attitudes.

The variation across areas of interfaces produce differing conditions affecting the macaques, and this affects how humans impact the macaques. Variation occurs as the result of a group's nationality, ethnicity, religion, socio-economic status, urban or rural lifestyle, etc. (Fuentes in press). Taking an ethnoprimatological approach to all of these different situations allows us to include anthropological factors into the ecological equation of understanding primate behavior and evolution at areas of interface (Fuentes and Hockings, 2010). The chapters in this book (see Part II) indicate some of these patterns and suggest that they are tied to human-directed changes to the ecological conditions, the behavior of humans and macaques, and the extent to which macaques fit within a local culture's attitudes and beliefs (Fuentes *et al.*, 2005). By focusing on human perceptions, patterns of behavior, and ecological contexts of the human-macaque interface, we gain a better understanding of how the characteristics of the interface itself affect the behavior of humans and macaques that are living together (Fuentes and Hockings, 2010; Fuentes, 2006a; Riley, 2007; Wolfe and Fuentes, 2007; Riley and Fuentes, 2010).

The evolutionary biological approach

The ethnoprimatological approach focuses on understanding human-macaque relationships by exploring the role of macaques in the local people's culture and how the two species interact. Its main focus is at the proximate level of behavior, as it explores how macaques and humans interact, and how people think and feel about the primates they are living with. While very useful for characterizing human-macaque interfaces zones, ethnoprimatologists have not yet directly investigated how the two species have evolved together, and specifically how they have influenced each other's reproductive biology (although they recognize and suggest this is happening). Of particular importance will be to increase efforts to understand the impact that human activity and the anthropogenic environment have on the genetic evolution of synathropic macaque populations. To do this, we will need to begin conducting more detailed studies on how the anthropogenic environment affects the mortality, reproductive success, and population genetics of macaques living beside humans. An evolutionary biological approach will provide a fuller picture of synanthropic macaques and will also allow us to potentially determine key traits involved in their ability to living in and adjusting to human-affected environments.

When primates first encounter humans, they quickly flee and it takes time for them to grow used to being near people. Across primate species there are very large differences in the speed they will accustom, or habituate, to being near people, and this can take a few weeks to several years (Williamson and Feistner, 2003). A key measurable factor in habituation is the distance at which the animal flees from humans. Researchers of canine domestication refer to this range as the animal's "flight distance" (Coppinger and Coppinger, 2001), a distance that lowers as the animal becomes less fearful of human presence. Long-tailed macaques go through the process of habituating to people when human communities develop on forest edges, and over the developmental process of lowering flight distance the macaques adjust to living more within the anthropogenic landscape as they become more comfortable and move closer into proximity with their neighboring humans. Aspects of this change are developmental and learned, but there will also be components influencing a species flight distance and ability to habituate to humans that are heritable, as indicated by the large degree of variation in the habituation process across species. This means that heritable aspects providing a greater ability for habituation and lowering flight distance around humans could be selected for if living near humans has reproductive advantages.

Thus far, we have no studies on the development of human-macaque interfaces, and can only speculate on the time frame and processes that occur as the two species increase in overlap. We can observe the end result of the process though, and it is clear macaques in interface zones do have lower flight distances to people. This presents a significant change in the biotic environment and thus presents a new exertion of selective pressures facing the macaques. Reproductive costs will be incurred based on the degree to which people harm, trap, kill, or otherwise remove macaques from the breeding pool (e.g., take as pets or trade) (Eudey, 1994; Schillaci *et al.*, 2010; Aggimarangsee, 1992; Louden *et al.*, 2006; Eudey, 2008; Lane *et al.*, 2010). Moreover, exposure to infectious agents through contact with humans (Engel and Jones-Engel, Chapter 7) can place new selective pressures on their immune systems. Selective advantages may also occur and these will be based on the amount of support, care, and provisioning local communities provide to macaques. Human support can increase survivorship and allow individuals to survive that might not have done so in less anthropogenic-influenced environments (see Box 7.1). The degree of human impact on macaque survival will vary across regions and will depend on many of the factors studied by ethnoprimatologists, such as local cultures, religions, and attitudes towards animals. Therefore, differing human communities will exert different selective pressures on the macaques they live with.

The important question in the evolutionary biological approach to human-macaque sympatry is whether there are any common traits shared by the individual macaques that are selected for or against by human activity. For example, do macaques removed by humans (i.e., culled, killed, or otherwise removed from breeding) tend to be less tolerant of humans and thus enter into greater levels of overt conflict (i.e., aggression) with their human neighbors than other individuals that are not removed? Are animals that or more passive and relaxed around their human neighbors under less threat to be killed by their human neighbors? Or perhaps, animals with temperaments that people prefer (e.g., complacent and docile) are removed to be taken as pets. If we do find any local directional selection for or against any particular behavioral traits, human influence would act as a selection pressure for that population of macaques. Over time, the traits better fitted for not being removed and thus better fitted for the human-inhabited environment would propagate, and the macaques would begin to biologically adapt to the anthropogenic ecology by which they are affected. The traits most likely to be affected would be those related to direct interactions with people, such as their temperamental tendencies to show fear, submission, aggression, and interest towards humans.

A model for comparison is the domestication of the dog, which has been the result of an increasing proximity and interaction between humans and wolves over at least the last 14,000 years (Clutton-Brock, 1995; Coppinger and Shneider, 1995; Coppinger and Coppinger, 2001). According to some scientists studying canine evolution, the basic transition occurred with wolves (*Canis lupus*) beginning to live near humans, as they scavenge litter in garbage dumps and hunted livestock. They suggest that humans attempted to repel wolves, which would have selected against the most aggressive and dangerous individuals. Overall, this created a directional selection pressure that left a higher proportion of submissive and tame individuals living on the fringes of human settlement to reproduce. Over time, through these processes and also the direct capture and breeding of pets, wolves were eventually selected to resemble village dogs (*Canis familiaris*). They evolved a higher frequency of tame traits adapted for living in a human environment and directly interacting with people (Coppinger and Schneider, 1995).

Studies on silvered foxes (*Vulpes vulpes*) have shown that with intense artificial selection against aggressive individuals, tamer populations can evolve within several generations (Belyaev, 1979; Trut, 1999). These changes in temperament also correlated with changes in morphology, such as floppy ears and patchy fur coloration (Trut, 1999), indicating there can be marked physical changes that co-evolve with behavioral traits. Further research on silver foxes has shown these behavioral adaptations are associated with changes in brain gene expression (Linberg *et al.*, 2005). Other research investigating tameness in rats (*Rattus norvegicus*) has shown clear genetic differences in rats which are more tolerant and comfortable around humans, which is related to changes in the genetic expression affecting hypo-pituitary-adrenal (HPA) axis and serotonin system activity (Albert *et al.*, 2008, 2009). Overall, this line of work suggests that human influence can potentially alter the behavioral and morphological evolution of local populations when there is sustained directional selection across only several generations. This work has all been artificially done in captivity, and thus it remains to be studied how much influence human activity could actually have on animal behavior in natural environments. The overlap of humans and macaques provides a scenario by which these types of questions can be readily investigated in natural settings.

No work to date has explored the influence of human activity on the evolution of synanthropic macaques. Given the speed in generation time that evolution appears to be able to act on temperaments in other mammals when artificially selected, it is plausible that such evolutionary forces might play a significant selective role in the behavioral evolution of macaques naturally interfacing with people. Species like long-tailed macaques may be in an evolutionary relationship with humans similar to what some experts suggest canines underwent. Evidence supports that the domestic dogs appear to be more temperamentally aligned with people and better adapted to human signals than wolves (Hare and Tomasello, 2005) (although there is dispute, see Udell *et al.*, 2008 and Hare *et al.*, 2010 for discussion). These adaptations seem associated with innate components affecting the time frame and receptivity for development of understanding human communication (Riedel *et al.*, 2008; Dorey *et al.*, 2010), and may result from the selection pressures exerted on them across their historical relationship with people (Coppinger and Coppinger, 2001).

Research on how human influence affects the evolution of macaque behavior might uncover relationships between the evolutionary impact of the local community's cultural practices on macaque reproductive biology and the intensity of conflict. For example, it is worth investigating whether the high level of feeding at tourist temple sites in Thailand and Bali, as well as the restrictions on harming macaques at these sites, may allow the most aggressive animals a reproductive advantage. Is it possible that the way the humans relate to the macaques at these temples allow highly aggressive animals to outcompete other macaques at clumped feedings, thus reproducing better and increasing the population of animals with aggressive/despotic temperaments? This might occur because the most aggressive animals will get more food from human sources without facing the selective pressure of being killed by humans when they become too aggressive, because they are specially protected at the temple. This is plausible, as it has been shown in a few studies that macaques face more violence from humans outside of temple areas than on the immediate protected grounds (Schillaci *et al.*, 2010; Aggimarangsee, 1991).

Interfaces at Balinese temples are reported to have very significant levels of aggressive contact against people (Fuentes, 2006a). In contrast, macaque-to-human aggressive contact is much rarer in Singapore, and their interface is considered the most benign in Southeast Asia (Fuentes *et al.*, 2008; Sha *et al.*, 2009a). One difference is that, at the tourist temple sites, macaques get far more human food resources (e.g., predictable mass feeding of large quantities of food every day), and thus there are much higher levels of contest competition over these highly clumped food resources. Contest over clumped resources leads to more agonistic or dominance-based relationships, where aggression would be competitively advantageous, and thus there would be selection for greater levels of despotism (Sterk *et al* 1997; van Schaik, 1989). A second difference is that in Singapore, regular trapping and removal of individuals occurs when conflict between humans and local communities escalates, while little or no such effort to remove potentially aggressive animals occurs at tourist temples sites. This raises an important question – Does the variation in the human-produced feeding ecology and the difference in macaque-to-human aggression (i.e., human killing of macaques) represent an important anthropogenic selective pressure on macaque behavior? (Figure 13.2).

Understanding how humans have affected the evolution of macaque behavior will be important not only to evolutionary biologists, but also to management planners for human-macaque conflict. Evolutionary-based research holds the potential for identifying breeding and population control programs that can reduce human-macaque conflict. An evolutionary approach also raises important considerations to organizations that are manipulating the reproductive biology of their macaques through sterilization (Martelli, 2009; Bunluesilp, 2009; Wong and Chow, 2004). At this point in time, we simply do not yet know how our activity affects the population genetics of macaques. Therefore, current management practices that utilize sterilization or culling are working blindly in terms of reproduction control. By taking an evolutionary biological approach to human–primate relationships, we could make significant improvements in our basic understanding of the important factors driving human-macaque sympatry, and use this information for better planning of management strategies related to reproductive control.

Long-term research and human-macaque communities

The importance of long-term research programs in supporting and developing sustainable human-macaque communities cannot be understated. Long-term research is known to support conservation and the sustainable development of communities by obtaining knowledge on the people, wildlife, and environment of the region. Moreover, the project personnel maintain a stable, long-term presence, and thus can develop strong relationships with the local communities and governments, playing an important role in educating and advising these communities (Wrangham and Ross, 2008). Research programs focusing on long-tailed macaques and their interfacing human communities have occurred in Bali, and have played important roles in learning about monkeys and how they interact with their communities (Wheatley and Harya Putra, 1994; Wheatley and Harya Putra, 1995; Wheatley, 1999; Fuentes *et al.*, 2005; Lane *et al.*, 2010). Moreover, these studies have investigated how the local communities relate to their macaques and the efforts they have taken to sustain them. These programs have established ties with local universities, governmental agencies, and NGO's, and have played a strong role in tying the human-macaque community together.

The type of information generated from research programs is an invaluable starting point for beginning to think through the strategies necessary to develop sustainable human-macaque communities. Research programs are also beginning in Singapore (Gumert *et al.*, 2009; Sha *et al.*, 2009a, 2009b;

Figure 13.2. Human-impact changes the ecology affecting long-tailed macaques. How humans feed them varies the distribution of food. (a) At Ulu Watu temple in Bali, macaques are provisioned daily on platforms with potatoes and other foods; (b) At an eco-lodge in Kalimantan, rice and trash are thrown out randomly for the macaques depending on activity at the lodge. Human aggression also changes selective pressures on macaque populations; (c) At a temple in Petchaburi Thailand, people repel monkeys by using sling shots, which can be damaging but not lethal. In contrast, at a small village in Kalimantan; (d) people trap macaques in nets and kill them. (Photographs curtesy of M. D. Gumert.)

Fuentes *et al.*, 2008), and if sustained in the long-term will be resources for the human-macaque communities there as well. Long-term research provides data on several factors: demographics, the behavior and temperaments of individual macaques, human-macaque interactions, and human attitudes and perceptions. Moreover, long-term research provides the only basis for evaluating the success of attempted conflict resolution strategies. Personnel from such research projects will also be important interlocutors between the macaques and their surrounding communities, playing an integral role in educating communities, advising governments, and enlisting and supervising the support of NGOs working on problems occurring in human-macaque interface zones.

Conclusion

Despite the ubiquity of long-tailed macaques across the region, we still have an impoverished understanding of the most common monkey of Southeast Asia. Future research directions will need to focus on better understanding

Figure 13.2. (*cont.*)

Figure 13.2. (*cont.*)

the population characteristics of long-tailed macaques. We will need to focus on questions such as, (1) How many macaques are there in Southeast Asia, and what proportion of their population interfaces with human societies?, and (2) what is the true extent of genetic variation across their vast geographical range, and how valid is the current subspecific classification system currently in use? We need to better study and consider what conservation measures may be needed for the various subspecies and for unique behavioral traditions that occur in small populations. Another critical aspect of research is to fully understand the criteria and conditions that allow long-tailed macaques to colonize islands where they have been released. This can be done by studying the few islands where macaques have colonized and then compare the degree of ability to propagate under varying ecological and anthropogenic conditions on each island. We also need to continue characterizing human-macaque interfaces using the ethnoprimatological approach, but we now need to begin investigating how variations in the anthropogenic-based ecological influences on macaques affects their evolution alongside human communities. Overall, long-tailed macaque population-level research will allow us to better understand the coevolution of human and non-human primate populations. Moreover, it will provide us with the information database necessary to support quality population monitoring and management programs aimed at alleviating human-macaque conflicts, controlling trade, and maintaining a sustainable long-tailed macaque population into the future.

References

Aggimarangsee, N. 1992. Survey for semi-tame colonies of macaques in Thailand. *Natural History Bulletin of the Siam Society* **40**: 103–166.

Albert, F. W., Shchepina, O., Winter, C. *et al.* 2008. Phenotypic differences in behavior, physiology and neurochemistry between rats selected for tameness and for defensive aggression towards humans. *Hormones and Behavior* **53**: 413–421.

Albert, F. W., Carlborg, O., Plyusnina, I. *et al.* 2009. Genetic architecture of tameness in a rat model of animal domestication. *Genetics* **182**: 541–554.

Belyaev, D. K. 1979. Destablizing selection as a factor in domestication. *Journal of Heredity* **70**: 301–308.

Berenstain, L. 1986. Responses of long-tailed macaques to drought and fire in Eastern Borneo: A preliminary report. *Biotropica* **18**: 257–262.

Blancher, A., Bonhomme, M., Crouau-Roy, B., Terao, K., Kitano, T. and Saitou, N. 2008. Mitochondrial DNA sequence phylogeny of 4 populations of the widely distributed cynomolgus macaque (*Macaca fascicularis fascicularis*). *Journal of Heredity* **99**: 254–264.

Bunluesilp, N. 2009. No monkey business: Thailand launches primate birth control. *Reuters Life*, August 21, 2009.

Burton, F. and Carroll, A. 2005. By-product mutualism: Conservation implications amongst monkeys, figs, humans, and their domesticants in Honduras. In *Commensalism and Conflict: The Primate-Human Interface.* J. D. Paterson and J. Wallis (eds.). Norman, OK: American Society of Primatology Publications.

Carpenter, A. 1887. Monkeys opening oysters. *Nature* **36**: 53.

Clutton-Brock, J. 1995. Origins of the dog: Domestication and early history. In *The Domestic Dog its Evolution, Behaviour and Interactions with People.* J. Serpell (ed.). Cambridge University Press.

Coppinger, R. and Coppinger, L. 2001. *Dogs: A Startling New Understanding of Canine Origin, Behaviour and Evolution.* New York: Scribner.

Coppinger, R. and Schneider, R. 1995. Evolution of working dogs. In *The Domestic Dog its Evolution, Behaviour and Interactions with People.* J. Serpell (ed.). Cambridge University Press.

Diamond, J. 1997. *Guns, Germs, and Steel: The Fates of Human Societies,* New York: W. W. Norton and Company.

Di Silva, R. 2008. SpotLight: Culling solution to macaque 'explosion'. *New Straits Times.* Kuala Lumpur. 11 February 2008.

Dorey, N. R., Udell, M. A. R., and Wyner, C. D. L. 2010. When do domestic dogs, *Canis familiaris*, start to understand human pointing? The role of ontogeny in the development of interspecies communication. *Animal Behaviour* **79**: 37–41.

Eudey, A. A. 1994. Temple and pet primates in Thailand. *Revue D'Ecologie (Terre et la Vie)* **49**: 273–280.

2008. The crab-eating macaque (*Macaca fascicularis*) widespread and rapidly declining. *Primate Conservatio* **23**: 129–132.

Fooden, J. 1995. Systematic review of Southeast Asian longtail macaques, *Macaca fascicularis* (Raffles, 1821). *Fieldiana: Zoology, n.s.* **81**: v + 206.

2006. Comparative review of fascicularis-group species of macaques (primates: *Macaca*). *Fieldiana: Zoology, n.s.* **107**: 1–43.

Froehlich, J., Schillaci, M., Jones-Engel, L., Froehlich, D., and Pullen, B. 2003. A Sulawesi beachead by longtail monkeys (*Macaca fascicularis*) on Kabaena Island, Indonesia. *Anthropologie* **41**: 76–74.

Fuentes, A. 2006a. Human-nonhuman primate interconnections and their relevance to anthropology. *Ecological and Environmental Anthropology* **2**: 1–11.

2006b. Human culture and monkey behavior: assessing the contexts of potential pathogen transmission between macaques and humans. *American Journal of Primatology* **68**: 880–896.

2007a. Social organization: Social systems and the complexities in understanding the evolution of primate behavior. In *Primates in Perspective.* C. Campbell, A. Fuentes, K. Mackinnon, M. Panger, and S. Bearder (eds.). Oxford University Press.

2007b. Monkey and human interconnections: The wild, the captive, and the in-between. In *Where the Wild Things are Now: Domestication Reconsidered.* R. Cassidy and M. Mullin (ed.). Oxford: Berg Publishers.

(in press) Being human and doing primatology: National, socioeconomic, and ethnic influences on primatological practice. *American Journal of Primatology.*

Fuentes, A. and Gamerl, S. 2005. Disproportionate participation by age/sex classes in aggressive interactions between long-tailed macaques (*Macaca fascicularis*) and human tourists at Padangtegal monkey forest, Bali, Indonesia. *American Journal of Primatology* **66**: 197–204.

Fuentes, A. and Hockings, K. 2010. The ethnoprimatological approach in primatology. *American Journal of Primatology* **71**: 1–7.

Fuentes, A. and Wolfe, L. D. (eds.) 2002. *Primates Face to Face: Conservation Implications of Human and Nonhuman Primate Interconnection.* Cambridge University Press.

Fuentes, A., Suaryana, K., Rompis, A. *et al.* 2002. Behavior and demography of a semi-free ranging population of long-tailed macaques (*Macaca fascicularis*) at Padangtegal, Bali, Indonesia. *American Journal of Physical Anthropology* **117**: 73.

Fuentes, A., Southern, M., and Suaryana, K. 2005. Monkey forests and human landscapes: Is extensive sympatry sustainable for *Homo sapiens* and *Macaca fascicularis* on Bali. In *Commensalism and Conflict: The Primate-Human Interface.* J. D. Patterson (ed.) Norman, OK: The American Society of Primatologists Publications.

Fuentes, A., Kalchik, S., Gettler, L., Kwiatt, A., Konecki, M., and Jones-Engel, L. 2008. Characterizing human-macaque interactions in Singapore. *American Journal of Primatology* **70**: 879–883.

Groves, C. P. 2001. *Primate Taxonomy*, Washington, DC: Smithsonian Institute Press.

Gumert, M., Sha, J., Lee, B. P. Y.-H., and Chan, S. 2009.) Factors influencing the interface between humans and long-tailed macaques in Singapore. *American Journal of Primatology* **71**: 56.

Gumert, M., Kluck, M., and Malaivijitnond, S. 2009a. Stone tool characteristics and usage patterns in long-tailed macaques from the Andaman Sea Region. *American Journal of Primatology* **71**: 91.

2009b. The physical characteristics and usage patterns of stone axe and pounding hammers used by long-tailed macaques in the Andaman Sea region of Thailand. *American Journal of Primatology* **71**: 594–608.

Gumert, M., Low, K., Tan, V., and Malaivijitnond, S. 2010. Sex differences, handedness, and selection preferences in the stone tool-use of Andaman long-tailed macaques (*Macaca fascicularis aurea*). 23rd Congress of the International Primatological Society, Kyoto.

Hare, B. and Tomasello, M., 2005. Human-like social skills in dogs? *Trends in Cognitive Sciences* **9**: 439–444.

Hare, B., Rosati, A., Kaminski, J., Bräuer, J., Call, J., and Tomasello, M. 2010. The domestication hypothesis for dogs' skills with human communication: A response to Udell *et al.* (2008) and Wynne *et al.* (2008). *Animal Behaviour* **79(2)**: e1–e6.

Harrison, T. 1996. The paleoarcheological context of Niah cave, Sarawak: Evidence from the primate fauna. *Indo-Pacific Prehistory Association Bulletin* **14**: 90–100.

Hart, D. and Sussman, R. W. 2008. *Man the Hunted: Primates, Predators, and Human Evolution* (expanded edition). New York: Westview Press.

Heinsohn, T. 2003. Animal translocation: Long-term human influences on the vertebrate zoogeography of Australasia (natural dispersal versus ethnophoresy). *The Australian Zoologist* **32**: 351–376.

Lane, K. E., Lute, M., Rompis, A. *et al.* 2010. Pests, pestilence, and people: The long-tailed macaque and its role in the cultural complexities of Bali. In *Indonesian Primates*. S. Gursky-Doyen, and J. Supriatna (eds.). Berlin: Springer Science.

Lindberg, J., Björnerfeldt, S., Saetre, P. *et al.* 2005. Selection for tameness has changed brain gene expression in silver foxes. *Current Biology* **15**: R915–R916.

Lorence, D. and Sussman, R. 1986. Exotic species invasion into Mauritius wet forest remnants. *Journal of Tropical Ecology* **2**: 1470–162.

Loudon, J. E., Howells, M. E., and Fuentes, A. 2006. The importance of integrative anthropology: A preliminary investigation employing primatological and cultural anthropological data collection methods in assessing human-monkey co-existence in Bali, Indonesia. *Ecological and Environmental Anthropology* **2**: 2–13.

Lowe, S., Browne, M., Boudjelas, S. and De Poorter, M. 2000. One hundred of the worst invasive species: A selection from the global invasive species database. *Aliens 12.*

Kemp, N. and Burnett, J. 2003. Final report: A biodiversity risk assessment and recommendations for risk management of long-tailed macaques (*Macaca fascicularis*) in New Guinea. Indo-Pacific Conservation Alliance and Universitas Cenderawasih.

2007. A non-native primate (*Macaca fascicularis*) in Papua: Implications for biodiversity. In *The Ecology of Papua: Part II*: A. J. Marshall B. M. and Beehler (eds.). Singapore: Periplus Editions Ltd.

Malaivijitnond, S., Lekprayoon, C., Tandavanittj, N., Panha, S., Cheewatham, C., and Hamada, Y. 2007. Stone-tool usage by Thai long-tailed macaques (*Macaca fascicularis*). *American Journal of Primatology* **69**: 227–233.

Martelli, P. 2009. *Endoscopic Tubectomy in Macaques* (DVD). Endoskope.

Menninger, K., Wieczorek, G., Riesen, S. *et al.* 2002. The origin of cynomolgous monkey affects the outcome of kidney allografts under Neoral immunosuppression. *Transplantation Proceedings* **34**: 2887–2888.

Mungroo, Y. and Tezoo, V. 1999. Control of invasive species in Mauritius. In *Invasive Species in Eastern Africa: Proceedings of a workshop held at ICIPE*, E. Lyons and S. Miller (eds.).

Ong, P. and Richardson, M. 2008. *Macaca fascicularis*. IUCN 2010: IUCN Red List of Threatened Species.

Patterson, J. D. and Wallis, J. (eds.) 2005. *Commensalism and Conflict: The Primate-Human Interface*. Norman, OK: The American Society of Primatologists Publications.

Poirier, F. E. and Smith, E. O. 1974. The crab-eating macaques (*Macaca fascicularis*) of Angaur Island, Palau, Micronesia. *Folia Primatologica* **22**: 258–306.

Riedel, J., Schumann, K., Kaminski, J., Call, J., and Tomasello, M. 2008. The early ontogeny of human-dog communication. *Animal Behaviour* **75**: 1003–1014.

Riley, E. 2007. The human-macaque interface: Conservation implications of current and future overlap and conflict in Lore Lindu National Park, Sulawesi, Indonesia *American Anthropologist* **109**: 473–484.

Riley, E. P. and Fuentes, A. 2010. Conserving social-ecological systems in Indonesia: human-nonhuman primate interconnections in Bali and Sulawesi. *American Journal of Primatology* **71**: 1–13.

van Schaik, C. P. 1989. The ecology of social relationships amongst female primates. In *Comparative Socioecology: The Behavioural Ecology of Humans and Other Mammals*. V. Standen, R. and Foley (eds.) Boston: Blackwell Science.

2002. Fragility of traditions: the disturbance hypothesis for the loss of local traditions in orangutans. *International Journal of Primatology* **23**: 527–538.

Schillaci, M. A., Engel, G. A., Fuentes, A. *et al.* 2010. The not-so-sacred monkeys of Bali: A radiographic study of human-primate commensalism. In *Indonesian Primates*. S. Gursky-Doyen and J. Supriatna (eds.). New York: Springer.

Sha, J., Gumert, M., Lee, B., *et al.* 2009a. Macaque-human interactions and the societal perceptions of macaques in Singapore. *American Journal of Primatology* **71**: 825–839.

Sha, J., Gumert, M., Lee, B., *et al.* 2009b. Status of the long-tailed macaque *Macaca fascicularis* in Singapore and implications for management. *Biodiversity and Conservation* **18**: 2909–2926.

Shepherd 2010. Illegal primate trade in Indonesia exemplified by surveys carried out over a decade in North Sumatra. *Endangered Species Research* **11**: 201–205.

Sponsel, L. E. 1997. The human niche in Amazonia: Explorations in ethnoprimatology. In *New World Primates: Ecology, Evolution, and Behavior*. W. G. Kinzey (ed.) Hawthorne, NY: Aldine de Gruyter.

Sterck, E. H. M., Watts, D. P., and Van Schaik, C. P. 1997. The evolution of female social relationships in nonhuman primates. *Behavioral Ecology and Sociobiology* **41**: 291–309.

Stewart, A.-M., Gordon, C., Wich, S., Schroor, P., and Meijaard, E. 2008. Fishing in *Macaca fascicularis*: A rarely observed innovative behavior. *International Journal of Primatology* **29**: 543–548.

Strier, K. B. 2006. *Primate Behavioral Ecology*, 3rd edn. Needam, MA: Allyn and Bacon.

Sussman, R. W. and Tattersall, I. 1981. Behavior and ecology of *Macaca fascicularis* in Mauritius: A preliminary study. *Primates* **22**: 192–205.

Tenaza, R. R. 1996. Kupang's cave monkeys. *Wildlife Conservation* **99**: 48–53.

Trut, L. N. 1999. Early canid domestication: The farm-fox experiment. *American Scientist* **87**: 160–169.

Twigg, I. C. and Nijman, V. 2008. Export of wild-caught long-tailed macaque from Southeast Asia. 22nd Congress of the International Primatological Society, Edinburgh, UK.

Udell, M. A. R., Dorey, N. R., and Wynne, C. D. L. 2008. Wolves outperform dogs in following human social cues. *Animal Behaviour* **76**: 1767–1773.

Waite, T., Chhangani, A., Campbell, L., Rajpurohit, L. and Mohnot, S. 2007. Sanctuary in the city: Urban monkeys buffered against catastrophic die-off during ENSO-related drought. *EcoHealth* **4**: 278–286.

Watanabe, K., Urasopon, N., and Malaivijitnond, S. 2007. Long-tailed macaques use human hair as dental floss. *American Journal of Primatology* **69**: 940–944.

Wheatley, B. 1999. *The Sacred Monkeys of Bali*. Prospect Heights, IL: Waveland Press, Inc.

Wheatley, B. and Harya Putra, D. K. 1994. Biting the hand that feeds you: Monkeys and tourists in Balinese monkey forests. *Tropical Biodiversity* **2**: 317–327.

1994. The effects of tourism on conservation at the monkey forest in Ubud, Bali. *Revue D'Ecologie (Terre et la Vie)* **49**: 245–257.

1995. Hanuman, the monkey god, leads conservation efforts in Balinese Monkey Forest at Ubud, Indonesia. *Prim. Rep.* **41**: 55–64.

Wheatley, B., Stephenson, R., and Kurashina, H. 1999. The effects of hunting on the longtailed macaques of Ngeaur Island, Palau. In and *The Nonhuman Primates*. P. Dolhinow and A. Fuentes (eds.). Mountain View, CA: Mayfield Publishing Company.

Wheatley, B., Stephenson, R., Kurashina, H., and Marsh-Kautz, K. 2002. A cultural primatological study of *Macaca fascicularis* on Ngeaur Island, Republic of Palau. In and *Primates Face-to-Face: Conservation Implications of Human and Nonhuman Primate Interconnections*. A. Fuentes and L. Wolfe. (eds.). Cambridge University Press.

Williamson, E. A. and Feistner, A. T. C. 2003. Habituating primates: Processes, techniques, variables and ethics. In *Field and Laboratory Methods in Primatology: A Practical Guide*. J. M. Setchell and D. J. Curtis (eds.). Cambridge University Press.

Wolfe, L. D. and Fuentes, A. 2007. Ethnoprimatology: Contextualizing human/primate interactions. In *Primates in Perspective*. C. Campbell, A. Fuentes, K. Mackinnon, M. Panger, and S. K. Bearder (eds.). Oxford University Press.

Wong, C. L. and Chow, G. 2004. Preliminary results of trial contraceptive treatment with SpayVac™ on wild monkeys in Hong Kong. *Hong Kong Biodiversity: AFCD Newsletter* **6**: 13–16.

Wrangham, R. W. and Ross, E. 2008. *Science and Conservation in African Forests: The Benefits of Long-term Research*. Cambridge University Press.

Yanuar, A., Bekti, D., and Saleh, C. 1993. The status of the Karimata primates *Presbytis rubicunda carimatae* and *Macaca fascicularis carimatensis* in Karimata Island, Indonesia. *Tropical Biodiversity* **1**: 157–162.

Index

Printed in the United States
by Bookmasters

Printed in the United States
By Bookmasters